Animal Health and Welfare in Livestock Production Enterprises

Richard Skiba

AFTER MIDNIGHT
PUBLISHING

Skiba, Richard (author)

Animal Health and Welfare in Livestock Production Enterprises

Non-fiction

Contents

CHAPTER ONE

Introduction

In the context of Livestock Production Enterprises, the interdependence of animal welfare and animal health is critical for achieving successful and sustainable livestock management. Animal welfare encompasses the physical and psychological well-being of livestock, emphasizing their ability to live in an environment that minimizes stress, suffering, and discomfort. Key components of animal welfare include adequate shelter, nutrition, and water, as well as freedom from pain, injury, and disease. Furthermore, allowing animals to express natural behaviours—such as grazing and socializing—contributes significantly to their welfare [1, 2]. The ethical treatment of animals, as outlined by various regulatory bodies, also plays a vital role in promoting high welfare standards, which in turn enhances productivity and efficiency within livestock enterprises [3, 4].

The agreed international definition of animal welfare from the World Organisation for Animal Health (OIE) states [5]:

Animal welfare means the physical and mental state of an animal in relation to the conditions in which it lives and dies. An animal experiences good welfare if the animal is healthy, comfortable, well nourished, safe, is not suffering from unpleasant states such as pain, fear and distress, and is able to express behaviours that are important for its physical and mental state. Good animal welfare requires disease prevention and appropriate veterinary care, shelter, management and nutrition, a stimulating and safe environment, humane handling and humane slaughter or killing. While animal welfare refers to the state of the animal, the treatment that an animal receives is covered by other terms such as animal care, animal husbandry, and humane treatment.

Animal health is equally crucial, focusing on the prevention, management, and treatment of diseases that can adversely affect livestock productivity and welfare. Regular veterinary care, vaccinations, and biosecurity measures are essential to maintain animal health and prevent the spread of infections [4]. Additionally, proper nutrition is fundamental for promoting immunity and growth, while hygiene and sanitation in living environments help mitigate health risks [4]. Stressors such as overcrowding and environmental extremes must also be managed to safeguard animal health, as stressed animals are more susceptible to disease and less productive overall [6, 7].

The synergy between animal welfare and health directly influences the productivity of livestock enterprises. Healthier animals that are well-cared-for tend to produce higher-quality outputs, such as meat, milk, and wool, thereby enhancing the overall profitability of the operation [2, 8]. Studies have shown that improving animal welfare not only benefits the animals themselves but also leads to better economic outcomes for livestock stakeholders, as higher welfare standards often correlate with improved product quality and marketability [2, 9]. Thus, prioritizing both animal welfare and health is essential for the long-term sustainability and success of livestock production systems.

Livestock Production Enterprises, as referred to throughout this book, are agricultural businesses focused on the breeding, raising, and management of animals to produce various products and services. These enterprises play a crucial role in the agricultural industry by providing essential goods such as meat, milk, eggs, wool, and leather. Beyond product generation, livestock enterprises also contribute to land management, cultural practices, and the economies of many rural areas.

These enterprises manage a wide range of livestock, including cattle for beef and dairy, sheep for wool and meat, goats, pigs, and poultry such as chickens and turkeys. Some operations even specialize in raising animals like horses, rabbits, and alpacas. The type of livestock raised typically depends on market demand, climate, and available resources.

Figure 1: Example of a Livestock Production Enterprise. CCO, via Flickr.

Livestock production is often categorized based on its primary focus. Meat production is a significant part of the industry, with livestock like cattle, pigs, sheep, and poultry raised for human consumption. Dairy production involves cows, goats, and sheep, which are used to produce milk and dairy products like cheese, butter, and yogurt. Egg production primarily comes from poultry, especially chickens. Fibre production focuses on animals like sheep, goats, and alpacas, providing wool and other fibres for textiles. Some enterprises also specialize in breeding livestock, providing animals for other farms or specific markets.

Farm management in these enterprises requires meticulous planning and the management of resources such as feed, water, shelter, veterinary care, and animal handling practices. Animal welfare and health are critical factors in maintaining high-quality and safe products, ensuring the well-being of livestock.

Livestock production enterprises have a substantial environmental and economic impact. They contribute significantly to global economies, especially in rural areas, by creating jobs, ensuring food security, and producing goods for international trade. However, managing their environmental footprint, including land use, waste management, and greenhouse gas emissions, is an ongoing challenge that these enterprises must address.

Many livestock production enterprises are also adopting new technologies and innovations to improve productivity. These include precision feeding systems, genetic selection for breeding, health monitoring, and automation for managing large-scale operations. These advancements help increase efficiency and maintain sustainability in livestock farming.

Livestock production enterprises are a vital component of the agricultural sector. They contribute to global food systems and economies while being governed by standards that promote sustainable, humane, and efficient production practices.

Poor livestock welfare is a significant concern in animal husbandry, often arising from inadequate care, poor living conditions, and inhumane treatment. Various factors contribute to this issue, including overcrowding, malnutrition, lack of veterinary care, inappropriate shelter, inhumane handling, lack of cleanliness, failure to provide enrichment, improper use of growth hormones and antibiotics, neglect of pregnant or nursing animals, and excessive isolation. Each of these factors can severely impact the physical and psychological well-being of livestock.

Overcrowding is a prevalent issue in livestock management, where animals are kept in cramped conditions that restrict their movement and natural behaviours. Such environments can lead to increased stress, aggression, injuries, and the rapid spread of diseases among animals. Research indicates that inadequate space can exacerbate these problems, leading to significant welfare concerns [10, 11]. The lack of proper space not only affects the physical health of the animals but also their mental well-being, as they are unable to engage in natural behaviours such as grazing and socializing [11].

Malnutrition or inadequate feeding is another critical aspect of poor livestock welfare. Providing insufficient or low-quality feed can lead to malnutrition, which compromises the immune system, stunts growth, and reduces overall productivity [11]. Additionally, failure to provide adequate water exacerbates these issues, leading to further health complications [11]. Studies have shown that livestock performance is significantly affected by the quality and availability of feed, highlighting the importance of proper nutrition in maintaining animal welfare [11].

The lack of veterinary care is a significant contributor to poor livestock welfare. In many regions, inadequate access to veterinary services results in untreated injuries and diseases, leading to prolonged suffering and decreased productivity [12, 13]. The absence of regular health check-ups and vaccinations can have dire consequences for livestock health, as diseases can spread rapidly in populations that lack proper medical attention

[12]. Furthermore, the reliance on informal veterinary services often leads to substandard care, further compromising animal welfare [13].

Inappropriate shelter exposes livestock to extreme weather conditions, which can lead to serious health issues such as heat stress, frostbite, and hypothermia [10, 11]. Animals require adequate shelter to protect them from harsh environmental factors, and failure to provide such protection can lead to significant welfare challenges [11]. Similarly, inhumane handling and transport practices contribute to poor welfare outcomes. Rough handling during loading and transportation, along with long transport times without breaks for food and water, can cause significant distress and suffering in animals [14].

Lack of cleanliness and hygiene in livestock facilities is another critical factor affecting animal welfare. Unsanitary conditions can lead to infections, respiratory issues, and the spread of parasites and diseases [11]. Maintaining a clean environment is essential for preventing health problems and ensuring the overall well-being of livestock [11]. Moreover, the failure to provide enrichment opportunities denies animals the chance to engage in natural behaviours, leading to boredom and stress, which can manifest in abnormal behaviours such as pacing or self-mutilation [11].

Improper use of growth hormones and antibiotics poses additional risks to livestock welfare. The overuse of these substances can lead to health complications and contribute to antibiotic resistance, which is a growing concern in both veterinary and human medicine [15]. Furthermore, neglecting the specific needs of pregnant or nursing animals can result in complications for both the mother and offspring, further highlighting the importance of attentive care in livestock management [11].

Excessive isolation or lack of socialization can lead to significant psychological distress in livestock, as these animals are inherently social beings [11]. Keeping them isolated for extended periods can result in anxiety and behavioural issues, underscoring the need for social interaction in maintaining their welfare [11].

Addressing the various factors contributing to poor livestock welfare is essential for ensuring the health and productivity of these animals. Comprehensive strategies that encompass adequate space, nutrition, veterinary care, shelter, humane handling, hygiene, enrichment, responsible use of medications, and socialization are crucial for maintaining high welfare standards in livestock management.

Poor Animal Welfare and Production Losses

When animal welfare is compromised, livestock producers often see a decline in production. Research has shown that animals suffering from poor physical or mental health exhibit lower rates of growth, reproduction, and overall productivity. For example, livestock housed in inadequate environments may experience stress, poor nutrition, and discomfort, forcing their bodies to divert energy from growth and reproduction to simply maintaining internal balance. As a result, poor welfare conditions directly impact profitability in the livestock industry [16].

Producers may notice that animals experiencing welfare issues show changes in behaviour, such as increased aggression or lethargy, or a drop in production levels. While many producers are aware of the importance of providing adequate housing, food, water, and social contact, they may not always recognize the impact that poor handling practices have on animal welfare. Studies have shown that poor handling can lead to livestock developing a high fear of humans, which raises stress levels, further diminishing growth, production, and reproductive success. In pigs, for instance, research conducted in both Australia and the United States showed that high fear levels can reduce growth rates by 6-7% and reproduction rates by 4-7% [17].

CHAPTER TWO

Animal Welfare Regulations and Codes of Practice

Knowing the Animal Welfare Regulations and Codes of Practice in your jurisdiction is important for several reasons, especially for those involved in livestock production, animal care, and related industries. Compliance with these regulations ensures that operations are conducted legally, ethically, and sustainably, benefiting both the animals and the businesses involved.

First and foremost, legal compliance is a critical aspect of adhering to animal welfare regulations. These laws are legally binding, and failure to comply can lead to serious consequences, including fines, penalties, and even the closure of operations. Understanding the legal framework helps individuals and businesses ensure they are operating within the boundaries of the law, avoiding potential legal issues and maintaining their business operations smoothly.

Beyond legal obligations, there is an ethical responsibility to ensure animals are treated humanely. By understanding welfare regulations, businesses can ensure that animals are cared for in a way that prevents unnecessary suffering. Following codes of practice promotes good animal husbandry, aligning with the moral duty to support the physical and psychological well-being of animals under their care.

Another key reason for understanding animal welfare regulations is the importance of public and consumer trust. Consumers are increasingly concerned about how animals are treated in the production of food and other products. Adhering to welfare standards demonstrates a commitment to humane practices, which builds trust with the public, enhances brand reputation, and improves marketability. This trust can be a significant differentiator in competitive markets.

From a productivity perspective, animal health and productivity are closely linked to good welfare practices. Healthier animals are more productive, and poor welfare conditions can lead to stress, injuries, and diseases, reducing both productivity and profitability. Knowing and following welfare regulations helps producers maintain the well-being of their livestock, resulting in higher-quality yields and reduced veterinary costs.

In terms of risk management, non-compliance with animal welfare laws can expose businesses to risks such as legal action, loss of licenses, or consumer boycotts. Adhering to regulations minimizes these risks, ensures safer operations, and reduces liabilities, thereby protecting the business from potential threats.

Additionally, some codes of practice also address environmental concerns. This enhances environmental and community relations by ensuring that issues such as waste management and land use are properly managed. Compliance with welfare and environmental standards helps maintain positive relationships with local communities and regulatory authorities, preventing conflicts over animal treatment or environmental impact.

Knowledge of welfare regulations is also essential for market access and certifications. Many markets, particularly international ones, require proof of adherence to strict welfare standards. Being well-versed in these regulations can open up new market opportunities and increase product value, especially for certifications like organic or animal-friendly labels.

Finally, understanding animal welfare regulations supports improved staff training and management. By creating clear guidelines and training programs, businesses can ensure that staff involved in animal care and handling follow the best practices. This reduces the likelihood of welfare breaches and ensures proper care is given to animals at all stages of the production process.

Knowing and adhering to animal welfare regulations and codes of practice is essential not only for legal and ethical reasons but also for improving productivity, gaining consumer trust, managing risks, and accessing valuable markets. These practices form the foundation of a sustainable and responsible livestock operation.

Animal Welfare Regulations and Codes of Practice vary around the world, reflecting differing cultural, economic, and environmental contexts. However, many countries and regions share a common commitment to ensuring that animals are treated humanely, with laws and guidelines that aim to protect their well-being.

Specific Regions

Below is an overview of how different regions approach animal welfare regulations and codes of practice.

1. European Union (EU)

The **European Union** has some of the strictest and most comprehensive animal welfare laws globally, applying both to farm animals and animals used for research or as pets.

- **Regulations**: The EU has enacted several directives, including:

  - **Council Directive 98/58/EC**: Outlines general animal welfare standards for animals kept for farming purposes, focusing on proper care, housing, feeding, and health management.

  - **Council Directive 2007/43/EC**: Specifically focuses on broiler chickens, establishing minimum welfare standards such as stocking density and lighting conditions.

  - **Regulation (EC) No 1099/2009**: Regulates animal welfare at the time of slaughter, ensuring humane treatment during slaughtering procedures.

  - **Regulation (EC) No 1/2005**: Governs the welfare of animals during transportation, emphasizing rest periods, feeding, and water access.

- **Codes of Practice**: Each EU country may also implement national codes of practice that offer guidance for industries on how to meet legal welfare standards. These can include guidelines on handling, housing, and humane euthanasia.

The European Union (EU) has established harmonized rules that cover a wide range of animal species and welfare-related issues, reflecting the EU's strong commitment to animal welfare. These rules are in place to ensure that all animals, particularly farm animals, are protected under uniform standards across member states, addressing various aspects such as housing, feeding, transport, and slaughter. These regulations aim to prevent unnecessary suffering and promote humane treatment [18].

A central piece of legislation is Council Directive 98/58/EC, which lays down the minimum standards for the protection of all farmed animals. This directive provides general guidelines for ensuring the well-being of animals across all species raised for farming purposes. It covers essential welfare aspects like the provision of adequate food and water, sufficient space, and appropriate living conditions. The directive is intended to ensure that animals are free from hunger, thirst, discomfort, and pain, while also enabling them to express natural behaviours [18].

In addition to this broad directive, the EU has enacted specific directives aimed at the protection of individual animal species. These targeted laws provide more detailed welfare requirements for animals like pigs, poultry, and cattle, ensuring that each species' particular needs are addressed. For example, certain species may require more specific housing or handling practices to meet welfare standards effectively [18].

The EU also sets comprehensive welfare standards for the transport and slaughter of animals. Transport regulations aim to minimize stress and suffering during the movement of animals, ensuring that they have access to water, rest, and are transported in suitable vehicles. Similarly, welfare standards for stunning and slaughter ensure that animals are handled and killed in a manner that minimizes pain and distress, reinforcing the EU's commitment to humane practices throughout an animal's life [18].

Legislation concerning wild animals and animals used for scientific purposes falls under the responsibility of the Directorate-General for Environment (DG Environment). This division oversees the implementation of regulations that govern the welfare of animals in non-farmed contexts, such as those in research or wildlife conservation, ensuring that their treatment aligns with environmental and ethical standards [18].

On 20 May 2020, the European Commission announced its intention to revise the existing EU animal welfare legislation to bring it into alignment with the latest scientific evidence. This decision recognizes the evolving understanding of animal welfare and the need for regulations that reflect current knowledge. The proposed updates aim to enhance protections for animals across various sectors, including farming, transport, and scientific research [18].

As part of this process, on 7 December 2023, the Commission adopted a proposal for the revision of the Regulation on the protection of animals during transport. This revision seeks to improve the welfare conditions during transportation, addressing concerns such as overcrowding, insufficient ventilation, and long transport times, which have been shown to cause significant stress to animals. Additionally, the Commission adopted a proposal for a new regulation on the welfare of dogs and cats, focusing on their traceability. This new regulation aims to combat illegal breeding and trading, ensuring that domestic animals are raised and sold under humane conditions [18].

The work is ongoing for other planned legislative proposals related to animal welfare, as the EU continues to refine and strengthen its regulations. These efforts highlight the EU's dedication to advancing animal welfare, ensuring that the latest scientific insights and ethical considerations are incorporated into law to improve the lives of animals across the Union [18].

2. United States

In the **United States**, animal welfare laws are a mix of federal, state, and local regulations, with some standards set by industry-specific codes.

- **Federal Regulations**:

  - **Animal Welfare Act (AWA)**: Enacted in 1966, it regulates the treatment of animals in research, exhibition, transport, and by dealers. However, the AWA does not cover farm animals raised for food production.

  - **Humane Slaughter Act**: Requires that animals be rendered insensible to pain before being slaughtered.

  - **Food Safety and Inspection Service (FSIS)**: Oversees humane handling of livestock before and during slaughter, particularly under USDA regulations.

- **Industry Codes of Practice**: Although farm animal welfare is less regulated at the federal level, the agricultural industry often adopts **voluntary codes of practice**. For example, the **National Chicken Council** and **National Pork Board** provide welfare guidelines for producers.

- **State Regulations**: Some U.S. states, such as California, have passed specific animal welfare laws, like Proposition 12, which requires more space for farm animals like chickens and pigs.

In the United States, animal protection laws are created and enforced at every level of government, making the legal landscape complex and multifaceted. The majority of animal protection legislation is enacted at the state level, but there are also federal laws and local ordinances passed by cities and counties. This layered system makes it crucial for animal advocates to engage with lawmakers across all levels of government, as each level has the power to create meaningful change in animal welfare [19].

Federal Animal Protection Laws: While most animal protection laws are at the state level, the U.S. does have a few important federal laws that provide some level of protection to animals [19].

1. **The Animal Welfare Act (AWA)**, signed into law in 1966, is the primary federal law governing animal welfare. It regulates the treatment of animals in research labs, zoos, and those bred commercially for sale, such as those in puppy mills. The AWA directs the Secretary of the Department of Agriculture to set minimum standards for the handling, care, treatment, and transportation of these animals. It also prohibits dogfighting and cockfighting, provided these activities cross state lines. However, the AWA has been criticized for allowing some inhumane practices to continue unchecked due to loopholes and inconsistent enforcement.

2. The **28 Hour Law**, enacted in 1873, mandates that animals being transported to slaughter must be given a break every 28 hours for exercise, food, and water. However, birds such as chickens and turkeys, which are the most-farmed animals in the U.S., are exempt from this law, limiting its impact on large-scale agricultural operations.

3. The **Humane Slaughter Act**, passed in 1958 and amended in 1978, requires that livestock be rendered unconscious before slaughter to minimize pain. Unfortunately, like the 28 Hour Law, this act excludes birds, meaning chickens and turkeys are not afforded the same protection. Moreover, enforcement has been inconsistent, as noted by government inspectors.

4. The **Endangered Species Act (ESA)**, enacted in 1973, protects species classified as threatened or endangered in the U.S. and internationally. The ESA is administered by the U.S. Fish and Wildlife Service and the National Oceanic and Atmospheric Administration Fisheries Service. The law outlines procedures for protecting these species and imposes penalties for violations, but it focuses mainly on wildlife conservation rather than animal welfare per se.

5. The **Preventing Animal Cruelty and Torture (PACT) Act**, signed into

law in 2019, made it a federal crime to commit extreme forms of animal cruelty—such as crushing, burning, drowning, or impaling—in cases that affect interstate commerce or occur within U.S. territories. The PACT Act builds upon the 2010 **Animal Crush Video Prohibition Act**, which banned the production and distribution of videos showing small animals being tortured or killed for entertainment.

6. The **Lacey Act**, enacted in 1900, prohibits wildlife trafficking and the sale of plants and animals taken illegally. It was the first federal law aimed at protecting wild animals and also penalizes those who falsify documents about the transport and sale of wildlife.

7. The **Swine Health Protection Act (SHPA)**, passed in 1980, focuses on public health by regulating the feeding of "garbage" to pigs, as garbage may contain harmful pathogens. States can implement their own programs under SHPA, provided they meet federal guidelines.

State and Local Animal Protection Laws: The vast majority of animal protection laws are created and enforced at the state level. Each state has its own laws and regulations that define what constitutes animal cruelty, with penalties that vary depending on the severity of the offense. Every state now has felony animal cruelty laws, although the specifics differ widely [19].

Most state laws offer the strongest protections to companion animals, typically dogs and cats, though some also extend these protections to birds, horses, and other animals. Many states have passed laws regulating the care and treatment of companion animals, such as mandating how long shelters must hold stray animals before adoption or euthanasia. Laws requiring vaccinations, such as for rabies, and measures like "hot car laws," which make it illegal to leave pets in dangerously hot vehicles, are also common. Some states also have anti-tethering laws that restrict how long pets can be tied up outside, especially in extreme weather conditions. In recent years, states like California and Maryland have passed retail pet sale bans, which prohibit pet stores from selling animals from commercial breeders, requiring them to source pets from shelters and rescue groups instead [19].

Wildlife protections also exist at the state level, including laws governing hunting and fishing seasons and the humane treatment of animals used in entertainment. Some states have passed bans on the use of wild animals in performances, such as Illinois and New York, which were the first to ban the use of elephants in circuses [19].

Farmed animals, however, often receive less protection under state laws, with many farm practices exempt from general animal cruelty statutes. Some states have started to address this by enacting laws against intensive confinement in agriculture, which refers to practices like keeping hens in battery cages or pigs in gestation crates, where they are unable to move freely. States like California have led the way in passing laws to phase out these practices [19].

Local laws at the city and county level also play a critical role in animal protection. For example, many cities and counties have passed their own versions of retail pet sale bans, anti-tethering laws, and bans on wild animal performances. These local ordinances often pave the way for broader state or even federal regulations, as they reflect growing public concern for animal welfare [19].

In summary, animal protection laws in the United States exist at multiple levels of government, from federal laws like the Animal Welfare Act and the Endangered Species Act to state and local ordinances that address specific concerns within communities. While there are significant protections for some animals, particularly companion animals, farmed animals and those used in laboratories often receive far fewer protections, and enforcement of existing laws is sometimes inconsistent. Advocating for stronger and more consistent animal protection laws at all levels of government is critical to improving the welfare of animals in the U.S.

3. Australia

Australia is known for its detailed animal welfare standards, particularly regarding livestock production and transport.

- **Regulations**:

    - **Australian Animal Welfare Standards and Guidelines**: Developed in consultation with stakeholders, these standards apply to the welfare of livestock across different stages of production. The standards are legally enforceable in most states and cover issues such as proper shelter, nutrition, handling, and humane slaughter.

    - **Prevention of Cruelty to Animals Acts**: Enacted in each state or territory, these laws prohibit cruelty to all animals, ensuring their humane treatment in various industries.

- **Codes of Practice**: Australia has specific **Model Codes of Practice for the Welfare of Animals**, which serve as guidelines for specific species (cattle, sheep, pigs, etc.) and production systems. These codes detail acceptable management practices, housing, feeding, and healthcare.

While there are no national laws governing animal welfare in Australia, each state and territory has its own legislation that regulates the treatment of animals. These laws set out the obligations of animal owners and handlers to prevent cruelty and ensure humane care. The legislation includes [16]:

- Animal Welfare Act 1992 (ACT)

- Prevention of Cruelty to Animals Act 1979 (NSW)

- Animal Welfare Act (NT)

- Animal Care and Protection Act 2001 (QLD)

- Animal Welfare Act 1985 (SA)

- Animal Welfare Act 1993 (TAS)

- Prevention of Cruelty to Animals Act 1986 (VIC)

- Animal Welfare Act 2002 (WA)

These acts are enforced by respective government departments, such as Agriculture Victoria in Victoria or the Department of Primary Industries and Regional Development in Western Australia. Each state's legislation aims to prevent cruelty to animals and ensure that their welfare needs are met through proper care, nutrition, housing, and medical attention.

At the local government level, regulations focus on the management of companion animals, such as pet ownership and public health concerns. Local governments also play a role in promoting responsible pet ownership, ensuring that companion animals do not cause harm to people, other animals, or the environment.

The Australian Government has specific responsibilities related to international trade and agreements that impact animal welfare. This includes overseeing the welfare of animals involved in the live animal export trade and animals processed at export-registered slaughter facilities. The government also regulates the commercial killing of kangaroos through State Management Plans approved by the Department of Environment and Energy. Additionally, the Australian Government manages introduced animal populations under National Threat Abatement Plans, monitors animal research conducted on federal lands, and provides input into international animal welfare negotiations and treaties, such as the Convention on International Trade in Endangered Species of Wild Fauna and Flora (CITES).

Australia's three tiers of government – federal, state/territory, and local – share responsibilities for animal welfare. The state and territory governments take the lead, preparing and enforcing the majority of animal welfare laws, but all levels collaborate to ensure consistency in legislation, scientific research, and enforcement across the country. The Australian Government plays a significant role in international negotiations, trade, and treaties that have implications for animal welfare.

Animal Welfare Responsibilities

Everyone who interacts with animals in Australia has a responsibility to protect their welfare, ensuring they are adequately cared for. This duty applies to individual owners, animal industry groups, and community animal welfare organizations, each with specific roles in promoting animal welfare practices.

- Individual owners are expected to fulfill their duty of care, ensuring that animals in their charge are treated humanely. This includes following best practices from relevant legislation, codes of practice, and guidelines, and preventing their animals from harming others.

- Animal industry groups must represent their members' interests in animal welfare discussions, provide factual information, promote best practices, and participate in developing policies and codes of practice. They are also responsible for sponsoring or conducting research that advances the knowledge and application of animal welfare standards.

- Community animal welfare groups advocate for animal welfare, raise public awareness, and work with the government to develop and enforce animal welfare legislation. These groups also play a key role in promoting best practices and advancing scientific understanding of animal care.

Government's Role

The state and territory governments are responsible for the creation, enforcement, and promotion of animal welfare laws within their respective regions. They also collaborate with the Australian Government to work towards a consistent approach to legislation across the country. The Australian Government oversees animal welfare in relation to trade, international agreements, and certain national issues such as wildlife management and animal research.

Governments at all levels are tasked with:

- Developing scientifically backed policies and legislation to improve animal welfare.

- Promoting consistent animal welfare standards.

- Enforcing regulations and encouraging best practices across different industries.

- Consulting with stakeholders, including industry, animal welfare groups, and the public, to refine and develop policies that ensure the humane treatment of animals.

The Australian Animal Welfare Standards and Guidelines for Cattle

The Australian Animal Welfare Standards and Guidelines for Cattle were agreed upon by state and territory governments in 2016. These standards aim to provide clear and consistent guidelines for cattle welfare, ensuring humane treatment across different states. However, the timeline and method of implementation vary between states [20].

- In South Australia, the standards became legally enforceable on 15th April 2017.

- The Northern Territory has yet to implement the standards, but compliance will be overseen by Veterinary and Livestock Biosecurity Officers once they are adopted.

- In New South Wales, the standards are prescribed under the Prevention of Cruelty to Animals Act 1979, meaning they serve as guidelines that can be used as evidence in legal proceedings. These came into effect on 15th December 2017.

- Queensland supports the adoption of these standards but has not yet finalized their implementation.

- Western Australia has not regulated these standards yet but plans to do so after amendments to the Animal Welfare Act 2002 are completed.

- Victoria intends to implement the standards through a new Animal Welfare Act, currently under development.

- Tasmania is in the process of legislating parts of the standards, and it is expected that they will become law within 12 months.

- The Australian Capital Territory is reviewing the standards and is expected to implement them within 18 months as a code of practice.

Australia's approach to animal welfare is complex, involving multiple levels of government and various stakeholders. State and territory governments hold the primary responsibility for legislation and enforcement, while the Australian Government focuses on areas related to trade, international agreements, and national wildlife management. Animal welfare in Australia is supported by a comprehensive framework of laws, guidelines, and codes of practice that ensure the humane treatment of animals across a variety of industries and settings. Ongoing efforts to align standards across the country, particularly in livestock industries, highlight the commitment to improving and maintaining high welfare standards.

4. Canada

In **Canada**, animal welfare laws are primarily provincial, with federal regulations focused on specific areas such as transport and slaughter.

- **Regulations**:

    - **Health of Animals Act**: Federally regulates the humane treatment of animals during transport and at slaughter.

    - **Provincial Regulations**: Each province has its own laws governing animal welfare. For example, Ontario's **Provincial Animal Welfare Services (PAWS) Act** sets out standards for the care of all animals.

- **Codes of Practice**: Canada uses **National Farm Animal Care Council (NFACC) Codes of Practice**, which provide guidelines for the care and handling of farm animals. These codes cover housing, nutrition, health management, and transportation and are often referenced in provincial regulations.

In Canada, the responsibility for animal welfare primarily falls on the provinces and territories. Each region has its own legislation to ensure the protection and humane treatment of both farm and companion animals. These laws are typically broad in scope, covering various aspects of animal welfare, although some provinces have specific regulations addressing particular species or issues [21].

The Canadian Food Inspection Agency (CFIA) has a more limited role, primarily focusing on ensuring the humane transport of animals and the humane treatment of food animals in federally registered abattoirs. While animal care on farms is the responsibility of the producers and provincial or territorial authorities, the CFIA works closely with stakeholders to set standards of care and biosecurity. It also verifies that humane transport and slaughter requirements are respected in all federal slaughter facilities. In addition, the Criminal Code of Canada prohibits wilful acts of animal cruelty, including causing unnecessary suffering through neglect, pain, or injury. This law is enforced by police, provincial SPCA organizations, and ministries of agriculture [21].

Alberta

In Alberta, the Animal Protection Act and related regulations ensure the welfare of animals during all activities, excluding generally accepted practices like animal management or slaughter. Peace Officers, along with the RCMP and municipal police, enforce these laws. The Meat Inspection Act ensures that animals are slaughtered humanely when their meat is intended for human consumption. The Livestock Industry Diversification Act sets standards for farmed cervids and velvet antler removal, with oversight from the Ministry of Agriculture and Irrigation.

British Columbia

In British Columbia (BC), the Animal Care Codes of Practice Regulation sets the National Farm Animal Care Council (NFACC) Codes of Practice as the standard for farm animal welfare. The Prevention of Cruelty to Animals Act protects animals from distress during activities like farming. The Milk Industry Act and Food Safety Act ensure humane handling of dairy animals and farm animals during slaughter, respectively. Enforcement is carried out by the Ministry of Agriculture and Food, police, and the BC SPCA.

Manitoba

Manitoba's Animal Care Act governs the welfare of farm animals, including during transportation, and is enforced by the Office of the Chief Veterinarian and special animal protection officers. The Act covers most farm animals, excluding generally accepted practices in husbandry and management.

New Brunswick

In New Brunswick, the Society for the Prevention of Cruelty to Animals Act provides protection for farm animals, with enforcement by the provincial SPCA and police. The Poultry Health Protection Act ensures the welfare of chicks during housing, with oversight from inspectors appointed by the Minister of Agriculture.

Newfoundland and Labrador

In Newfoundland and Labrador, the Animal Health and Protection Act and related regulations adopt codes of practice to ensure the welfare of domestic animals, including farm animals. These laws are enforced by appointed inspectors and police officers. The Meat Inspection Act governs the humane treatment of animals during slaughter, enforced by the Department of Government Services.

Nova Scotia

Nova Scotia enforces animal welfare through the Animal Health and Protection Act, which regulates disease control, quarantine, and vaccination. The Animal Protection Act safeguards animals during activities like slaughter, and the Meat Inspection Act ensures humane treatment during slaughter. Inspectors from the Ministry of Agriculture enforce these laws.

Northwest Territories and Nunavut

In both the Northwest Territories and Nunavut, the Herd and Fencing Act protects farm animals during housing, with enforcement by officers appointed by the Commissioner.

Ontario

Ontario's Provincial Animal Welfare Services Act establishes inspection powers and mandates standards of care for farm animals. The Bill 156, Security from Trespass and Protecting Food Safety Act and the Animals for Research Act regulate the treatment of farm animals in research and testing facilities. Various acts such as the Food Safety and Quality Act and the Livestock Community Sales Act cover the humane transport, slaughter, and sale of farm animals. Enforcement is carried out by inspectors appointed by the Ministry of Agriculture, Food, and Rural Affairs.

Prince Edward Island

In Prince Edward Island, the Animal Welfare Act protects farm animals from distress during any activity, with enforcement by inspectors appointed by the Ministry of Agriculture.

Québec

In Québec, the Animal Welfare and Safety Act ensures the protection and welfare of farm and companion animals, including standards for euthanasia and slaughter. The Food Products Act regulates the humane treatment of animals during transportation and slaughter. Enforcement is carried out by inspectors and veterinarians appointed by the Québec Ministry of Agriculture, Fisheries, and Food.

Saskatchewan

Saskatchewan's Animal Protection Act protects all animals from distress and establishes duties for animal care. The Animal Health Act focuses on disease prevention and biosecurity. Other laws, like the Stray Animals Act, deal with animal impoundment and care. These acts are enforced by various agencies, including animal protection officers and local governments.

Yukon

In the Yukon, the Animal Protection and Control Act mandates that all animal owners provide adequate care and control, referencing national codes of practice for most livestock species. The Act is enforced by officers from the Yukon Government and the RCMP.

Across Canada, animal welfare is regulated through a combination of provincial and territorial legislation, each addressing specific issues related to farm and companion animals. The Canadian Food Inspection Agency (CFIA) plays a crucial role in ensuring humane transport and slaughter, while the Criminal Code of Canada prohibits animal cruelty at a federal level. These laws and regulations ensure that animals are protected from unnecessary suffering, with enforcement carried out by various government agencies, police, and animal protection organizations [21].

5. United Kingdom

The **UK** has a long-standing commitment to animal welfare, with stringent laws that apply to both pets and farm animals.

- **Regulations**:

  - **Animal Welfare Act 2006**: This is the core piece of legislation for protecting animals in England and Wales. It mandates that owners meet the needs of their animals, including providing proper nutrition, environment, and medical care.

  - **Welfare of Farmed Animals (England) Regulations 2007**: This outlines the standards for farm animal welfare, including housing, feeding, handling, and disease prevention.

  - **Welfare of Animals at Slaughter or Killing (WASK) Regulations 1995**: These regulations ensure that animals are slaughtered humanely.

- **Codes of Practice**: The UK government has also published **codes of practice** for various species, providing guidance for farmers on meeting the legal requirements of welfare standards.

The Animal Welfare Act 2006 is the cornerstone of animal welfare legislation in the UK, providing protection for all vertebrate animals. It establishes a framework under which animal owners and keepers have a duty of care to ensure the well-being of animals under their responsibility. The Act outlines specific needs that must be met, including providing a suitable environment, proper diet, and opportunities for animals to exhibit normal behaviour. It also specifies requirements for animals to be housed with or apart from other animals, depending on their needs, and mandates protection from pain, injury, suffering, and disease [22].

The Act strictly prohibits any form of animal cruelty, which includes actions that cause unnecessary suffering, mutilation, or poisoning of animals. Failure to comply with these regulations can result in severe penalties, including being banned from owning animals, facing unlimited fines, or even imprisonment for up to five years. This underscores the serious legal and ethical responsibilities placed on those who own or care for animals [22].

Specific Regulations Under the Act

Several regulations supplement the Animal Welfare Act 2006, including the Welfare of Farmed Animals (England) Regulations 2007, which set minimum welfare standards for all farm animals. These regulations outline basic conditions under which farm animals must be kept, with species-specific standards provided for different types of animals. For instance, Schedules 2 to 9 of these regulations give detailed requirements for the care of specific species, such as pigs, cattle, poultry, and sheep. These standards ensure that farm animals are provided with adequate space, nutrition, and health care to maintain good welfare [22].

Additionally, the Mutilations (Permitted Procedures) (England) Regulations 2007 make it an offense to carry out procedures that interfere with the sensitive tissues or bone structure of an animal, except in specific circumstances where exemptions are allowed. These exemptions apply to procedures such as ear tagging for identification or castration for reproduction control, provided certain conditions are met. This regulation aims to prevent unnecessary harm while allowing for essential procedures that support animal management and welfare [22].

On-Farm Animal Welfare Codes of Practice

In addition to legislation, farmers and animal keepers must also adhere to species-specific codes of practice. These welfare codes provide detailed guidance on how to meet the requirements of the legislation. While they are not legally binding, the codes play a crucial role in promoting good standards of stockmanship and animal care. They outline best practices for caring for farm animals, ensuring that their welfare needs are met in accordance with both ethical standards and legal requirements [22].

Although these codes of practice are not law, failure to follow them can be used as evidence in court to demonstrate breaches of the Animal Welfare Act or other relevant legislation. This makes it essential for anyone working with animals to familiarize themselves with the codes and incorporate them into their daily management practices. The codes cover a wide range of species, including laying hens, broiler chickens, ducks, turkeys, cattle, pigs, sheep, goats, deer, rabbits, and gamebirds reared for sport [22].

6. New Zealand

New Zealand has some of the most progressive animal welfare laws, focusing on treating animals as sentient beings with specific needs.

- **Regulations**:

  - **Animal Welfare Act 1999**: This is the primary piece of legislation in New Zealand. It requires all those responsible for animals to meet their physical, health, and behavioural needs, ensuring protection from unnecessary harm.

  - **Animal Products Act 1999**: This act addresses the humane treatment of animals in the meat production industry.

- **Codes of Welfare**: New Zealand's **Codes of Welfare**, issued by the Ministry for Primary Industries, set out minimum standards and best practices for different species, covering areas like housing, health, and humane euthanasia.

In New Zealand, the management and care of farm animals are regulated by several key laws and codes of practice, all aimed at ensuring the well-being of animals. These regulations are designed to provide guidelines that not only protect the welfare of animals but also ensure that they are cared for in a way that meets their physical, behavioural, and health needs. New Zealand is recognized for its strong commitment to animal welfare, and the comprehensive legal framework reflects this [23].

The five essential laws and regulations governing farm animal care and management in New Zealand are: the Animal Welfare Act 1999, the Code of Welfare for Dairy Cattle, the Code of Welfare for Pigs, the Code of Welfare for Poultry, and the Code of Animal

Health and Welfare. Together, these legal instruments cover the wide spectrum of animal welfare concerns, from general principles to species-specific guidelines.

The Animal Welfare Act 1999

The Animal Welfare Act 1999 is the cornerstone of animal welfare legislation in New Zealand. This Act establishes the general principles of animal care and ensures that all animals are treated with respect and dignity. The Act defines the responsibilities of animal owners and handlers to meet the needs of their animals, ensuring they have proper food, water, shelter, and care. It also prohibits the cruel treatment of animals, including neglect, abuse, and unnecessary suffering [23].

A key feature of the Animal Welfare Act is the creation of the National Animal Welfare Advisory Committee (NAWAC), which advises the government on animal welfare matters, including the development of codes of practice. The Act's strong emphasis on preventing cruelty and ensuring that animals' welfare needs are met provides a foundation for the specific welfare codes that follow [23].

The Code of Welfare for Dairy Cattle

The Code of Welfare for Dairy Cattle is a specific code of practice that outlines the minimum standards of care for dairy cattle. This code covers every aspect of dairy cattle husbandry, from housing and feeding to health care and handling. It includes guidelines for milking practices, ensuring that animals are not overmilked or handled in ways that could cause stress or injury. The code also sets out recommendations for maintaining proper nutrition and hygiene, as well as ensuring that cows have access to clean water and comfortable shelter [23].

The Dairy Cattle Code ensures that dairy cows are managed in a way that promotes their health and welfare, taking into account their behavioural needs. By setting out these standards, the code helps dairy farmers maintain high welfare standards while optimizing productivity in the industry.

The Code of Welfare for Pigs

The Code of Welfare for Pigs provides the minimum standards of care required for the management of pigs. This code addresses various aspects of pig farming, including housing, nutrition, health management, and welfare practices. The welfare of pigs is a critical concern in intensive farming systems, and this code is designed to ensure that pigs are not subjected to conditions that could cause unnecessary stress or suffering.

The Pig Welfare Code includes guidelines for housing systems, such as ensuring that pens are large enough to allow pigs to move freely and express natural behaviours. It also covers health care, emphasizing the importance of regular health checks, proper disease management, and appropriate veterinary care. By adhering to this code, pig farmers can ensure they meet both the legal and ethical requirements for pig welfare [23].

The Code of Welfare for Poultry

The Code of Welfare for Poultry sets out the minimum standards for the care of poultry, including chickens, turkeys, and other birds raised for eggs or meat production. This code addresses the specific needs of poultry, from their housing and feeding arrangements to their health and overall welfare. It also includes standards for stocking density, ensuring that poultry have sufficient space to move around and exhibit natural behaviours.

The Poultry Welfare Code covers all aspects of husbandry, including ventilation, lighting, and hygiene in poultry houses, as well as health care and biosecurity measures to prevent disease outbreaks. This code ensures that poultry are raised in conditions that meet their physical and behavioural needs, contributing to both their welfare and productivity.

The Code of Animal Health and Welfare

The Code of Animal Health and Welfare is a broad code of practice that applies to all animals, regardless of species. It sets out minimum standards for animal husbandry, including housing, feeding, health care, and overall welfare. This code is designed to ensure that animals are managed in a way that promotes their health, prevents suffering, and allows them to live in a suitable environment.

The Animal Health and Welfare Code applies to a wide range of farm animals, ensuring that they are provided with the care necessary to keep them healthy and productive. It includes guidelines for disease prevention, proper feeding practices, and maintaining suitable living conditions for all farm animals. By following this code, farmers and animal handlers can ensure they meet the overall welfare requirements for the animals in their care [23].

7. Brazil

In **Brazil**, animal welfare is increasingly gaining attention due to the country's large livestock industry.

- **Regulations:**

  - **Federal Law 9,605/98** (Environmental Crimes Law): Although not specific to animals, it prohibits mistreatment of animals and provides penalties for violations.

- ○ **Normative Instructions**: Brazil has issued various normative instructions focused on animal welfare, especially concerning transport and slaughter.

- **Codes of Practice**: The Brazilian government has developed **guidelines** in collaboration with industry organizations for cattle, poultry, and other livestock, aiming to improve the welfare of animals in production systems.

Brazil's approach to animal welfare has a long history, beginning with the 1924 decree that prohibited any behaviour or activity causing suffering to animals. This early law laid the foundation for Brazil's ongoing commitment to protecting animals from cruelty. It prohibited harmful recreational activities involving animals and established basic protections for their welfare. A more comprehensive decree followed in 1934, stipulating that animals must not be overworked or kept in environments where they cannot breathe, move, or rest. The decree also mandated that animals be provided with air, light, and a quick, humane death, whether or not they were destined for human consumption. It also prohibited the abandonment of sick, injured, or mutilated animals and required owners to provide necessary care, including veterinary assistance. This decree was notable for allowing lawyers from the Public Ministry or animal protection organizations to represent animals in court, a unique legal provision that granted animals a degree of legal standing not found in many countries, such as the United States [24, 25].

Brazil's 1988 constitution further strengthened animal protection by mandating that the government must actively protect animals from cruelty. This constitutional protection was invoked in 1997 in a landmark decision by the Brazilian Supreme Court to uphold a ban on the Farra do Boi, a festival in Santa Catarina where bulls and oxen were tortured and killed. This ruling was significant as it set a legal precedent, reinforcing that cultural practices cannot justify cruelty to animals, and affirming the constitutional mandate to protect animals from mistreatment [24, 25].

In 1998, Brazil passed a law that expanded upon the protections established by the 1934 decree by imposing harsher penalties for cruelty to domestic and wild animals. The law set a punishment of 3 months to a year of imprisonment and a fine, with the sentence increased by one-sixth to one-third if the animal died as a result of cruelty. A further update to the Penal Code in 2012 increased the penalties, raising imprisonment to 1 to 4 years, and up to 6 years if the animal was killed. These escalating penalties reflect Brazil's growing commitment to punishing acts of cruelty more severely and its intention to deter such behaviours [24, 25].

However, the 1998 law did not specifically address farm animals, which are a significant part of Brazil's agricultural economy. In 2000, the government introduced a Normative Instruction to regulate pre-slaughter handling and slaughter methods, ensuring that handling minimized stress and banning the use of aggressive instruments during slaughter. In 2008, additional instructions were issued, focusing on procedures for rearing and transporting animals and encouraging the adoption of Manuals of Good Practice, although these manuals are voluntary. That same year, the Ministry of Agriculture, Livestock, and Food Supply established the Permanent Technical Commission of Animal Welfare, which works to promote animal welfare through education, training, and the dissemination of technical materials related to livestock [24, 25].

One major shortfall in Brazilian animal welfare law is the exclusion of birds—which represent the majority of land animals slaughtered for food—from laws governing transport or export related to animal welfare. This gap highlights an area where further protections are needed [24, 25].

Brazil made significant strides in regulating the use of animals in scientific research with the passage of its first federal law on the subject in 2008. This law created a National Animal Control and Experimentation Committee and institutional Ethical Committees on Animal Use, requiring adherence to the principles of the Three Rs: replacement of animals with non-animal methods, reduction in the number of animals used, and refinement of techniques to minimize suffering. This law marked an important development in the ethical treatment of animals used for research purposes, aligning Brazil with global trends in reducing reliance on animal testing [24, 25].

In 2014, Brazil took another step forward by banning most animal testing for cosmetics, although exceptions remain for testing ingredients with unknown effects and for testing conducted in other countries, provided the products are sold in Brazil. While this was a positive move, the continued allowance of some forms of testing shows that more progress is needed to fully eliminate cosmetic testing on animals [24, 25].

Despite these advancements, Brazil's animal welfare system has room for improvement. In 2014, the country received a C rating on the World Animal Protection's Animal Protection Index, indicating that while progress had been made, Brazil still lags behind in certain areas, especially compared to countries with more stringent animal protection laws [24, 25].

In 2020, under President Jair Bolsonaro, Brazil further toughened its stance on animal cruelty with a law specifically targeting the mistreatment of dogs and cats. This law imposes harsher penalties, with those found guilty of mistreatment facing 2 to 5 years of imprisonment and a fine. Additionally, convicted individuals are banned from owning animals, marking a significant step toward addressing cruelty toward pets [24, 25].

8. South Africa

In **South Africa**, animal welfare is regulated through laws primarily aimed at preventing cruelty.

- **Regulations**:

  - **Animals Protection Act of 1962**: Prohibits the abuse or neglect of animals and provides penalties for violations.

  - **Meat Safety Act**: Ensures humane treatment of animals during slaughter.

- **Codes of Practice**: South Africa has developed **best practice guidelines** for different livestock sectors, often aligned with international welfare standards.

In South Africa, animal welfare is regulated through a series of laws and regulations primarily focused on preventing cruelty and ensuring the humane treatment of animals. The legislative framework covers various aspects of animal care, including protection from abuse and neglect, as well as standards for humane slaughter. These laws reflect South Africa's commitment to safeguarding animals, both domestic and farmed, from mistreatment, and they align with global trends toward ethical treatment of animals.

Animals Protection Act of 1962

The cornerstone of South Africa's animal welfare legislation is the Animals Protection Act of 1962. This law prohibits any form of abuse or neglect toward animals, making it a criminal offense to inflict unnecessary suffering, harm, or failure to provide proper care for animals. The Act applies to all animals, including farm animals, pets, and working animals. Under this law, those found guilty of animal cruelty can face penalties such as fines, imprisonment, or both, depending on the severity of the offense. The Act is critical in addressing intentional acts of cruelty as well as cases of neglect, where animals are deprived of basic necessities such as food, water, and shelter.

The Animals Protection Act serves as the primary legal instrument for preventing animal cruelty in South Africa and is widely enforced by various animal welfare organizations and law enforcement agencies. The law ensures that individuals who own or work with animals are held accountable for their treatment, promoting a legal obligation to care for and protect animals under human control.

The primary law, the Animal Protection Act of 1962, prohibits cruelty to animals and lays out specific offenses related to animal suffering. While the law provides for the protection of a wide range of animals, including domestic animals, birds, reptiles, and certain wild animals under human control, it does not explicitly recognize animals as sentient beings [26].

The Animal Protection Act is a key piece of legislation that prohibits abuse, neglect, and cruelty toward animals. The Act covers a broad spectrum of offenses, including overloading animals, keeping them in unsanitary or overcrowded conditions, denying them food or water, and failing to provide necessary veterinary care. It also criminalizes inflicting psychological suffering, such as infuriating or terrifying animals, which acknowledges the mental distress animals may experience [26].

However, the Act's limitations are notable. It does not extend to fish or wild animals in their natural environments, focusing mainly on animals under human control. Additionally, the law does not explicitly define animals as sentient beings, meaning their capacity to experience emotions and pain is not formally acknowledged in legal terms, although some recognition of sentience can be inferred through the Act's prohibitions on physical and psychological suffering [26].

The enforcement of the Animal Protection Act is carried out primarily by the Societies for the Prevention of Cruelty to Animals (SPCAs) and the police. The Minister of Justice has the overall responsibility for ensuring compliance with the law. Offenders found guilty of animal cruelty can face fines, imprisonment, confiscation of animals, or a ban on animal ownership. Additionally, organizing animal fighting can lead to more severe penalties, including imprisonment for up to two years [26].

However, there are concerns about the effectiveness of enforcement. It is unclear whether the government provides consistent financial support to the SPCAs for enforcing the Act, and there are indications that enforcement is under-resourced. Reports suggest that public authorities often struggle with limited capacity, a lack of training, and inadequate penalties for offenders. Moreover, while the Animals Matter Amendment Act No. 42 of 1993 allows the Minister of Justice to allocate funds to animal welfare organizations, this funding appears to be discretionary rather than a formal, budgeted contribution, raising concerns about the sustainability of enforcement efforts [26].

South Africa has developed several Codes of Conduct and Standards for different species, including dairy cattle, pigs, and poultry. These standards, produced by the South African Bureau of Standards (SABS) and the Livestock Welfare Coordinating Committee

(LWCC), are based on best practices for animal welfare and often align with international guidelines. However, these codes are largely voluntary and non-binding, meaning they rely on industry self-regulation rather than government enforcement [26].

While these codes offer detailed guidance on humane treatment, handling, transportation, and slaughter, they are not legally enforceable. This voluntary nature poses challenges in maintaining consistent animal welfare standards across the country. For instance, the public must purchase these codes from the SABS, which limits access and awareness among the broader public and stakeholders, further hindering widespread adoption of best practices.

Meat Safety Act

The Meat Safety Act is another significant piece of legislation that plays a key role in ensuring the welfare of animals, specifically during the process of slaughter. This Act establishes regulations to ensure that animals are treated humanely and experience minimal stress during pre-slaughter handling and the slaughtering process itself. The humane treatment of animals is especially important in the livestock industry, and the Meat Safety Act outlines protocols for handling, transporting, and slaughtering animals to minimize their suffering. These protocols include requirements for stunning animals before slaughter to prevent pain, as well as guidelines on the handling of animals in slaughterhouses.

The **Meat Safety Act** is critical for ensuring that animal welfare standards are maintained throughout the meat production process. The Act also contributes to food safety by regulating the conditions under which animals are slaughtered, thereby ensuring that meat products are safe for human consumption. By enforcing humane slaughter practices, South Africa aligns itself with international welfare standards, promoting ethical practices in the livestock sector.

Codes of Practice

In addition to legislative measures, South Africa has developed a range of Codes of Practice that provide best practice guidelines for different livestock sectors. These codes are not legally binding but are widely adopted by farmers, producers, and industry stakeholders as standards for ensuring high levels of animal welfare. The guidelines cover various aspects of animal care, including housing, feeding, transportation, and health management, with the aim of minimizing stress and promoting the well-being of livestock.

Many of these Codes of Practice are aligned with international welfare standards, ensuring that South Africa's animal welfare practices are consistent with global norms. By adopting these best practices, the livestock industry is able to improve animal health, reduce suffering, and enhance productivity. These codes also help businesses meet the expectations of consumers, who are increasingly concerned with the ethical treatment of animals in food production.

9. India

India has growing awareness around animal welfare, with strong legal protections, especially for animals considered sacred.

- **Regulations**:

    - **Prevention of Cruelty to Animals Act, 1960**: This is the main law that protects animals from abuse, cruelty, and neglect.

    - **Transport and Slaughter Laws**: India also has specific regulations regarding the humane treatment of animals during transport and slaughter.

- **Codes of Practice**: India is developing animal welfare codes, often informed by international standards, especially for cattle and poultry.

India, as one of the most bio-diverse regions in the world, is home to a vast array of species, including iconic animals such as the Bengal Tigers and the Great Indian Rhinoceros. Over recent years, animal protection and welfare have gained significant prominence within the country, with laws like the Prevention of Cruelty to Animals Act of 1960 and the Wildlife Protection Act of 1972 providing a legal framework for the protection of animals. Additionally, various states have enacted their own legislation, such as cattle protection and cow slaughter prohibition laws, to address region-specific animal welfare concerns. These legislative efforts demonstrate India's commitment to safeguarding its diverse wildlife and ensuring the welfare of domestic animals.

Animal welfare is enshrined in the Indian Constitution as a fundamental duty of every citizen. Article 48A, introduced by the 42nd Amendment in 1976, mandates that the State shall strive to protect and improve the environment, including safeguarding the country's forests and wildlife. While these constitutional provisions are not directly enforceable in courts, they lay the groundwork for various animal protection laws and policies at both the central and state levels. Moreover, courts have taken an expansive interpretation of these provisions, incorporating them into the fundamental right to life and liberty under Article 21, making them judicially enforceable. This development underscores the importance India places on the protection of animals, both in terms of legislation and constitutional duties.

In India, the primary sources of law include the Constitution, statutes, customary law, and case law. The country operates as a federal union, with 28 States and 8 Union Territories, each governed by its own legislative body. Parliament at the central level enacts laws for the entire country, while state legislatures pass laws for their respective states. These laws are supplemented by subordinate legislation, such as rules and regulations, enacted by both central and state governments. The common law system—inherited from British colonial rule—forms the backbone of the legal system, with significant reliance on judicial precedents. This system allows higher court decisions, such as those from the Supreme Court of India and High Courts, to shape legal interpretations and guide lower courts.

The Indian Penal Code (IPC) of 1860 also plays an important role in animal welfare. Sections 428 and 429 of the IPC provide penalties for acts of cruelty against animals, such as killing, poisoning, or maiming. These provisions complement specific animal welfare statutes, providing further legal protection to animals across the country.

The Prevention of Cruelty to Animals Act of 1960 is the cornerstone of India's animal cruelty laws. Its primary goal is to prevent the infliction of unnecessary pain or suffering on animals. The Act defines "animal" as any living creature other than a human being and outlines a wide range of offenses under Section 11. These offenses include beating, kicking, overloading, and abandoning animals, as well as keeping animals in cages or chains that restrict their movement. Additionally, the Act prohibits using animals as bait for other animals and organizing animal fights for entertainment.

To further enforce animal welfare, the Act established the Animal Welfare Board of India (AWBI), responsible for advising the government on animal welfare policies, including the humane treatment of animals during transport and captivity. The AWBI also promotes public education on animal welfare and oversees the registration of animals used in entertainment.

However, the Act provides some exemptions. Practices such as dehorning and castration of cattle, destruction of stray dogs in lethal chambers, and the killing of animals for religious purposes are not considered cruelty. These provisions reflect the diverse cultural and religious practices in India, balancing animal welfare with traditional customs.

Animal protection is also embedded in India's Directive Principles of State Policy. Article 48A emphasizes the state's responsibility to protect and improve the environment, which includes the preservation of wildlife. While these principles are not directly enforceable, they have paved the way for comprehensive animal protection laws at both the central and state levels. Over time, Indian courts have developed an extensive legal jurisprudence in animal law, interpreting the Right to Life under Article 21 to include the protection of animals from cruelty.

India's federal structure divides legislative powers between the central and state governments. Article 245 of the Indian Constitution allows Parliament to make laws for the entire country, while Article 246 allocates specific subjects to Union, State, and Concurrent lists. In the context of animal rights, the State List gives states the power to preserve and protect livestock and prevent animal diseases, while the Concurrent List allows both central and state governments to legislate on the prevention of cruelty to animals and the protection of wild animals and birds. This system ensures that animal welfare is addressed at both national and regional levels.

Despite India's extensive legislative framework, there are several challenges in effectively protecting animals. Penalties under the Prevention of Cruelty to Animals Act are often considered too lenient to deter cruelty. For example, the fine for first-time offenders can be as low as Rs. 10, while subsequent offenses carry a maximum fine of Rs. 100 or three months imprisonment. This has led to calls for stricter penalties and more robust enforcement mechanisms.

Another challenge lies in the exemptions provided by the Act, particularly regarding religious practices and the destruction of stray dogs. While these exemptions aim to respect cultural diversity, they also create loopholes that can be exploited to evade accountability for animal cruelty.

Global Cooperation and Standards

Internationally, several organizations contribute to setting global standards for animal welfare:

- **World Organisation for Animal Health (OIE)**: Provides international guidelines on animal welfare, including transportation, slaughter, and disease prevention.

- **Food and Agriculture Organization (FAO)**: Works on improving global food systems and includes animal welfare as part of its sustainable agriculture goals.

Across the globe, animal welfare regulations and codes of practice are becoming more stringent and increasingly aligned with ethical, environmental, and economic considerations. While the specific details of these regulations vary by country and region, the general principles of humane treatment, proper care, and welfare management are shared universally. Understanding these regulations and codes is essential for individuals and businesses to ensure legal compliance, promote ethical standards, and meet consumer expectations.

Applying Regulations and Codes of Practice

Livestock production enterprises use Animal Welfare Regulations and Codes of Practice as a foundational framework to guide and shape their activities and practices. These regulations not only ensure compliance with legal obligations but also promote ethical, humane treatment of animals, which is essential for maintaining public trust, improving productivity, and accessing certain markets. Here's a detailed look at how these enterprises integrate these standards into their operations:

1. Regulatory Compliance and Legal Obligations

One of the primary reasons livestock production enterprises adhere to **animal welfare regulations** is to remain in compliance with national and regional laws. Regulatory bodies, such as the European Union (EU) and the U.S. Department of Agriculture (USDA), establish minimum standards of care for livestock, covering aspects like housing, feeding, health management, transportation, and slaughter.

For example, the Animal Welfare Act in the U.S. sets standards for animals in commercial use, while the EU's Council Directive 98/58/EC outlines the protection of all farmed animals. These laws mandate practices that reduce animal suffering, ensure humane treatment, and avoid penalties. Non-compliance can lead to legal consequences, including fines, penalties, and reputational damage. Thus, adherence to welfare laws is an integral part of daily operations for these enterprises, guiding how animals are housed, fed, and transported.

2. Ethical Treatment and Industry Best Practices

Beyond legal requirements, livestock enterprises use Codes of Practice to inform ethical and humane treatment of animals. These codes, which are often developed by industry bodies, governments, or animal welfare organizations, provide practical guidance on animal care. They address species-specific needs and outline best practices for various aspects of animal management, including nutrition, behaviour, and veterinary care.

For instance, Australia's Model Codes of Practice for the Welfare of Animals provide detailed recommendations for livestock handling, transport, and husbandry that go beyond the legal minimum. By following these guidelines, enterprises ensure that animals are treated with dignity and respect, which is essential for maintaining ethical standards and enhancing the reputation of the industry.

3. Improving Animal Health and Productivity

Good animal welfare directly correlates with better animal health and increased productivity. Livestock production enterprises understand that animals in poor health or living in stressful conditions are less productive and more prone to disease, which can affect meat, milk, and egg quality. By adhering to welfare regulations and codes, enterprises ensure that animals are kept in environments that reduce stress and improve overall health.

For example, ensuring that animals have access to adequate space, ventilation, and proper nutrition as per welfare guidelines helps to prevent common health issues like respiratory problems, malnutrition, and injuries from overcrowding. Healthier animals require less veterinary intervention, which reduces operational costs and improves yield. In this way, adherence to welfare codes becomes an economically beneficial practice.

4. Shaping Livestock Housing and Infrastructure

Animal housing and infrastructure are crucial aspects of livestock management, and animal welfare regulations play a pivotal role in shaping these elements. Welfare standards often specify the minimum space requirements, bedding types, ventilation, lighting, and temperature control needed to maintain animal health and well-being.

For example, EU welfare regulations dictate that certain species, such as pigs and chickens, must be kept in enclosures that allow them to engage in natural behaviours like turning around, stretching, and lying down. In intensive farming settings, this means enterprises must design their facilities to prevent overcrowding and promote comfort. Codes of practice may also recommend features like windows for natural light, slatted floors for waste management, or automated feeding systems to ensure animals have consistent access to food and water.

5. Transport and Slaughter Practices

Transportation and slaughter are two critical areas where welfare regulations are closely followed by livestock production enterprises. Welfare standards typically require that animals be transported with minimal stress and handled humanely during the process. For example, the EU's Regulation (EC) No 1/2005 sets out detailed requirements for the transport of livestock, including rest periods, food, and water access, as well as limits on transport times.

In terms of slaughter, regulations such as the Humane Slaughter Act in the U.S. and the Humane Methods of Slaughter Act in many countries mandate that animals be stunned before slaughter to minimize pain and suffering. Livestock enterprises must invest in equipment and train staff to ensure compliance with these laws, which are critical to maintaining animal welfare and avoiding penalties.

6. Staff Training and Animal Handling

Proper animal handling is essential for maintaining welfare standards, and enterprises often rely on welfare regulations and codes of practice to inform staff training programs. Welfare codes provide guidance on how to handle animals during everyday activities, such as feeding, moving, and caring for sick animals. Enterprises must train their employees to ensure that they handle animals with care, reducing stress and preventing injury.

For instance, in Australia, codes of practice emphasize the importance of training staff in low-stress handling techniques, which are designed to calm animals and prevent aggressive behaviour during handling or transportation. By providing staff with the knowledge and skills to handle animals according to welfare standards, livestock enterprises improve animal welfare outcomes and reduce the risk of accidents or injuries.

7. Access to Markets and Certifications

Many export markets and retailers demand that livestock products come from farms that adhere to high welfare standards. Compliance with animal welfare regulations and codes of practice is often a prerequisite for obtaining certifications like organic, free-range, or animal-friendly labels, which are increasingly important to consumers. Meeting these welfare standards allows enterprises to access premium markets and increases the value of their products.

For example, Certified Humane and Animal Welfare Approved are labels that indicate livestock have been raised according to stringent welfare guidelines. Adhering to these standards can differentiate products in competitive markets and appeal to consumers who prioritize animal welfare in their purchasing decisions.

8. Responding to Public and Consumer Pressure

As consumer awareness of animal welfare issues grows, livestock production enterprises are increasingly influenced by public pressure to improve their practices. Regulations and welfare codes often reflect societal expectations about how animals should be treated. By adhering to these guidelines, enterprises can maintain consumer trust and avoid public backlash.

In some cases, enterprises may go beyond the minimum legal requirements to adopt higher welfare standards in response to consumer demand. For example, companies may phase out intensive confinement systems like gestation crates for pigs or battery cages for hens, even if these practices are still legal. This proactive approach not only enhances animal welfare but also improves the enterprise's public image.

Livestock production enterprises use Animal Welfare Regulations and Codes of Practice as essential tools to shape their daily operations, from housing and feeding to transport and slaughter. These standards ensure legal compliance, promote ethical treatment, improve animal health and productivity, guide infrastructure design, and help meet market demands. By following these regulations and codes, enterprises can enhance their reputation, gain access to premium markets, and align with societal expectations for humane treatment of animals.

CHAPTER THREE

Animal Welfare Policies, Procedures and Programs

In livestock production enterprises, animal welfare policies, procedures, and programs form the foundation of ethical and humane treatment of animals throughout their lifecycle. These frameworks help businesses adhere to legal requirements while also improving productivity, enhancing marketability, and gaining consumer trust. Below is a detailed explanation of the key elements and how these policies are implemented in practice.

Animal Welfare Policies

Animal Welfare Policies represent formal commitments by livestock production enterprises to prioritize the well-being of animals. These policies often align with global standards, such as those from the **World Organisation for Animal Health (OIE)** or regional frameworks like the **EU Animal Welfare Standards**.

- **Ethical Treatment**: Policies explicitly state that animals will be treated ethically, free from unnecessary stress, pain, injury, and disease. Ensuring their well-being is central to daily operations.

- **Compliance**: Enterprises commit to following local, national, and international animal welfare regulations, such as the Animal Protection Act or Prevention of Cruelty to Animals Act, ensuring that legal obligations are met at all times.

- **Training**: Regular training programs are included in these policies to ensure that employees are well-versed in proper animal handling and welfare standards, providing consistency and high levels of care across operations.

Sample Animal Welfare Policy for Livestock Production Enterprises

Introduction

At [Company Name], we are committed to the highest standards of animal welfare. We recognize that ensuring the well-being of our livestock is not only a moral and ethical obligation but also essential to maintaining the productivity and sustainability of our operations. This policy outlines our formal commitment to animal welfare, detailing how we will ensure ethical treatment, compliance with relevant laws, and proper training for all employees involved in the care of animals.

1. Ethical Treatment of Animals

We prioritize the ethical treatment of all animals under our care, ensuring they are treated humanely at every stage of their lifecycle. Our commitment to animal welfare includes:

- **Minimizing Stress**: Animals will be raised, handled, and transported in ways that minimize unnecessary stress and anxiety. Housing, handling, and transportation practices will be continually reviewed to ensure they align with best practices.

- **Freedom from Pain and Suffering**: Animals will be protected from unnecessary pain, injury, and suffering. In situations where medical intervention is required, veterinary care will be provided promptly to relieve suffering and treat injuries or illnesses.

- **Natural Behaviour**: We provide environments that allow animals to express natural behaviours. Livestock will have adequate space, proper shelter, and access to nutrition that supports their physical and psychological health.

2. Compliance with Animal Welfare Regulations

We are fully committed to complying with all applicable animal welfare regulations, including:

- **Local and National Legislation**: Our operations comply with all local and national animal welfare laws, such as the **Animal Protection Act, Prevention of Cruelty to Animals Act**, and other applicable regulations governing the treatment of livestock.

- **International Standards**: We align our practices with global standards, such as those established by the **World Organisation for Animal Health (OIE)**, ensuring that our operations meet or exceed internationally recognized guidelines for animal welfare.

- **Auditing and Monitoring**: We will conduct regular internal audits and work with third-party auditors to assess our compliance with both legal obligations and ethical standards of animal care.

3. Training and Education

To ensure the highest level of animal welfare in our operations, we are committed to providing ongoing training for all staff involved in animal care. This training will include:

- **Animal Handling Practices**: Staff will receive training on proper handling techniques to minimize stress and ensure the safe movement of animals within our facilities and during transportation.

- **Health and Welfare Monitoring**: Employees will be trained to identify signs of illness, injury, or distress in animals. They will be instructed on how to respond to these situations, ensuring that animals receive prompt veterinary care when needed.

- **Emergency Protocols**: Staff will be trained on emergency procedures to handle crises such as disease outbreaks, natural disasters, or equipment malfunctions, ensuring the continued well-being of animals in unforeseen circumstances.

4. Continuous Improvement

At [Company Name], we believe in continuous improvement of our animal welfare practices. We commit to:

- **Review and Update**: Regularly reviewing and updating our policies and procedures to reflect the latest research and advancements in animal welfare science.

- **Stakeholder Engagement**: Engaging with industry experts, animal welfare organizations, and regulatory bodies to ensure we stay informed about emerging trends and best practices in animal welfare.

- **Feedback**: Encouraging feedback from employees, customers, and stakeholders to identify areas for improvement and implement changes that enhance the well-being of our animals.

5. Accountability and Reporting

We hold ourselves accountable for maintaining the highest standards of animal welfare. To support this commitment:

- **Monitoring and Reporting**: We will maintain detailed records of our animal welfare practices, including health monitoring, feeding, housing conditions, and veterinary interventions. These records will be available for review during internal audits or third-party assessments.

- **Transparency**: We are committed to transparency with our stakeholders, providing regular updates on our animal welfare practices and reporting any significant issues or improvements in line with our continuous improvement goals.

Conclusion

At [Company Name], animal welfare is a core value and central to our operations. By adhering to the highest standards of ethical treatment, compliance, and staff training, we ensure that our livestock production practices contribute to the well-being of animals, the satisfaction of our customers, and the success of our business. We are committed to reviewing and improving our policies regularly to uphold the trust placed in us by our stakeholders and the wider community.

Approved by: [Name]

Position: [Position]

Date: [Date]

Next Review Date: [Review Date]

Animal Welfare Procedures

These are the detailed operational steps that enterprises follow to implement the goals outlined in their welfare policies. Procedures are crucial in maintaining welfare standards and ensuring daily activities are aligned with best practices.

- **Housing and Shelter**: Livestock must be provided with appropriate shelter that protects them from extreme weather conditions, allowing them to exhibit natural behaviours such as lying down, moving freely, and socializing. Housing design must also reduce the risk of injury, and overcrowding is avoided to minimize stress.

- **Feeding and Nutrition**: Animals need access to clean water and a balanced diet, tailored to their species, age, and production system. Nutrition programs are developed based on recognized guidelines, ensuring animals are properly fed, promoting health and productivity.

- **Health Monitoring**: Regular health checks and access to veterinary care are essential. Preventive measures like vaccinations and parasite control are implemented to minimize disease risks. Sick or injured animals are promptly treated or humanely euthanized to prevent suffering.

- **Transportation**: Proper transportation procedures are in place to minimize stress during transit. This includes providing adequate space, ventilation, and trained handlers who follow established guidelines for loading, unloading, and securing livestock. Transportation time is also minimized to reduce fatigue.

- **Slaughter Practices**: Humane slaughter procedures are mandatory, requiring animals to be stunned before slaughter to ensure they are unconscious and free from pain. Regular inspections of slaughterhouses are conducted to ensure compliance with welfare regulations.

Sample Animal Welfare Procedure for Livestock Production Enterprises

Objective

This procedure outlines the operational steps required to implement our Animal Welfare Policy effectively. The goal is to maintain high standards of animal welfare in daily operations, ensuring livestock are housed, fed, monitored, transported, and slaughtered in a humane and ethical manner.

1. Housing and Shelter Procedure

Purpose:

To ensure that all livestock are provided with appropriate shelter that protects them from extreme weather, allows them to express natural behaviours, and minimizes the risk of injury or stress.

Procedure:

- **Shelter Design and Maintenance:**

    - All livestock shelters must be designed to protect animals from extreme weather conditions, including heat, cold, rain, and wind.

    - Ventilation systems should be in place to ensure adequate airflow and temperature control.

    - Shelters must have sufficient space for animals to lie down, stand, and move freely without risk of overcrowding. The size of enclosures will vary depending on the species, age, and number of animals.

    - Shelters must be inspected daily for hazards such as sharp edges, broken equipment, or structural weaknesses that could cause injury.

- **Natural Behaviours:**

    - Ensure that the design of the shelter allows livestock to exhibit natural behaviours such as lying down, standing, walking, and socializing.

    - Enclosures must include soft bedding materials where appropriate, and regular cleaning schedules will be maintained to ensure hygienic conditions.

- **Overcrowding Prevention:**

    - Stocking density in shelters must adhere to guidelines for each species, ensuring animals have enough space to avoid stress from overcrowding.

    - Overstocking is strictly prohibited, and animal numbers will be reviewed regularly to ensure compliance.

2. Feeding and Nutrition Procedure

Purpose:

To provide livestock with access to clean water and a balanced diet that meets their nutritional needs, supporting their health, growth, and productivity.

Procedure:

- **Water Access:**

  - Animals must have continuous access to fresh, clean water. Water troughs or dispensers will be cleaned regularly to prevent contamination.

  - Water availability must be checked at least twice daily, particularly during periods of high heat or when animals are in production.

- **Diet and Feeding Plans:**

  - A tailored feeding plan will be developed for each species, age group, and production system, following recognized nutritional guidelines.

  - Feeding schedules will ensure animals receive adequate and consistent nutrition, with feed distribution times planned to avoid excessive competition among animals.

  - Feed storage facilities must be clean and free from pests, and all feed must be checked regularly for freshness and contamination.

- **Monitoring Nutrition:**

  - Animal body condition will be monitored regularly to ensure they are maintaining healthy weight and growth patterns. Adjustments to diet will be made based on these observations and veterinary advice.

  - Special dietary requirements, such as for pregnant, lactating, or recovering animals, will be addressed with specific feeding programs.

3. Health Monitoring Procedure

Purpose:

To ensure the ongoing health and well-being of all livestock through regular health checks, preventive care, and timely treatment of illness or injury.

Procedure:

- **Daily Health Checks:**

- Stockpeople will conduct daily health inspections, looking for signs of illness, injury, stress, or abnormal behaviour. Any concerns must be reported immediately to the farm manager or veterinarian.

- Observations will include checking for signs of lameness, respiratory distress, skin conditions, or changes in appetite or behaviour.

- **Preventive Measures:**

  - Vaccination schedules will be maintained for each species according to veterinary advice to prevent common diseases.

  - A parasite control program, including deworming and pest management, will be implemented regularly.

  - Livestock pens and living areas will be cleaned regularly to minimize the spread of disease and reduce stress.

- **Veterinary Care:**

  - Sick or injured animals must receive prompt veterinary attention. If treatment is not viable or if the animal's suffering cannot be alleviated, humane euthanasia must be performed following veterinary advice.

  - Health records for each animal, including treatments, vaccinations, and health incidents, must be updated and reviewed regularly to ensure ongoing health monitoring.

4. Transportation Procedure

Purpose:

To ensure that livestock are transported with minimal stress, using appropriate vehicles and handling techniques that prioritize their welfare during transit.

Procedure:

- **Preparation for Transport:**

  - Livestock must be assessed prior to transport to ensure they are fit for travel. Animals showing signs of illness or injury will not be transported unless veterinary approval is provided.

- Transport vehicles must be checked for cleanliness, ventilation, and proper loading capacity. Overloading is strictly prohibited.

- **Loading and Unloading:**

  - Animals will be handled gently during loading and unloading, with the use of ramps or chutes to prevent injury or stress. Handlers must use non-aggressive methods and avoid shouting, striking, or startling the animals.

  - Animals must be secured in the vehicle to prevent overcrowding and allow them to stand or lie down comfortably.

- **During Transport:**

  - Ensure that transport routes are planned to minimize travel time, with stops made to check on the animals' welfare as necessary. Long journeys should include rest breaks where animals can access water and food.

  - Proper ventilation must be maintained throughout the journey to prevent overheating or suffocation.

5. Slaughter Practices Procedure

Purpose:

To ensure that slaughter practices are carried out humanely, minimizing pain and distress for the animals in accordance with welfare regulations.

Procedure:

- **Pre-Slaughter Handling:**

  - Animals must be handled calmly and without unnecessary force prior to slaughter. Handlers must follow approved methods to guide animals into the slaughter area, ensuring minimal stress.

  - Any animal showing signs of illness or injury prior to slaughter must be assessed by a veterinarian.

- **Stunning Process:**

  - All animals must be stunned before slaughter to ensure they are unconscious and unable to feel pain. Only approved stunning methods, such as captive bolt, electrical stunning, or gas stunning, may be used.

- The effectiveness of the stunning process will be verified by checking for unconsciousness before proceeding with slaughter.

- **Slaughter Inspection:**

  - Regular inspections of slaughterhouse operations will be conducted to ensure compliance with welfare regulations. These inspections will include verifying proper equipment maintenance and that staff are following humane handling and stunning procedures.

Conclusion

The implementation of these procedures ensures that [Company Name] adheres to the highest animal welfare standards in all aspects of livestock management. These operational guidelines are integral to maintaining the health, productivity, and ethical treatment of our animals while ensuring compliance with local, national, and international regulations. Regular training, audits, and reviews will be conducted to ensure continuous improvement in our animal welfare practices.

Animal Welfare Programs

Animal welfare programs are systematic approaches that aim to implement, monitor, and continuously improve welfare practices within livestock production enterprises. These programs often include audits, certifications, and ongoing assessments to ensure welfare standards are maintained.

- **Good Agricultural Practices (GAP)**: These programs focus on sustainable and ethical production methods, including maintaining high standards of animal welfare. Farms certified under GAP undergo regular inspections to verify compliance with welfare guidelines.

- **Animal Welfare Audits**: Regular audits, often conducted by third-party organizations, assess compliance with established welfare standards. These audits help identify areas for improvement and ensure that livestock operations maintain their commitments to animal welfare.

- **Certification Programs**: Enterprises can pursue certifications from organizations such as **RSPCA Assured**, **Global Animal Partnership (GAP)**, or **Certified Humane**. These certifications indicate that farms meet specific welfare standards, enhancing marketability and consumer confidence in the welfare of the livestock products they purchase.

- **Employee Training Programs**: Ongoing training ensures that farm staff are equipped to handle animals humanely, recognize signs of stress or illness, and follow the correct procedures for care, transportation, and slaughter. Training also covers emergency protocols for handling disease outbreaks or natural disasters.

- **Technology and Innovation**: Advanced technologies such as **precision livestock farming** are increasingly used to monitor animal welfare indicators, including activity levels, feeding behaviour, and stress markers. These tools provide real-time data that enable proactive measures to address welfare issues before they escalate.

Ensuring good welfare for livestock is crucial for maintaining their health, productivity, and overall well-being. To achieve this, producers should follow a set of guidelines that cover regular surveillance, housing and husbandry practices, and ways to reduce the impact of negative husbandry procedures. Below are detailed explanations of how these guidelines contribute to better animal welfare .

One of the most critical aspects of ensuring good welfare is the regular surveillance of livestock. Stock should be inspected frequently to detect any early signs of distress or changes in behaviour that might indicate compromised welfare. For intensively housed animals, it is recommended to inspect them at least 1-2 times per day, while for extensively housed livestock, inspections should occur every 1-2 days. Familiarizing yourself with your stock's normal behaviour makes it easier to identify when something is amiss, such as lethargy, agitation, or abnormal movement, which could signal health or welfare issues.

Behavioural changes are often the first signs that an animal's well-being is being compromised. When unusual behaviour is observed, further investigation, such as approaching or handling the affected animals, is necessary to diagnose the issue and take appropriate action. Additionally, regular surveillance helps to build trust between the animals and the handler, reducing their fear of humans. This can make future handling and routine procedures easier and less stressful for both animals and handlers, especially in extensive grazing systems where livestock may not interact with humans as frequently.

Regular checks also help detect potential predator problems by allowing stockpeople to notice signs such as carcasses or signs of unease among the animals. Ensuring access to adequate feed, water, and shelter should also be part of routine inspections to maintain good welfare and prevent malnutrition or dehydration. Many industries provide auditing guidelines to help producers monitor their stock's welfare and ensure the production of high-quality products.

Proper housing and husbandry practices are essential for maintaining good welfare. Each livestock industry has its own Codes of Practice, which set out minimum requirements for housing, care, and handling. These codes are not legally binding but can help producers meet industry standards for good animal welfare. For example, the codes provide guidance on the appropriate living conditions for livestock, such as providing adequate space, ventilation, bedding, and protection from the elements.

While these minimum standards help prevent harm, they do not always ensure optimal welfare. Producers should aim to exceed these standards by consulting industry experts or Animal Health Officers on ways to improve housing and husbandry conditions. Good housing should enable animals to exhibit their natural behaviours, reduce stress, and minimize the risk of injury. Overcrowding should be avoided, and facilities should be designed to promote comfort and ease of movement.

Negative or invasive husbandry procedures, such as castration, dehorning, or branding, can cause stress and discomfort to animals. However, following the Code of Practice can minimize their negative impact. The codes provide guidelines on the appropriate age for performing such procedures and the necessary care to ensure animals are comfortable during and after the process. Recognizing the signs of pain or distress, such as vocalization, restlessness, or decreased appetite, is crucial for managing animal welfare.

One effective way to reduce the stress associated with handling is through habituation. By regularly interacting with your stock and moving quietly among them, you can help them become accustomed to human presence, reducing the fear and anxiety associated with handling and procedures. This is particularly beneficial for young animals or animals new to the operation. Running animals through the yard or crush without performing any procedures can help them become familiar with the handling process, making future procedures less stressful and more efficient.

In addition, providing positive reinforcement around the time of a husbandry procedure, such as feeding them a favourite food, can help reduce the negative association between the procedure and the stockperson. Research has shown that animals can distinguish between familiar and unfamiliar handlers, as well as handlers wearing distinctive clothing. Thus, changing into a different set of clothing or hiring a contractor to perform aversive procedures can prevent animals from associating their regular caretakers with discomfort or pain.

In environments where animals lack complexity, such as intensive pig farming, positive human-animal interactions can be particularly beneficial. These interactions may include allowing animals to engage with humans in a way that stimulates their curiosity and reduces stress, thereby improving their overall well-being.

Good livestock welfare is a combination of consistent care, proper housing, and humane handling practices. By conducting regular surveillance, providing suitable housing, and mitigating the impact of necessary but invasive husbandry procedures, producers can ensure that their animals remain healthy and productive. Exceeding the minimum requirements of welfare codes and adapting handling techniques to reduce stress are essential components of maintaining high welfare standards. Additionally, fostering positive human-animal relationships can further improve animal welfare and enhance productivity in livestock production enterprises.

To effectively integrate the principles of ensuring good welfare for livestock into an **animal welfare program**, the program must focus on systematic approaches that address regular surveillance, proper housing and husbandry, and mitigating the impact of negative procedures. Below is a step-by-step guide to implementing these elements within a comprehensive animal welfare program.

1. Regular Surveillance of Livestock

Incorporating regular surveillance into an animal welfare program ensures that animals are consistently monitored, and any signs of distress or compromised welfare are detected early.

- **Daily Inspection Routines**: Establish a formal schedule where stock handlers inspect intensively housed livestock at least 1-2 times per day and extensively housed livestock at least once every 1-2 days. This can be part of a broader welfare checklist to ensure thorough evaluation during each inspection.

- **Behavioural Monitoring Protocols**: Develop protocols for identifying early behavioural changes (e.g., lethargy, abnormal movement) as key indicators of compromised welfare. Train staff on behavioural observation techniques to detect these signs and take immediate action if needed, such as further investigation or veterinary intervention.

- **Technology Integration**: Use technology such as cameras or sensors to monitor animals, especially in extensive farming systems where human inspection may be more challenging. These tools can detect movement patterns or changes in behaviour that may not be visible during a physical check.

- **Familiarization Programs**: Encourage handlers to spend time interacting with livestock, helping them become accustomed to human presence. This can reduce fear and anxiety, making handling easier during necessary procedures.

2. Housing and Husbandry Practices

Providing adequate housing and appropriate husbandry is essential for maintaining good welfare and minimizing stress. The animal welfare program should detail specific standards and operational procedures for housing and care.

- **Compliance with Codes of Practice**: Ensure that the program aligns with existing Codes of Practice specific to each livestock sector. Although these codes set minimum requirements, the program should aim to exceed them where possible by consulting industry experts or Animal Health Officers.

- **Housing Design Standards**: Outline housing standards that provide appropriate space, ventilation, bedding, and shelter from extreme weather. The program should also avoid overcrowding, which can cause stress and health issues.

- **Animal Comfort and Safety**: Develop guidelines to ensure that housing enables animals to exhibit natural behaviours (e.g., grazing, socializing) and minimizes injury risks. Regular audits should be conducted to assess housing conditions and ensure they meet these standards.

- **Continuous Improvement**: Incorporate a feedback loop within the welfare program, allowing for constant evaluation and upgrades to housing and husbandry practices as new technologies or research findings emerge.

3. Mitigating the Impact of Negative Husbandry Procedures

Many livestock enterprises must conduct invasive or negative husbandry procedures such as castration, dehorning, or branding. The welfare program should implement strategies to reduce the stress and discomfort associated with these procedures.

- **Age-Appropriate Procedures**: Follow guidelines from the relevant Codes of Practice that specify the appropriate age for conducting invasive procedures. This helps minimize the pain and stress that may be inflicted on animals.

- **Pain Management**: Integrate pain relief strategies (e.g., anaesthesia or analgesics) for more invasive procedures. Ensure that veterinarians or trained staff are available to administer these treatments.

- **Habituation Practices**: Incorporate habituation into the welfare program by allowing livestock to become accustomed to handling and routine procedures through regular interaction. Running animals through handling facilities without performing a procedure helps them become more familiar with the environment and reduces stress during actual procedures.

- **Positive Reinforcement**: Use positive reinforcement, such as offering food rewards or gentle handling, during and after negative procedures to create a positive association with the experience. This approach can reduce the animals' fear and anxiety, leading to more efficient handling and better welfare outcomes.

- **Handler Training**: Ensure all handlers are trained in gentle handling techniques and can recognize signs of stress or pain in animals. Specific clothing or uniforms for handlers during negative procedures can help animals differentiate between routine and potentially stressful events.

4. Incorporating Audits, Certification, and Employee Training

To ensure continuous improvement, animal welfare programs should include regular audits, employee training, and certifications.

- **Animal Welfare Audits**: Implement regular welfare audits, ideally by third-party organizations, to assess compliance with welfare standards. These audits should cover housing conditions, feeding practices, health checks, and procedures like transportation and slaughter. Audits can help identify areas for improvement and ensure the enterprise maintains its welfare commitments.

- **Certification Programs**: Seek certifications such as RSPCA Assured, Global

Animal Partnership (GAP), or Certified Humane, which verify that the farm complies with high welfare standards. Certification adds credibility to the welfare program and assures consumers that animals are raised ethically.

- **Ongoing Employee Training**: Ensure that employees undergo continuous training to handle animals humanely and understand welfare protocols. Training should also cover how to detect signs of distress or illness, follow transportation guidelines, and use pain management techniques during husbandry procedures.

- **Emergency Preparedness**: Include protocols for emergencies such as disease outbreaks or natural disasters. Train employees on how to manage these situations to ensure that animal welfare is maintained even in difficult conditions.

5. Utilizing Technology for Welfare Monitoring

Incorporating technology and innovation into an animal welfare program enhances the ability to monitor livestock welfare continuously.

- **Precision Livestock Farming (PLF)**: Use sensors and monitoring devices that track livestock activity, feeding patterns, and health markers in real-time. PLF technologies allow farmers to proactively address welfare issues before they escalate by detecting early signs of illness or stress.

- **Data Analytics**: Integrate data analysis tools to monitor trends in animal health, behaviour, and productivity. This information can be used to refine welfare practices, improve housing conditions, and reduce the impact of husbandry procedures.

Sample Animal Welfare Program for Livestock Production Enterprises
Objective:

The purpose of this Animal Welfare Program is to implement, monitor, and continuously improve animal welfare practices within our livestock production enterprise. The program is structured to ensure compliance with welfare standards, ethical treatment of animals, and efficient farm management. This will be achieved through regular audits, certifications, employee training, and the use of innovative technologies.

1. Good Agricultural Practices (GAP)

Purpose:

To maintain high standards of sustainable and ethical production methods, focusing on animal welfare as a core principle of farm management.

Program Details:

- **GAP Certification:**

    - Our enterprise will participate in a recognized Good Agricultural Practices (GAP) certification program to ensure compliance with the latest animal welfare standards.

    - GAP certification will involve regular inspections by certified auditors who will assess animal housing, feeding, health care, and general management practices to verify that welfare standards are consistently met.

    - Certification ensures access to premium markets and consumer trust in the welfare standards followed on the farm.

- **Sustainable Practices:**

    - We commit to incorporating sustainable practices, such as water conservation, waste management, and biodiversity protection, as part of our GAP program. These practices will also contribute to improving animal welfare by promoting a healthier, low-stress environment for livestock.

2. Animal Welfare Audits

Purpose:

To ensure that all animal welfare standards are adhered to and continuously monitored through systematic and regular assessments.

Program Details:

- **Third-Party Audits:**

    - Our enterprise will engage third-party auditors from reputable organizations to conduct regular audits of our animal welfare practices. These audits will assess compliance with local, national, and international animal welfare regulations, as well as internal company standards.

    - The audits will evaluate key areas such as housing conditions, feeding and nutrition, transportation, health monitoring, and humane slaughter practices.

- **Internal Audits:**

  - In addition to third-party audits, we will conduct internal audits every quarter to proactively identify areas for improvement.

  - Findings from these audits will be documented, and corrective actions will be implemented to address any deficiencies found. This ensures a continual improvement process within our animal welfare program.

3. Certification Programs

Purpose:

To achieve and maintain certifications that verify compliance with animal welfare standards, thereby enhancing our marketability and consumer trust.

Program Details:

- **Certification Process:**

  - Our farm will pursue certification from leading organizations, including RSPCA Assured, Global Animal Partnership (GAP), and Certified Humane.

  - Each certification program has specific animal welfare standards that focus on housing, feeding, transportation, and humane slaughter. Compliance with these standards will be verified through regular inspections and audits.

- **Market Benefits:**

  - Achieving certification ensures that our products meet the welfare expectations of consumers and retailers, enhancing the marketability of our livestock products.

  - Certification logos will be displayed on product packaging, assuring consumers that the meat, dairy, or other livestock products they purchase come from farms adhering to high animal welfare standards.

4. Employee Training Programs

Purpose:

To ensure that all farm staff are equipped with the knowledge and skills to handle animals humanely and follow correct welfare procedures.

Program Details:

- **Initial Training:**

  - All new employees will receive comprehensive training on proper animal handling techniques, the recognition of stress and illness, and emergency protocols. This training will cover every aspect of livestock care, from feeding and housing to transportation and humane slaughter.

  - Employees will also be educated on company policies related to animal welfare and the specific welfare standards associated with our certification programs.

- **Ongoing Training:**

  - Regular refresher courses will be provided for all employees to ensure they stay up-to-date with the latest welfare practices and regulations.

  - Specialized training will be offered on emergency protocols, including handling disease outbreaks, responding to extreme weather, and implementing biosecurity measures to prevent the spread of infectious diseases.

- **Performance Monitoring:**

  - Employee performance related to animal handling will be monitored and evaluated regularly. Staff who excel in implementing welfare practices will be recognized, while additional training will be provided to those who require improvement.

5. Technology and Innovation

Purpose:

To leverage advanced technologies to monitor animal welfare indicators and make informed decisions for the proactive management of livestock.

Program Details:

- **Precision Livestock Farming (PLF):**

- Our farm will integrate Precision Livestock Farming (PLF) technologies to monitor key welfare indicators such as activity levels, feeding behaviour, and stress markers. This data-driven approach will allow us to make real-time decisions that enhance the well-being of animals.

- Sensors and automated monitoring systems will be installed in housing units to track temperature, humidity, and animal movement, ensuring that environmental conditions are optimal for animal welfare.

- **Health Monitoring Systems:**

  - We will use health monitoring technologies such as wearable devices or smart collars that track vital signs and alert staff to potential health issues before they escalate. This will help ensure early detection and treatment of illness, reducing suffering and improving overall livestock health.

- **Data Analysis and Reporting:**

  - The data collected from PLF technologies will be analysed and reported regularly to management. This analysis will provide insights into areas for improvement and allow us to address welfare issues before they become critical.

  - Reports will be shared with auditors and certification bodies as part of our commitment to transparency and accountability.

6. Continuous Improvement and Review

Purpose:

To ensure the program remains up-to-date with evolving welfare standards and continually improves animal welfare practices.

Program Details:

- **Annual Program Review:**

  - Our Animal Welfare Program will be **reviewed annually** to ensure that it reflects current best practices and complies with any changes to welfare regulations.

  - Input from third-party audits, internal audits, employee feedback, and technological data will be incorporated into the review process to identify areas for improvement.

- **Stakeholder Engagement:**

  - We will **engage with stakeholders**, including veterinarians, welfare experts, certification bodies, and industry partners, to stay informed about emerging trends and innovations in animal welfare.

  - Feedback from consumers and retailers will also be considered in our efforts to align our practices with market expectations.

Conclusion

This Animal Welfare Program is designed to implement and maintain high welfare standards throughout our livestock production enterprise. Through the use of Good Agricultural Practices, regular welfare audits, certifications, comprehensive training, and innovative technology, we will ensure the ethical treatment of animals, while improving productivity and marketability. Continuous improvement and a commitment to proactive welfare management will remain central to our operations.

Approved by: [Name]
Position: [Position]
Date: [Date]
Next Review Date: [Review Date]

Benefits of Implementing Animal Welfare Policies and Programs

- **Enhanced Productivity**: Well-treated animals are healthier, more productive, and tend to yield higher-quality products, whether it's milk, meat, or reproduction rates. Ensuring welfare reduces stress-related impacts, leading to better overall performance.

- **Market Access**: Many premium markets, particularly in Europe and North America, require adherence to strict animal welfare standards. Livestock enterprises that comply with these standards gain access to these markets, attracting consumers who prioritize ethical animal treatment.

- **Risk Management**: Adhering to proper welfare practices reduces the risk of disease outbreaks, injuries, and accidents. This leads to lower veterinary costs and operational risks, making businesses more resilient.

- **Consumer Trust**: As consumers become more concerned about the ethical treatment of animals, demonstrating a commitment to high welfare standards enhances brand reputation. Consumers are more likely to trust and remain loyal to companies that prioritize animal welfare.

- **Sustainability**: Good animal welfare practices are closely linked to sustainable farming. They encourage the responsible use of resources, reduction of waste, and promote eco-friendly practices, which are increasingly important in the global marketplace.

Animal Welfare Policies, Procedures, and Programs in livestock production enterprises are essential for ensuring the ethical and humane treatment of animals. These frameworks not only help enterprises comply with legal obligations but also improve productivity, open up premium markets, and build consumer trust. They are integral to the success and sustainability of livestock farming operations, ensuring that animals are cared for throughout their lifecycle in a way that benefits both the business and society.

Stockperson Competency

Stockpeople play a vital role in livestock production enterprises, taking on diverse responsibilities that ensure the care, health, and overall well-being of the animals. Their tasks are crucial for maintaining animal welfare, productivity, and efficiency within the livestock production process. Below is a detailed overview of the key duties of stockpeople in these enterprises.

Stockpeople are primarily responsible for the routine care of livestock, making sure animals are well-fed, have access to clean water, and live in a clean, safe environment. This includes providing the right feed according to each animal's dietary needs and ensuring a constant supply of fresh water. Additionally, they are tasked with maintaining the cleanliness of housing areas to prevent the buildup of waste, which could lead to disease or stress for the animals. Monitoring livestock behavior is another critical part of their role, as they observe animals for signs of illness, injury, or distress, and take appropriate action to address these issues.

An essential aspect of stockpeople's duties is the regular monitoring of animal health. This involves identifying early signs of disease or injury, such as lethargy, loss of appetite, or unusual behaviour. Stockpeople are often trained to administer treatments, including medications, vaccinations, and other necessary interventions, under the guidance of veterinarians. They also assist veterinarians during examinations and procedures, ensuring that the animals receive the best possible care.

Stockpeople are responsible for the safe handling and movement of livestock within the farm or during transportation. They utilize low-stress handling techniques to minimize fear and anxiety in animals, as high levels of stress can negatively impact both welfare and productivity. In transportation, stockpeople ensure that animals are safely loaded and unloaded from vehicles and that they remain comfortable during transit, with proper space, ventilation, and handling practices.

In livestock production, managing breeding activities is a significant responsibility of stockpeople. They monitor and manage breeding cycles by identifying animals in heat and ensuring that mating or artificial insemination is carried out effectively. Stockpeople also provide care for pregnant animals throughout gestation and assist with the birthing process. After birth, they ensure that newborn animals receive proper care, including feeding colostrum and conducting health checks to safeguard their early development.

In cases where animals are severely injured or suffering from untreatable diseases, stockpeople may be tasked with humane euthanasia. This is often carried out by trained personnel to ensure that the procedure is done ethically and with minimal stress or pain to the animal, in line with welfare standards.

Effective livestock management relies on accurate record-keeping, and stockpeople are responsible for maintaining comprehensive records related to animal health, feeding, growth, and reproduction. This data includes logging treatments, vaccinations, and health checks, tracking weight and feed consumption, and maintaining detailed reproductive records such as breeding schedules, pregnancies, and births. These records are essential for tracking the progress of livestock and ensuring compliance with industry standards.

Stockpeople ensure that all tasks align with the livestock enterprise's animal welfare policies and comply with relevant legislation. They adhere to national or industry welfare standards, such as those established by the World Organisation for Animal Health (OIE) or local animal protection laws. By maintaining a low-stress environment and implementing humane handling and management practices, stockpeople play a key role in ensuring animal welfare and enhancing productivity.

To provide a safe and comfortable environment for livestock, stockpeople are responsible for the maintenance of housing and equipment. This includes regular inspections to ensure barns, pens, and paddocks are in good condition, free from hazards, and offer adequate space for animals. They also maintain feeding and water systems, ensuring that automated equipment functions properly. Additionally, stockpeople ensure that animal handling equipment, such as chutes and crushes, is safe and well-maintained.

Experienced stockpeople often take on the role of training new staff, providing instruction on proper animal handling, welfare standards, and safety procedures. This ensures that all employees are equipped to manage livestock efficiently and humanely, maintaining high standards of care across the enterprise.

Stockpeople are critical in maintaining biosecurity measures to prevent the spread of disease within the farm. This involves implementing quarantine measures for new or sick animals to prevent the transmission of infectious diseases. They also ensure that facilities, equipment, and staff adhere to stringent hygiene practices, minimizing the risk of disease outbreaks.

Stockpeople are the foundation of livestock production enterprises, responsible for ensuring the health, welfare, and productivity of animals. From daily care and health monitoring to implementing animal welfare standards and maintaining biosecurity, their role is essential to the success of livestock operations. Their expertise, attention to detail, and commitment to ethical treatment are key to ensuring that livestock production is conducted in compliance with legal and welfare standards, while also promoting sustainable and humane practices.

Figure 2: Stockperson mustering cattle. CSIRO, CC BY 3.0, via Wikimedia Commons.

The competency of stockpeople is one of the most critical factors in ensuring good animal welfare, particularly in livestock operations such as pig farming. Under state government regulations in Australia for example, it is mandated that all stockpeople working with pigs must either be fully competent in maintaining the health and welfare of animals or work under the direct supervision of someone who is. Competency is typically gained through training and experience and is essential for maintaining not only the health of the livestock but also for promoting high welfare standards across the farm [27].

Stockpeople are expected to participate in training in various areas related to pig farming and animal welfare. These areas include [27]:

Implementing and Monitoring Workplace Procedures for Compliance with Animal Welfare Procedures

1. **Move and Handle Pigs**: This involves understanding the natural behaviour of pigs and using low-stress handling techniques to move them in a calm and safe manner.

2. **Care for Health and Welfare of Pigs**: Stockpeople must be trained to recognize signs of illness, injury, and discomfort in pigs, ensuring timely interventions.

3. **Comply with Industry Quality Assurance Requirements**: Depending on the enterprise, stockpeople may need to follow specific quality assurance procedures that align with industry standards.

4. **Animal Industry Welfare Requirements**: Knowledge of and compliance with animal welfare regulations is mandatory to ensure legal obligations are met.

5. **Occupational Health and Safety (OHS)**: Stockpeople must contribute to safe working conditions, recognizing that animal welfare and human safety are interconnected.

6. **Administer Medication**: Being competent in administering medication is crucial for treating sick or injured animals, following veterinary advice.

7. **Implement Health Control Programs**: This includes carrying out preventative health measures such as vaccinations and disease control protocols.

8. **Euthanize Livestock**: While optional, at least one person on each farm must be qualified to humanely euthanize livestock, ensuring the welfare of animals in severe health crises.

Stockpersonship and Its Importance in Animal Welfare

Stockpersonship refers to the attitude, behaviour, skills, and knowledge of individuals responsible for managing and handling farm animals. It plays a crucial role in determining the welfare and overall well-being of livestock. Good stockpersonship is built on a foundation of knowledge, caring, and low-stress animal management. When consistently practiced, it ensures that animals are treated with respect and that their physical and emotional needs are met. Stockpeople are at the heart of maintaining positive animal welfare, and their interactions with livestock directly affect the animals' quality of life.

The technical skills and knowledge of stockpeople are vital to effective animal management. This makes it essential to carefully select and train individuals who are not only proficient in handling animals but also possess the right attitude toward animal care. Stockpeople must be able to recognize both positive and negative indicators of animal welfare. Regular monitoring of animals' appearance, behaviour, and vocalizations allows stockpeople to detect changes that might signal distress or poor health. For example, observing whether animals are feeding and drinking normally, maintaining calm behaviour, and showing positive or neutral responses to the stockperson are key indicators of good welfare. The absence of abnormal behaviours and the proper use of environmental enrichment also serve as indicators of healthy, well-adjusted animals.

In addition to monitoring animal behaviour, stockpeople are responsible for ensuring that the animals live in an optimal environment. This requires a solid working knowledge of the specific husbandry system in place and an understanding of the animals' needs within that system. Stockpeople need to recognize signs of disease, injury, or distress early on to provide timely intervention. Maintaining animal welfare also involves paying attention to body condition, as this can be an early indicator of underlying health issues. By being proactive, stockpeople can prevent minor issues from escalating into significant health or welfare concerns.

The attitudes and competence of stockpeople have a direct impact on the animals they care for. The way animals perceive and respond to humans is shaped by their interactions with stockpeople. Animals that are handled with care and treated gently are less likely to experience fear, which improves their overall well-being. In contrast, animals that are handled roughly or with disregard for their needs may develop fear and stress, which not only affects their welfare but can also reduce their productivity and meat quality. Positive human-animal relationships are therefore critical for maintaining high standards of animal welfare in livestock production.

Animal Welfare Audits

Animal welfare audits are an essential tool for maintaining and improving the welfare of livestock. These audits are designed in collaboration with industry bodies and animal welfare research scientists to ensure that they are based on sound scientific principles and industry best practices. Auditing materials provide farmers and stockpeople with detailed, step-by-step guidelines for monitoring the welfare of their animals, identifying areas for improvement, and developing strategies to address any issues.

The primary goal of animal welfare audits is to promote productivity, reproductive success, and product quality by ensuring that animals are raised in environments that support their health and well-being. Auditing materials cover various aspects of animal management, including housing, feeding, handling, and transportation. These materials offer a structured approach to assessing animal welfare, making it easier for stockpeople to detect problems and take corrective action before they negatively affect the animals' well-being.

Many livestock industries have already developed or are in the process of developing animal welfare auditing programs for farms, transporters, and abattoirs. These audits help standardize welfare practices across the industry and provide a benchmark for evaluating the success of welfare initiatives. Farmers and stockpeople should seek out animal welfare auditing materials from their respective industry bodies to ensure they are meeting the highest welfare standards. By adhering to these standards, producers can enhance the welfare of their livestock while also meeting consumer demand for ethically produced animal products.

Using an Animal Welfare Checklist

An Animal Welfare Compliance Checklist can be a valuable tool used by several key individuals and teams within a livestock production enterprise environment. Its purpose is to ensure that the operation adheres to required animal welfare standards, improves the well-being of livestock, and maintains high levels of productivity and efficiency. Here's how and by whom the checklist can be used:

1. Farm or Livestock Managers

Farm managers are typically responsible for overseeing daily farm operations, including animal welfare standards. They can use the checklist to:

- **Monitor Compliance**: Ensure that the farm complies with local, national, and international animal welfare regulations and industry standards.

- **Assess Operations**: Review current practices in housing, feeding, health care, and animal handling to identify gaps and areas for improvement.

- **Prepare for Audits**: Ensure the farm is ready for internal or third-party audits by keeping records updated and ensuring all practices meet welfare requirements.

- **Track Progress**: Use the checklist to track and document improvements over time in terms of animal welfare, productivity, and operational efficiency.

2. Stockpeople (Animal Handlers)

Stockpeople are directly responsible for the daily care and handling of animals. They can use the checklist to:

- **Follow Standard Procedures**: Ensure that they are following proper protocols in animal handling, feeding, health monitoring, and other daily activities.

- **Identify Animal Welfare Issues**: Spot potential welfare issues early, such as signs of illness, stress, or poor housing conditions, and take immediate action to correct them.

- **Ensure Proper Animal Care**: Consistently apply procedures related to feeding, health care, transportation, and housing to maintain animal well-being.

3. Quality Assurance Teams

Quality assurance personnel play a critical role in ensuring that the farm's production practices meet internal and external standards. They can use the checklist to:

- **Conduct Audits**: Perform regular audits of animal welfare practices and verify that the farm meets all required standards and regulations.

- **Verify Compliance**: Review all necessary documentation, records, and procedures to confirm that animal welfare guidelines are being followed.

- **Recommend Improvements**: Use the checklist as a basis for recommending changes or improvements in animal welfare procedures to enhance overall compliance.

4. Veterinarians

Veterinarians are often called upon to oversee the health and welfare of livestock. They can use the checklist to:

- **Assess Health Practices**: Evaluate the health monitoring, vaccination, and treatment procedures followed by farm staff to ensure that they are appropriate and effective.

- **Provide Feedback**: Recommend adjustments to procedures related to health care, biosecurity, or other practices that affect animal welfare.

- **Guide Training**: Assist in training stockpeople and farm staff on best practices related to animal health and welfare.

5. Animal Welfare Auditors (Third-Party or Internal)

Auditors who assess compliance with welfare standards can use the checklist to:

- **Conduct Inspections**: Systematically review the farm's operations, from housing and feeding to transportation and slaughter, to ensure compliance with animal welfare policies and regulations.

- **Provide Certification**: Verify that the enterprise meets the necessary requirements for animal welfare certifications from organizations such as RSPCA Assured, Global Animal Partnership (GAP), or other certifications.

- **Create Action Plans**: Highlight areas for improvement and work with farm management to implement corrective actions and ensure continued compliance.

6. Training Coordinators

Training coordinators responsible for educating farm staff can use the checklist to:

- **Develop Training Programs**: Identify specific areas where additional training is needed, such as handling practices, health monitoring, or the use of equipment.

- **Track Employee Competency**: Ensure that employees are trained on all aspects of animal welfare and are competent in carrying out their responsibilities effectively.

- **Monitor Compliance Post-Training**: Follow up after training sessions to ensure that staff are applying their knowledge and maintaining welfare standards.

7. Government or Regulatory Authorities

In certain instances, government bodies or regulatory authorities may use the checklist to:

- **Enforce Compliance**: Ensure that livestock enterprises are complying with animal welfare laws and regulations.

- **Conduct Routine Inspections**: Use the checklist during official inspections to ensure that all legal and welfare obligations are being met.

- **Issue Penalties**: Identify violations of welfare standards and recommend penalties or corrective actions for non-compliance.

Benefits of the Compliance Checklist

Implementing an animal welfare compliance checklist plays a crucial role in ensuring that animals are treated humanely, which significantly reduces stress, injury, and illness. By adhering to established welfare standards, livestock production enterprises can ensure better care for their animals, leading to healthier and more content livestock. This not only promotes ethical practices but also directly contributes to the animals' well-being.

The health and welfare of livestock are directly linked to their productivity. Healthier animals are generally more productive, whether in terms of milk production, meat quality, wool yield, or reproduction rates. A comprehensive welfare checklist ensures that the animals' needs are consistently met, contributing to higher yields and overall profitability. Proper care minimizes the risks of disease, injury, or stress, which otherwise could negatively impact productivity.

Ensuring legal compliance is another key benefit of using a compliance checklist. By following welfare standards, livestock enterprises can ensure they are adhering to local, national, and international animal welfare laws. This helps avoid penalties, fines, or legal actions related to non-compliance, providing peace of mind and ensuring the enterprise operates within the regulatory framework.

A focus on animal welfare can also improve marketability. Many consumers are becoming increasingly conscious of the ethical treatment of animals and prefer to purchase products from sources that adhere to strict welfare standards. Compliance with these standards makes products more attractive to such consumers, enhancing the brand's reputation and opening doors to premium markets that value ethical production practices.

Finally, regularly using the checklist contributes to more streamlined operations. It helps standardize procedures across the enterprise, ensuring consistency in the way animals are handled, cared for, and managed. This, in turn, improves overall efficiency, reduces the likelihood of operational errors, and enhances the effectiveness of daily farm management activities.

Sample Animal Welfare Compliance Checklist

This Animal Welfare Compliance Checklist is designed to help ensure that your livestock production enterprise adheres to the highest standards of animal welfare. Use this checklist as a guide to monitor your compliance with animal welfare regulations, industry standards, and best practices.

1. Animal Housing and Shelter

Criteria	Yes	No	Comments/Notes
Adequate shelter is provided to protect livestock from extreme weather conditions (heat, cold, rain, wind)			
Housing allows animals to exhibit natural behaviours (e.g., lying down, moving freely)			
Adequate space is available to avoid overcrowding and reduce stress			
Clean, dry bedding is provided and changed regularly			
Housing is free from hazards that could cause injury			
Proper ventilation is maintained to ensure air quality			
Lighting is sufficient and appropriate for the livestock			

2. Feed and Water

Criteria	Yes	No	Comments/Notes
Animals have access to clean, fresh water at all times			
Feeding systems provide adequate nutrition suited to species, age, and production stage			
Feed is free from contaminants (e.g., mold, spoilage)			
Proper feeding schedules are followed			
Feed and water systems are regularly inspected and maintained			

3. Health Monitoring and Veterinary Care

Criteria	Yes	No	Comments/Notes
Animals are regularly inspected for signs of disease, injury, or distress			
Sick or injured animals are promptly treated or humanely euthanized if necessary			
Preventive measures (vaccinations, parasite control) are in place			
Veterinary care is readily available for emergencies			
Medical records for each animal are maintained and up to date			

4. Animal Handling and Movement

Criteria	Yes	No	Comments/Notes
Low-stress handling techniques are used to minimize fear and anxiety in livestock			
Stockpeople are trained in proper handling methods			
Animals are moved safely, with appropriate equipment (e.g., handling chutes, ramps)			
Procedures are in place to ensure the safe loading and unloading of livestock during transportation			

5. Transportation

Criteria	Yes	No	Comments/Notes
Transport vehicles are well-ventilated and provide adequate space for livestock			
Animals are transported for the shortest time possible to minimize stress			
Drivers and handlers are trained in safe transport procedures			
Animals are regularly checked during transit to ensure their well-being			

6. Breeding and Reproduction Management

Criteria	Yes	No	Comments/Notes
Breeding programs are managed to ensure the health and welfare of animals			
Pregnant animals are provided with appropriate care and nutrition			
Newborn animals receive adequate care (e.g., colostrum feeding, health checks)			

7. Euthanasia (When Necessary)

Criteria	Yes	No	Comments/Notes
Humane euthanasia procedures are in place and followed when necessary			
At least one trained person is responsible for carrying out euthanasia on the farm			

8. Biosecurity and Hygiene

Criteria	Yes	No	Comments/Notes
Biosecurity protocols are in place to prevent the spread of disease			
New or sick animals are quarantined to prevent contamination			
Hygiene practices are followed to maintain cleanliness in animal housing and handling areas			

9. Animal Welfare Audits and Compliance

Criteria	Yes	No	Comments/Notes
Regular animal welfare audits are conducted by internal or third-party auditors			
Corrective actions are taken in response to audit findings			
Welfare compliance documentation is maintained and easily accessible			

10. Employee Training

Criteria	Yes	No	Comments/Notes
Employees are trained in animal welfare standards and handling techniques			
New employees undergo animal welfare training as part of the induction process			
Refresher courses on animal welfare are regularly conducted			

11. Record Keeping

Criteria	Yes	No	Comments/Notes
Health, feeding, and welfare records are maintained for all animals			
Records of veterinary visits, treatments, and medications are up to date			

This checklist can be tailored to your specific enterprise needs, and is a useful tool for ensuring that your livestock operation adheres to all required animal welfare standards.

Reporting Animal Welfare Concerns, Grievances and Complaints

Reporting animal welfare concerns, grievances, and complaints is a critical part of ensuring the ethical treatment of animals in livestock production enterprises. Around the world, various systems, procedures, and frameworks have been put in place to address these concerns, ensuring transparency and accountability within the animal husbandry sector. The procedures typically focus on creating clear reporting mechanisms for workers, implementing swift and effective action plans, and providing follow-up to ensure issues are resolved and compliance is maintained.

1. Reporting Mechanisms

In many countries, reporting animal welfare concerns follows a structured process designed to make it accessible and straightforward for workers. This often includes:

- **Anonymous Reporting Systems**: Some regions have adopted anonymous reporting systems, allowing workers to raise concerns without fear of retaliation. These systems are often available through hotlines, online portals, or designated welfare officers within the organization.

- **Internal Reporting**: Workers are typically encouraged to report any signs of mistreatment or welfare issues to their direct supervisors or designated animal welfare officers. The internal reporting system should be clearly communicated during training and induction, so all staff are aware of their responsibilities and the procedures they must follow.

- **External Reporting**: In cases where internal reporting does not lead to resolution, or if workers feel unsafe reporting within their organization, many countries have external channels. For example, in the UK, workers can contact the Royal Society for the Prevention of Cruelty to Animals (RSPCA), while in the United States, they can reach out to the USDA or local animal welfare bodies. In Europe, the European Commission also provides reporting frameworks under the EU Animal Welfare legislation.

2. Handling Grievances and Complaints

Once a concern or complaint is reported, a formal investigation process should follow. This typically includes:

- **Acknowledge Receipt**: Once a complaint or concern is lodged, it should be acknowledged promptly by the organization. Clear communication is critical, and the complainant should be informed of the next steps.

- **Initial Assessment**: The designated welfare officer or compliance team conducts an initial assessment to determine the severity of the complaint. For more serious cases, such as neglect or abuse, immediate action may be required, including separating the involved animals or workers from the situation.

- **Investigation**: A thorough investigation is conducted to verify the validity of the complaint. This includes reviewing any surveillance footage, interviewing relevant workers, and examining the conditions of the animals involved. The process should be objective, ensuring that all evidence is properly collected and analysed.

- **Third-Party Auditors**: In some cases, external or third-party auditors may be involved to ensure impartiality. For example, organizations like the RSPCA or the Global Animal Partnership (GAP) may be called to review the case and recommend further action.

3. Actions Taken to Address Concerns

Once a valid animal welfare concern or complaint is identified, swift corrective actions are implemented. These may include:

- **Immediate Corrective Actions**: In serious cases of animal mistreatment or neglect, immediate corrective actions may be taken. This could involve removing the animals from harmful environments, providing necessary medical care, or, in severe cases, suspending or terminating the responsible individuals involved.

- **Training and Retraining**: One of the most common responses to identified welfare issues is additional training for the staff involved. This may focus on proper animal handling, low-stress practices, and legal compliance with welfare standards.

- **Infrastructure Changes**: In some cases, the grievance may be linked to inadequate facilities or resources. Corrective actions might involve updating or improving housing, feeding, or transportation systems to better meet welfare standards.

- **Policy Review**: If the investigation identifies gaps in the organization's animal welfare policies, those policies may be reviewed and revised. This could involve implementing stricter monitoring or refining procedures to prevent further welfare breaches.

4. Escalation and Legal Enforcement

In some cases, particularly if there is no appropriate resolution within the organization or the issue violates legal requirements, the concern may be escalated:

- **Regulatory Bodies**: Organizations can escalate unresolved concerns to national regulatory bodies. For instance, the USDA in the United States, DEFRA in the UK, and the European Commission in the EU are key regulatory bodies involved in enforcing animal welfare laws. These bodies have the authority to conduct further investigations, issue fines, and, if necessary, revoke permits or licenses.

- **Legal Action**: In severe cases, criminal charges can be pursued against individuals or enterprises found guilty of breaching animal welfare regulations. This often involves prosecution under national animal welfare legislation, such as the Animal Welfare Act in the US or the Prevention of Cruelty to Animals Act in India.

5. Follow-Up and Monitoring

After the initial actions are taken, follow-up is crucial to ensure that the changes are maintained and no further issues arise. This may include:

- **Regular Inspections**: Increased inspection frequency may be implemented to monitor the situation and ensure that the corrective actions have been effective.

- **Ongoing Audits**: Many enterprises establish regular welfare audits post-complaint to ensure continued compliance with welfare standards.

- **Worker Feedback**: Organizations may also implement systems that encourage ongoing feedback from workers on the welfare of animals, ensuring that any potential issues are caught early.

6. Global Variations in Reporting Procedures

Reporting animal welfare concerns may vary based on local regulations and frameworks:

- **European Union**: The EU places a strong emphasis on animal welfare, with various mechanisms in place to report and address grievances under its stringent legislation. EU member countries enforce compliance through agencies like the

European Food Safety Authority (EFSA).

- **United States**: In the US, the USDA and Animal and Plant Health Inspection Service (APHIS) oversee welfare regulations, particularly under the Animal Welfare Act. Workers can report concerns through designated channels, and the USDA conducts audits and investigations.

- **Australia**: Australia enforces animal welfare through state-level legislation, and reporting mechanisms often involve local animal welfare bodies or the RSPCA. Workers are encouraged to raise concerns both internally and, if needed, through external bodies.

Managing and Resolving Animal Welfare Concerns, Grievances and Complaints

Animal Welfare Issues – Australian Context

Animal welfare issues in Australian abattoirs and livestock industries are tightly regulated by a combination of national, state, and territory laws, designed to ensure the humane treatment of animals throughout their life cycle, including during transportation, handling, and slaughter. These guidelines focus on preventing unnecessary suffering, minimizing stress, and ensuring that practices align with ethical and legal standards. The key areas of concern include the welfare of animals in abattoirs, live export regulations, bobby calf welfare, camel culling, and specific concerns related to greyhounds, horses, kangaroo harvesting, and sheep mulesing.

Australian Abattoirs

Australian abattoirs are required to comply with strict animal welfare legislation that governs the humane slaughter of animals used for meat production. The Australian Standard for Hygienic Production and Transportation of Meat and Meat Products for Human Consumption outlines mandatory requirements for the slaughter process, which apply to both export and domestic abattoirs. The focus is on minimizing injury, pain, and suffering during slaughter, and this is closely monitored by veterinarians employed by the Australian Department of Agriculture in abattoirs licensed under the Export Control Act 1982. For abattoirs regulated by state and territory laws, similar animal welfare standards must be met. These standards are verified to ensure the humane treatment of animals at all stages of processing.

Live Export Industry

Australia's live export industry is subject to several government regulations that ensure the health and welfare of livestock being transported overseas. These regulations address the transportation conditions, health checks, and feeding schedules to ensure animals do not suffer from dehydration, malnutrition, or stress during transit. Government veterinarians and inspectors oversee the conditions in which animals are transported, and violations of these standards can lead to legal penalties.

Figure 3: Cows that arrived in live export from Australia, are in Zofar quarantine before being sent to feedlots and abattoirs. Anat Refuah, CC BY-SA 4.0, via Wikimedia Commons.

Bobby Calf Welfare

Bobby calves, which are young calves removed from their mothers in the dairy industry, are protected under the Australian Animal Welfare Standards and Guidelines for the Land Transport of Livestock. These guidelines stipulate that bobby calves must be in good health and adequately fed before transport. Transport must be completed within a specified time frame, and all calves must be tracked using an auditable system to ensure compliance. Industry partners have also committed to additional welfare standards, such as limiting time-off-feed to a maximum of 30 hours.

Camel Culling

Each state and territory in Australia is responsible for managing animal welfare within its jurisdiction. This includes camel culling programs, which are necessary to manage overpopulated and feral camel herds that pose environmental threats. While aerial culling is sometimes used due to the remote locations of feral camel populations, this method is carried out by accredited shooters aiming for humane kills. The intention is to prevent prolonged suffering, such as starvation, which is common during drought conditions.

Greyhound Welfare

The welfare of greyhounds, particularly in relation to their export for racing purposes, is a growing concern in Australia. While the Export Control Act 1982 regulates the export of greyhounds, requiring them to be certified as fit for travel, the welfare of these dogs once they reach their destination is governed by the importing country's laws. Greyhound Australasia, an industry body, works with other countries to address animal welfare issues and improve conditions for exported dogs.

Horse Welfare and Aerial Culling

Horses, particularly those used in thoroughbred racing, are protected by state and territory legislation enforced by the RSPCA or local departments of primary industries. These laws ensure the humane treatment of racing horses, and industry guidelines like the Australian Rules of Racing aim to minimize injury and protect the welfare of horses. In cases where wild or feral horses cause environmental damage, aerial culling is sometimes employed as a humane method of population control.

Kangaroo Harvesting

Kangaroo populations in Australia are managed under state and territory guidelines, which allow for the sustainable harvesting of kangaroos. Licensed commercial shooters must follow the National Code of Practice for the Humane Shooting of Kangaroos and Wallabies, which mandates humane killing methods. This code is regularly updated and endorsed by industry and animal welfare groups to ensure the ethical treatment of harvested animals.

Mulesing in Sheep

Mulesing, a procedure used to prevent flystrike in merino sheep, is a contentious practice in Australia. While mulesing remains the most effective method for preventing flystrike, the Australian wool industry is investing heavily in research to find alternatives, such as pain relief products, genetic breeding programs, and new non-surgical methods like liquid nitrogen treatments and laser therapy. These alternatives aim to improve animal welfare while maintaining flystrike prevention.

Mulesing is a surgical procedure performed on sheep, primarily Merino sheep, to prevent a parasitic condition called flystrike (or myiasis). Flystrike occurs when blowflies lay eggs in the moist, wool-covered skin folds of a sheep's rear end, leading to maggot infestation that can cause severe pain, infection, and even death. Mulesing involves removing strips of wool-bearing skin around the breech (the rear area) to create smooth, scarred skin that is less attractive to flies. While effective in preventing flystrike, mulesing is considered painful and has raised significant animal welfare concerns, particularly regarding the lack of pain relief used in some cases.

Steining as a Replacement for Mulesing

Steining (also known as intradermal injections) is one of the emerging alternatives to mulesing and involves using a chemical solution to cause the skin in the breech area to shrink and contract over time. This process eliminates skin folds in a way that mimics the outcome of mulesing but without the need for surgical intervention. Here's how steining works and why it's being considered a replacement for mulesing:

1. **Application of a Chemical Solution**: Steining involves injecting or applying a solution, typically a caustic chemical, to the skin around the breech area. This solution is absorbed into the skin, causing a reaction that leads to the skin shrinking and contracting over a period of time.

2. **Non-Invasive**: Unlike mulesing, which involves cutting away sections of skin, steining is non-surgical. There is no open wound, which reduces the risk of infection, bleeding, and prolonged recovery time. This significantly reduces the animal's pain and stress levels.

3. **Gradual Process**: The skin gradually contracts and smooths out over time, eliminating the wrinkles that attract blowflies. This gradual process is less traumatic for the animal than the immediate and invasive nature of mulesing.

4. **Less Pain and Stress**: One of the key advantages of steining is that it is less painful for the sheep compared to mulesing. Pain management, such as local anaesthetics or pain relief, can be more effectively used during the process, further reducing the discomfort to the animal.

5. **Improved Animal Welfare**: With growing concern about animal welfare and the ethical treatment of livestock, steining is seen as a more humane alternative to mulesing. It is becoming increasingly important to adopt methods that minimize pain and stress in farm animals, particularly as consumers become more aware of animal welfare issues.

6. **Effectiveness**: Although still in development and not yet as widely adopted as mulesing, steining has shown promise in preventing flystrike effectively. It reduces skin folds in a similar way to mulesing, thereby mitigating the risk of blowfly infestations.

While mulesing has been a traditional method for controlling flystrike in sheep, its invasive and painful nature has led to calls for more humane alternatives. Steining, as a chemical skin contraction process, presents a less invasive, less painful alternative, aligning with modern animal welfare standards. It's part of a broader movement in the livestock industry to adopt more ethical and sustainable practices, which not only benefit animal health and welfare but also meet increasing consumer demands for humane farming methods.

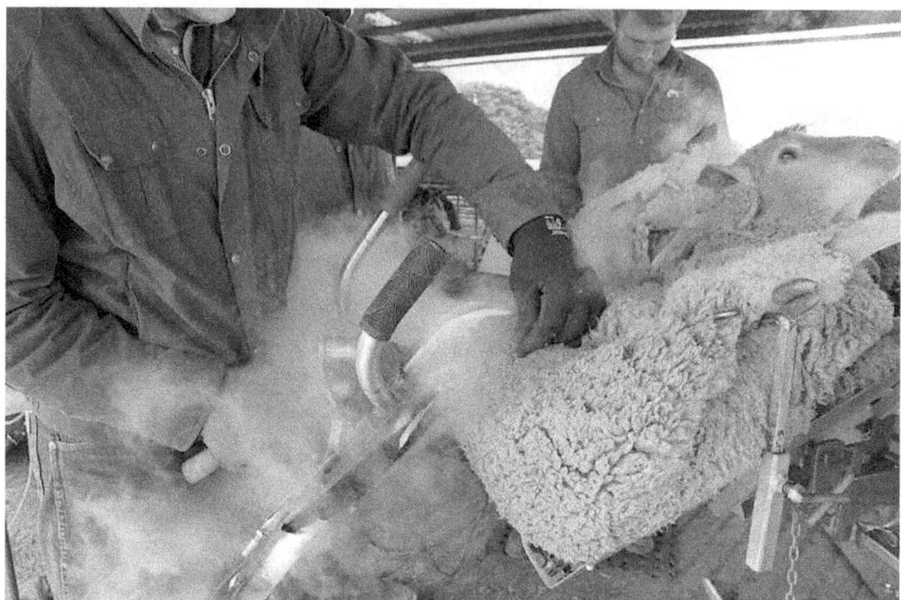

Figure 4: A lamb is having the hair around its tail and breech removed by a cryogenic process that destroys the growth cells in wool follicles. This is done to prevent maggot infestation (also called flystrike or breechstrike) by the Australian sheep blowfly. The process was invented by John Steinfort, an Australian veterinarian. Australian Public Broadcasting Agricultural Fair, Dubbo NSW, CC BY 4.0, via Wikimedia Commons.

National Animal Welfare Frameworks

Australia's livestock industry works closely with the government to improve animal welfare standards through collaborative research and development initiatives. The government provides funding to industry bodies and research centres, facilitating the development of best practices for animal husbandry. The Australian Animal Welfare Strategy plays a key role in the development of national welfare standards and guidelines, which are being gradually legislated across the country.

Pig Welfare

Pig welfare in Australia is guided by the Model Code of Practice for the Welfare of Animals: Pigs, which sets voluntary guidelines and enforceable standards to ensure humane treatment in pig farming. This includes restrictions on the use of individual gestation stalls for pregnant sows and specific guidelines for farrowing pens to protect piglets from being accidentally crushed by sows.

Poultry Welfare

Poultry welfare in Australia is regulated through the Model Code of Practice for the Welfare of Animals - Domestic Poultry, which outlines standards for housing systems like cages, barns, and free-range setups. While the government does not mandate specific farming methods, it does ensure that farmers adhere to the welfare code. Recent developments include improved egg labelling standards to help consumers make informed choices about free-range products.

Animal Welfare in Drought Conditions

During drought or poor seasonal conditions, livestock owners are responsible for ensuring that their animals' welfare is maintained, which includes providing adequate food and water. If this is not possible, owners must transport or sell animals while they are still fit for travel or, in extreme cases, humanely destroy unfit animals. Transporting animals that are in poor condition can result in prosecution under animal welfare legislation.

Animal welfare issues in Australia span a wide range of industries and activities, from abattoirs and live exports to specific practices like mulesing and kangaroo harvesting. National, state, and territory governments play a crucial role in regulating and enforcing welfare standards, while industries are increasingly adopting new technologies and research-based practices to improve animal welfare outcomes. These regulations and frameworks are vital for maintaining ethical practices in livestock production and ensuring compliance with both domestic and international welfare standards.

Animal Welfare Issues in New Zealand's Livestock Industries

Animal welfare in New Zealand's livestock industries is regulated by a combination of national laws and codes of practice designed to ensure the humane treatment of animals throughout their life cycle, including transportation, handling, and slaughter. These regulations aim to prevent unnecessary suffering, minimize stress, and promote ethical practices. Key areas of concern include the welfare of animals in slaughterhouses, live transport conditions, welfare standards for young animals like calves, and issues surrounding specific livestock sectors such as poultry, pigs, and cattle.

New Zealand Slaughterhouses

Slaughterhouses in New Zealand are required to comply with the Animal Welfare Act 1999, which mandates humane slaughter practices. This legislation requires that animals be stunned before slaughter to ensure they are unconscious and insensible to pain during the process. The Ministry for Primary Industries (MPI) oversees compliance with these regulations across domestic and export-oriented slaughterhouses. MPI inspectors regularly monitor slaughter practices to ensure the humane handling of animals and enforce penalties when violations occur, such as improper stunning or mishandling of livestock.

Live Animal Transport

The transport of livestock within New Zealand is governed by the Animal Welfare (Transport within New Zealand) Code of Welfare 2018. This code specifies conditions under which animals must be transported, including requirements for space, ventilation, access to water, and maximum journey times to minimize stress and fatigue. The code also states that animals must be fit for transport, and specific measures must be taken to ensure their welfare during transit. Violations of these transport standards can result in penalties, including fines or prosecution.

Calf Welfare

In New Zealand, calf welfare, particularly for dairy calves, is an important issue. The Code of Welfare for Dairy Cattle 2019 includes strict guidelines to ensure calves are treated humanely. These include requirements for adequate feeding, housing, and transportation. The Bobby Calf Welfare Regulations, introduced in 2016, ensure that young calves are properly fed and rested before transportation and that their time off-feed is limited to no more than 24 hours. These measures are in place to ensure the well-being of calves during critical stages of their development.

Wild Horse and Kaimanawa Horse Management

New Zealand's wild horse population, particularly the Kaimanawa horses, is managed under the Wild Animal Control Act 1977. The Department of Conservation (DOC) works to maintain the balance between the conservation of these horses and the preservation of native ecosystems. Regular roundups are conducted to control the population, and efforts are made to rehome as many horses as possible. Controversial practices such as aerial culling are avoided, and animal welfare groups work alongside the DOC to ensure humane treatment during these population control efforts.

Figure 5: Wild horses at Spirits Bay, North, New Zealand. Natalia Volna itravelNZ@ travel app, CC BY 2.0, via Wikimedia Commons.

Greyhound Welfare

The welfare of greyhounds used in racing is a growing concern in New Zealand. The Greyhound Racing New Zealand (GRNZ) has introduced welfare guidelines to ensure that racing dogs are treated humanely, with proper medical care, housing, and handling. However, concerns remain about the fate of greyhounds once their racing careers are over. Recent reforms have been aimed at improving the rehoming process for retired dogs and increasing transparency around greyhound welfare within the racing industry.

Horse Welfare in Racing and Euthanasia

Thoroughbred horse racing in New Zealand is governed by the New Zealand Thoroughbred Racing (NZTR), which enforces welfare standards for racehorses. These standards focus on the humane treatment of horses during their training, racing, and retirement. The use of veterinary oversight ensures that injured or suffering horses are euthanized humanely when necessary. In addition, the New Zealand Racing Integrity Unit (RIU) works to ensure that racing regulations are adhered to and that horse welfare is prioritized within the industry.

National Animal Welfare Frameworks

New Zealand's livestock industry works closely with the government to maintain and improve animal welfare standards through various regulations and welfare codes. The National Animal Welfare Advisory Committee (NAWAC) plays a crucial role in developing and updating these codes, which cover all aspects of livestock management, from husbandry and feeding to transportation and slaughter. Research and development efforts, supported by both industry and government, are aimed at promoting humane livestock management practices, such as the use of pain relief during invasive procedures and improving housing conditions for animals.

Pig Welfare

Pig welfare in New Zealand is guided by the Animal Welfare (Pigs) Code of Welfare 2018, which outlines standards for housing, feeding, and general care. Gestation crates for pregnant sows have been banned in New Zealand since 2015, with farmers adopting more humane alternatives such as group housing systems. The New Zealand Pork Industry Board supports these efforts by providing farmers with guidance on humane practices and auditing systems to ensure compliance with welfare standards. Euthanasia of pigs, when necessary, must be carried out according to strict guidelines to ensure it is done humanely.

Poultry Welfare

Poultry welfare is regulated under the Animal Welfare (Layer Hens) Code of Welfare 2018 and the Animal Welfare (Broiler Chickens) Code of Welfare 2018. These codes set out minimum standards for housing, feeding, and handling poultry in both cage and free-range systems. Free-range and organic certification schemes in New Zealand also include welfare requirements, such as access to outdoor spaces and lower stocking densities. Recent reforms in egg production have seen the phasing out of conventional battery cages, further improving the welfare of layer hens in the country.

Drought and Emergency Conditions

During periods of drought or other emergency conditions, New Zealand livestock producers are responsible for ensuring their animals receive adequate food and water. The MPI provides guidance and support to farmers during such times, including access to feed and water supplies. In extreme cases, farmers may be advised to reduce stock numbers or humanely euthanize animals that cannot be adequately cared for. Transporting animals in poor condition is prohibited, and non-compliance with welfare regulations can result in fines or legal action.

Animal welfare issues in New Zealand's livestock industries cover a wide range of sectors, from slaughterhouses and live transport to the treatment of horses, pigs, and poultry. The New Zealand government, along with industry bodies, plays a critical role in enforcing welfare standards and promoting humane treatment of animals across all stages of the livestock production process. Continuous research, development, and updates to welfare codes are essential for maintaining ethical practices in livestock production and ensuring compliance with both domestic and international standards.

Animal Welfare Issues – United States of America Context

Animal welfare issues in the United States' livestock industries are governed by a combination of federal and state laws designed to ensure the humane treatment of animals throughout their life cycle, including transportation, handling, and slaughter. These regulations aim to prevent unnecessary suffering, minimize stress, and promote ethical and legal practices. Key areas of concern include the welfare of animals in slaughterhouses, the conditions of live transport, welfare standards for young animals like veal calves, and issues surrounding specific livestock sectors such as poultry, pigs, and cattle.

U.S. Slaughterhouses

U.S. slaughterhouses must comply with federal animal welfare laws, particularly the Humane Methods of Slaughter Act (HMSA), which requires that animals be slaughtered in a manner that minimizes pain and suffering. The act mandates the stunning of animals before slaughter, ensuring that they are unconscious and insensible to pain during the process. The USDA's Food Safety and Inspection Service (FSIS) oversees compliance with these regulations in both domestic and export-oriented slaughterhouses. FSIS inspectors are responsible for ensuring humane handling and enforcing penalties when violations occur, such as improper stunning or excessive use of electrical prods.

Figure 6: USDA Agricultural Marketing Service (AMS)Â Commodity Graders (Red Meat) at work at Cargill Meat Solutions, Friona, Texas, Sept. 20, 2022. USDA/FPAC photos by Preston Keres, CCO, via Flickr.

Live Animal Transport

The transport of livestock within the U.S. is subject to Animal Welfare Act (AWA) regulations, which focus on the humane treatment of animals during transport. These rules specify the conditions under which animals can be transported, including requirements for ventilation, access to water, and limiting transport times to reduce stress and fatigue. The 28-Hour Law requires that animals transported across state lines for longer than 28 hours must be unloaded for rest, water, and feeding. Failure to comply with these transport regulations can lead to fines and penalties for transport companies.

Veal Calf Welfare

Young calves, particularly those raised for veal, are subject to specific welfare standards aimed at ensuring humane treatment. Federal and state laws address the conditions under which veal calves must be housed, fed, and transported. Many industry players adhere to the American Veal Association's welfare guidelines, which include prohibitions on extreme confinement and requirements for adequate nutrition and health monitoring. Additionally, some states have passed laws banning the use of restrictive veal crates, further promoting more humane farming practices.

Wild Horse and Burro Management

In the U.S., wild horse and burro populations are managed under the Wild Free-Roaming Horses and Burros Act of 1971, which aims to protect these animals from inhumane treatment and exploitation. The Bureau of Land Management (BLM) oversees wild horse populations, using techniques like helicopter roundups to manage herd sizes. While the BLM seeks to minimize stress during roundups, these operations are controversial, and animal welfare groups frequently call for more humane management practices, such as fertility control and relocation.

Greyhound Welfare

The welfare of greyhounds used in racing has been a significant concern in the U.S. In some states, such as Florida, greyhound racing has been banned due to concerns about the treatment of dogs, including injuries and euthanasia of unfit or retired dogs. The American Greyhound Council has established welfare guidelines for racing dogs, ensuring that they receive proper medical care, adequate housing, and humane handling. However, concerns remain about the conditions in some kennels and the fate of greyhounds once their racing careers are over.

Horse Welfare in Racing and Euthanasia

Horses used in racing and equestrian sports are protected under state laws that ensure their humane treatment. Regulations govern the conditions under which racehorses are kept, trained, and raced, with a focus on minimizing injury and stress. The Horseracing Integrity and Safety Act (HISA), passed in 2020, aims to create a national standard for horse welfare in racing, addressing issues such as doping, over-racing, and humane retirement. In situations where horses are severely injured or suffering, euthanasia is a necessary but carefully regulated procedure, carried out under veterinary supervision to ensure it is humane.

Mulesing in Sheep

Though mulesing is not commonly practiced in the U.S., the import of wool products from countries where this procedure is used, like Australia, raises animal welfare concerns. The U.S. government and wool industry stakeholders support research into alternatives to mulesing, such as selective breeding and pain relief treatments, to promote the welfare of sheep. U.S. consumers increasingly seek out wool products labelled as "mulesing-free," pushing international suppliers to adopt more humane practices.

National Animal Welfare Frameworks

The U.S. livestock industry works closely with federal agencies such as the USDA and the National Institute for Animal Agriculture (NIAA) to establish and maintain animal welfare standards. These standards include best practices for husbandry, feeding, transportation, and slaughter, all of which are designed to improve animal welfare across the livestock sector. The U.S. government also supports industry research and development programs that promote humane livestock management practices, including the use of pain relief for invasive procedures and improved housing conditions.

Pig Welfare

Pig welfare in the U.S. is guided by federal regulations under the AWA, as well as voluntary industry standards like the Pork Quality Assurance (PQA) program, which sets guidelines for the humane treatment of pigs throughout their lives. This includes restrictions on the use of gestation crates for pregnant sows and guidelines for humane euthanasia and transportation. The National Pork Board's Common Swine Industry Audit is another tool used by producers to ensure that animal welfare standards are met on pig farms.

Figure 7: Pigs in gestation crates. Humane Society of the United States, CC BY 3.0, via Wikimedia Commons.

Poultry Welfare

Poultry welfare in the U.S. is governed by the National Chicken Council's Animal Welfare Guidelines and federal regulations, including the Poultry Products Inspection Act. These guidelines establish standards for housing, feeding, and handling poultry in both conventional and free-range systems. Additionally, the USDA's Organic Certification Program includes specific welfare requirements for poultry labelled as organic, ensuring that animals have access to outdoor spaces and are raised under more humane conditions.

Drought and Emergency Conditions

During droughts or other emergency conditions, livestock producers in the U.S. are responsible for ensuring their animals' welfare by providing adequate food and water. The USDA offers assistance programs to help farmers manage livestock during droughts, including financial aid for feed and water supply. In cases where animals cannot be adequately cared for, producers are encouraged to sell or humanely euthanize them to prevent suffering. Transporting animals in poor condition can lead to prosecution under federal and state animal welfare laws.

Animal welfare issues in the U.S. livestock industry encompass a broad range of sectors and activities, from slaughterhouses and live transport to the treatment of racehorses, pigs, and poultry. Both federal and state governments play critical roles in enforcing welfare standards, while industries increasingly adopt new technologies and research-based practices to improve the treatment of animals. These regulations and frameworks are essential for maintaining ethical practices in livestock production and ensuring compliance with both domestic and international welfare standards.

Animal Welfare Issues in Canadian Livestock Industries

Animal welfare in Canadian livestock industries is governed by a combination of federal, provincial, and territorial laws, all aimed at ensuring the humane treatment of animals throughout their life cycle, from transportation and handling to slaughter. The primary goal of these regulations is to prevent unnecessary suffering, minimize stress, and promote ethical treatment of livestock. Key areas of concern in Canada include the welfare of animals in slaughterhouses, live animal transport conditions, welfare standards for young animals like veal calves, and industry-specific issues in sectors such as poultry, pigs, and cattle.

Canadian Slaughterhouses

Canadian slaughterhouses must adhere to federal animal welfare regulations, primarily under the Safe Food for Canadians Regulations (SFCR) and the Health of Animals Act. These laws ensure that animals are slaughtered humanely, requiring that animals be stunned before slaughter to minimize pain and suffering. The Canadian Food Inspection Agency (CFIA) oversees compliance with these regulations, monitoring both export-oriented and domestic slaughterhouses. CFIA inspectors ensure that proper stunning techniques are used and that animals are handled humanely during their final moments. Any violations, such as improper stunning or excessive use of electrical prods, result in penalties and corrective actions.

Live Animal Transport

The transport of livestock in Canada is regulated by the Health of Animals Regulations, which focus on the humane treatment of animals during transit. These rules outline conditions for ventilation, access to water, and requirements for limiting transport times to minimize stress and fatigue. Recent updates to the Health of Animals Regulations in 2020 reduced the allowable transport times for animals and set stricter guidelines for rest periods during long journeys. Livestock transported for more than 36 hours must be unloaded for rest, food, and water. Failure to comply with these regulations can result in fines and penalties for transport companies.

Veal Calf Welfare

In Canada, young veal calves are subject to specific welfare standards aimed at ensuring their humane treatment. Both federal and provincial regulations address the housing, feeding, and transport conditions for veal calves. The Canadian veal industry has implemented welfare guidelines that prohibit extreme confinement and promote adequate nutrition and health monitoring. Provincial governments, in some cases, have enacted laws that ban restrictive veal crates, promoting more humane farming practices across the country.

Wild Horse and Burro Management

In Canada, wild horse populations are primarily managed by provincial governments, particularly in areas like Alberta and British Columbia, where wild herds are more prevalent. These management practices focus on ensuring that wild horses are treated humanely, and techniques like fertility control or relocation are used to manage herd sizes. Provincial laws require that any interventions, such as roundups or relocations, are done with minimal stress to the animals. Public scrutiny and pressure from animal welfare groups ensure that humane treatment remains a priority in managing Canada's wild horse populations.

Greyhound Welfare

Greyhound racing is not a major industry in Canada, but the welfare of greyhounds is a concern where racing occurs, mainly in recreational or unofficial capacities. Provincial laws, alongside groups like the Canadian Federation of Humane Societies (CFHS), oversee the treatment of racing greyhounds, focusing on ensuring proper medical care, adequate housing, and humane handling. Animal welfare groups in Canada continue to advocate for strict regulations to ensure that greyhounds are treated ethically during and after their racing careers.

Horse Welfare in Racing and Euthanasia

Horses used in racing and equestrian sports in Canada are protected under provincial animal welfare laws, and the Canadian Pari-Mutuel Agency (CPMA) regulates aspects related to horse racing. These regulations ensure that horses are kept in humane conditions, with special attention to minimizing injury and stress during training and racing. The humane euthanasia of horses, when necessary, is a carefully regulated process, always carried out by veterinarians to ensure it is done ethically.

National Animal Welfare Frameworks

Canada's livestock industries work closely with federal agencies like the CFIA and provincial agricultural departments to establish and maintain animal welfare standards. These standards include best practices for feeding, transportation, housing, and slaughter, designed to improve welfare throughout the livestock sector. Canada's government supports research and development initiatives aimed at promoting humane management practices, such as the use of pain relief for invasive procedures and improvements to housing systems. The National Farm Animal Care Council (NFACC) plays a significant role in developing Codes of Practice, which outline national guidelines for the care and handling of various farm animals.

Pig Welfare

Pig welfare in Canada is guided by the Code of Practice for the Care and Handling of Pigs, developed by the NFACC. This code provides guidelines for the humane treatment of pigs, including restrictions on the use of gestation stalls for pregnant sows and standards for humane euthanasia and transport. Many producers voluntarily adopt these guidelines, and industry audits ensure compliance with welfare standards. The goal is to continuously improve pig welfare across farms through training and adherence to these best practices.

Poultry Welfare

Poultry welfare in Canada is regulated by the Code of Practice for the Care and Handling of Poultry, which sets standards for housing, feeding, and handling of birds in both conventional and free-range systems. Canadian poultry producers also follow specific requirements under the Canadian Organic Standards, which mandate more humane conditions for organic poultry production, including outdoor access. Provincial and federal agencies monitor compliance with these welfare standards, ensuring that poultry are treated ethically across all production systems.

Drought and Emergency Conditions

During droughts or other emergency conditions, Canadian livestock producers are responsible for maintaining animal welfare by ensuring adequate food, water, and shelter. The government, through various programs like the AgriRecovery framework, offers financial assistance to help farmers manage livestock during difficult conditions. If producers are unable to care for their animals, they are encouraged to sell or humanely euthanize them to prevent undue suffering. Transporting animals in poor condition is strictly regulated under Canadian animal welfare laws, with penalties for non-compliance.

Animal welfare in Canada's livestock industries covers a wide range of activities, from slaughterhouse practices and live transport to the treatment of horses, pigs, and poultry. Federal and provincial governments play a crucial role in enforcing welfare standards, while industry groups work to adopt new technologies and best practices to ensure the humane treatment of animals. These regulations and frameworks are essential for maintaining ethical standards in livestock production and ensuring compliance with both domestic and international animal welfare guidelines.

Animal Welfare Issues – United Kingdom Context

Animal welfare issues in the United Kingdom's livestock industries are governed by a combination of national laws and EU-derived standards designed to ensure the humane treatment of animals throughout their life cycle. This includes their care during transportation, handling, and slaughter. The regulations aim to prevent unnecessary suffering, reduce stress, and promote ethical farming practices. Key areas of concern include the welfare of animals in slaughterhouses, transport conditions, and specific welfare standards for sectors such as poultry, pigs, and cattle.

UK Slaughterhouses

Slaughterhouses in the UK are required to comply with strict animal welfare laws, particularly the Welfare of Animals at the Time of Killing (WATOK) Regulations. These regulations mandate the stunning of animals before slaughter to ensure they are unconscious and insensible to pain during the process. The Food Standards Agency (FSA) and Official Veterinarians (OVs) closely monitor these practices to ensure compliance with the law. Slaughterhouses that process meat for domestic consumption and export must meet stringent welfare standards, and penalties are enforced for non-compliance, such as improper stunning or unnecessary distress to animals.

Live Animal Transport

The transport of livestock in the UK is regulated under the Welfare of Animals (Transport) (England) Order 2006, which sets out clear guidelines for the humane transport of animals. These regulations cover aspects such as space, ventilation, rest periods, and access to food and water during journeys. For long-distance transport, especially across EU borders, the rules are even more stringent, with regular checks by authorities to ensure that animal welfare is upheld. The UK also adheres to the EU's Transport Regulation (EC) No 1/2005, which governs animal welfare during transport and limits the duration of journeys to reduce stress and fatigue.

Veal Calf Welfare

In the UK, welfare standards for veal calves are high, addressing concerns about housing, feeding, and transport conditions. The Welfare of Farmed Animals (England) Regulations 2007 specify that calves must be provided with bedding, and restrictive veal crates have been banned since 1990. Instead, calves must be housed in group pens with sufficient space to stand, turn around, and lie down. These measures ensure the ethical treatment of calves throughout their rearing and transportation, adhering to national and EU welfare standards.

Wild Horse and Feral Animal Management

In the UK, feral animal populations, such as wild ponies in Dartmoor or the New Forest, are carefully managed to prevent overpopulation and environmental damage. Welfare concerns during roundups and relocations are mitigated by strict adherence to guidelines laid out by the Wildlife and Countryside Act 1981 and additional local policies. In contrast to other countries, helicopter roundups and more invasive methods are generally not used, with management focused on conservation and minimal disruption to the animals.

Greyhound Welfare

The welfare of greyhounds used in racing is a significant concern in the UK. The industry is regulated by the Greyhound Board of Great Britain (GBGB), which ensures that racing dogs receive proper care during their racing careers and retirement. Greyhound kennels must meet welfare standards, including veterinary care, safe housing, and regular exercise. Retired greyhounds are rehomed through the Retired Greyhound Trust, addressing concerns about the welfare of dogs once their racing careers have ended.

Figure 8: Greyhound racing in the UK - Dorotas Wildcat, the English Greyhound Derby champion pictured in 2018. Dorotas Wildcat, CC BY-SA 4.0, via Wikimedia Commons.

Horse Welfare in Racing and Euthanasia

Horses used in equestrian sports and racing in the UK are protected by the Animal Welfare Act 2006, as well as industry-specific guidelines such as the Rules of Racing. These rules govern the care, training, and racing of horses, focusing on minimizing injury and stress. The British Horseracing Authority (BHA) oversees the implementation of these rules, ensuring the humane treatment of horses. In cases where a horse is severely injured or suffering, euthanasia is carried out by veterinary professionals to ensure the procedure is humane and pain-free.

Mulesing in Sheep

Although mulesing is not practiced in the UK, the import of wool products from countries like Australia, where mulesing is common, raises animal welfare concerns. UK consumers are increasingly seeking wool products that are certified as "mulesing-free," encouraging international wool suppliers to adopt more humane practices. The British Wool Marketing Board also supports research into non-invasive alternatives to mulesing, promoting higher welfare standards in the wool industry.

National Animal Welfare Frameworks

The UK livestock industry works closely with government bodies such as the Department for Environment, Food and Rural Affairs (DEFRA) to establish and maintain comprehensive animal welfare standards. These include the Code of Recommendations for the Welfare of Livestock, which sets out best practices for housing, feeding, and caring for animals. The UK government supports industry research and development initiatives that promote humane livestock management practices, such as the use of pain relief for invasive procedures and improving housing conditions.

Pig Welfare

Pig welfare in the UK is governed by both national and EU legislation. The Welfare of Farmed Animals (England) Regulations 2007 and the Pigs (Protection) Act 1994 outline the requirements for housing, feeding, and care, including restrictions on the use of farrowing crates for pregnant sows. The voluntary Red Tractor Assurance Scheme further ensures that farms meet higher welfare standards, focusing on humane transport, housing, and slaughter. This is complemented by government inspections to ensure compliance with welfare laws.

Poultry Welfare

Poultry welfare in the UK is regulated by the Code of Practice for the Welfare of Laying Hens and the Broiler Directive, which set specific standards for housing systems such as free-range, barn, and caged systems. The government ensures that poultry farmers comply with these codes, and the RSPCA Assured Scheme offers certification for farms that meet higher welfare standards. Improved egg labelling standards also allow consumers to make informed choices about free-range products, ensuring transparency in farming practices.

Figure 9: Free-range chickens at Balmuchy Farm. Sylvia Duckworth, CC BY-SA 2.0, via Geograph.

Drought and Emergency Conditions

In cases of drought or emergency conditions, livestock producers in the UK must ensure their animals have adequate food and water. DEFRA provides guidance and support during extreme weather events, including financial assistance through programs like the Farming Recovery Fund. If livestock cannot be adequately cared for, farmers are encouraged to sell or euthanize animals humanely to prevent suffering. Transporting animals in poor condition is prohibited under UK welfare laws, and violations can lead to prosecution.

Animal welfare in the UK livestock industry is regulated by a combination of national legislation and EU-derived standards, ensuring humane practices in transportation, handling, and slaughter. Both government bodies and industry stakeholders play key roles in maintaining and improving welfare standards, supported by consumer demand for ethically produced goods. These frameworks are vital for upholding ethical practices in livestock production and ensuring compliance with domestic and international welfare regulations.

Animal Welfare Issues in the European Union's Livestock Industries

Animal welfare in the European Union (EU) is governed by a comprehensive set of regulations aimed at ensuring the humane treatment of animals throughout their life cycle, including transportation, handling, and slaughter. The EU's stringent welfare standards are designed to prevent unnecessary suffering, reduce stress, and promote ethical practices across all livestock sectors. These regulations apply to the entire food supply chain and are enforced across all member states, with a particular focus on welfare in slaughterhouses, live animal transport, and sector-specific standards for young animals like veal calves, as well as poultry, pigs, and cattle.

EU Slaughterhouses

Slaughterhouses within the European Union must comply with the Council Regulation (EC) No 1099/2009 on the protection of animals at the time of killing, which mandates the humane treatment of animals during slaughter. The regulation requires that animals are stunned before slaughter to ensure that they are unconscious and do not experience pain. This applies to both domestic and export-oriented slaughterhouses. EU member states are responsible for enforcing these regulations through official veterinary inspectors who monitor slaughterhouse practices. Non-compliance, such as improper stunning methods or excessive use of electric prods, is met with penalties and corrective actions.

As a case study, the Alès Abattoir, located in southern France, gained international attention in 2015 when it was temporarily closed following the release of a disturbing undercover video by the animal rights group L214. The footage, filmed over several days, exposed severe animal welfare violations, showing horses, cows, pigs, and sheep being slaughtered in ways that violated legal standards. The video revealed instances where animals were not properly stunned and showed signs of consciousness while being slaughtered, which is against regulations aimed at minimizing pain and suffering during the slaughter process [28].

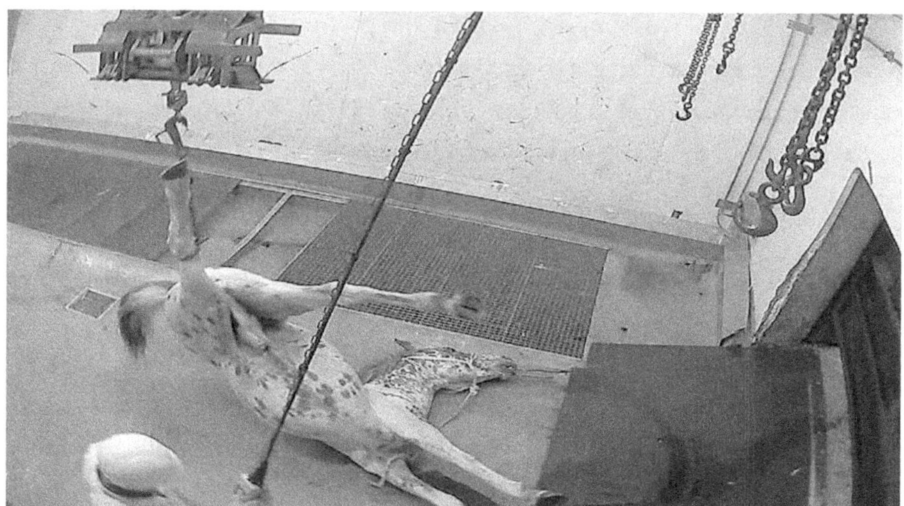

Figure 10: Abattoir Alès 2015. L214 - Ethique & Animaux, CC BY 3.0, via Wikimedia Commons.

The investigation led to public outrage and forced the local authorities, including the mayor of Alès, to temporarily shut down the facility while a thorough investigation took place. This incident brought attention to broader concerns about animal welfare in abattoirs across France and sparked debates about the enforcement of existing animal welfare laws in slaughterhouses.

Activists from L214 emphasized the need for stricter controls and greater transparency in the meat production industry. They called for reforms in slaughter practices and encouraged consumers to reflect on their choices regarding animal products.

Live Animal Transport

The transport of livestock within the EU is regulated by Council Regulation (EC) No 1/2005 on the protection of animals during transport, which outlines specific conditions for the humane transportation of animals. These conditions include adequate ventilation, access to water, rest intervals, and limits on transport times. For long journeys (over eight hours), additional rules apply, including the use of vehicles that meet higher welfare standards. Violations of transport regulations can result in substantial fines and penalties for transport operators. The regulation is strictly enforced by national authorities in collaboration with the European Commission to ensure animal welfare during cross-border and intra-EU transport.

Veal Calf Welfare

In the European Union, veal calves are protected by the Council Directive 2008/119/EC, which establishes minimum standards for the protection of calves, including rules on housing, feeding, and care. This directive prohibits the use of restrictive veal crates that were previously widespread and requires that calves are housed in group pens by a certain age. These standards ensure that calves have space to move, interact socially, and receive proper nutrition. Many EU countries have gone beyond these minimum standards, promoting higher welfare practices for veal production.

Wild Horse and Burro Management

Although wild horses are less common in Europe than in North America, the EU does have policies for managing semi-wild and feral animal populations in some member states. The management of these animals focuses on maintaining a balance between conservation and humane treatment, with strategies such as fertility control and relocation being prioritized to avoid unnecessary harm. EU animal welfare legislation ensures that any interventions, such as roundups or relocations, are carried out in a manner that minimizes stress and suffering.

Greyhound Welfare

Greyhound racing exists in some EU countries, but the welfare of racing dogs is a significant concern. EU member states have varying levels of regulation over the sport, with some banning it entirely due to welfare concerns. In countries where greyhound racing continues, welfare guidelines focus on ensuring adequate housing, medical care, and humane handling of the dogs. The Federation of European Greyhound Racers has set out guidelines to promote the welfare of racing dogs, but concerns persist over the treatment of retired greyhounds and the conditions in some racing kennels.

Horse Welfare in Racing and Euthanasia

Thoroughbred horse racing is popular in many EU countries, and horses involved in this sport are protected by both national and EU animal welfare laws. The focus is on minimizing injury and stress to the animals, with rules governing their training, racing, and retirement. In cases where horses are injured or suffering, euthanasia is carefully regulated and must be performed under veterinary supervision to ensure it is humane. EU member states are responsible for enforcing these regulations through national animal welfare bodies.

National Animal Welfare Frameworks

The European Union has established a robust framework for animal welfare, which is harmonized across member states. The EU Strategy for the Protection and Welfare of

Animals (2012-2015) laid the groundwork for the development of welfare standards in agriculture, transport, and slaughter. The EU works closely with national governments and the European Food Safety Authority (EFSA) to research and promote best practices for animal welfare. The EU also supports ongoing research into improving livestock management practices, including the use of pain relief for invasive procedures and advancements in animal housing conditions.

Pig Welfare

Pig welfare is a key concern in the EU, regulated under the Council Directive 2008/120/EC, which sets minimum standards for the welfare of pigs. This directive includes restrictions on the use of gestation crates, requiring that pregnant sows are given freedom of movement for most of their pregnancy. The directive also mandates humane handling practices, space requirements, and the use of enrichment materials to allow pigs to engage in natural behaviours. Enforcement of these standards is carried out by national authorities, and the Common Agricultural Policy (CAP) provides funding to support farms that adopt higher welfare standards.

Poultry Welfare

The welfare of poultry in the EU is governed by Council Directive 1999/74/EC, which sets standards for the protection of laying hens and Council Directive 2007/43/EC, which governs the welfare of broiler chickens. These directives mandate specific conditions for cage, barn, and free-range systems, ensuring that birds have space to move, perch, and nest. Additionally, organic certification schemes within the EU require even higher welfare standards, including access to outdoor areas. The EU has been at the forefront of promoting alternatives to conventional caged systems, with many member states phasing out battery cages altogether.

Drought and Emergency Conditions

In cases of drought or other emergency conditions, livestock producers in the EU must comply with strict welfare regulations that ensure animals have adequate food, water, and shelter. The EU offers financial support to farmers through programs like the European Agricultural Fund for Rural Development (EAFRD), which helps farmers manage livestock during challenging conditions. If farmers are unable to meet welfare standards, they are encouraged to sell or humanely euthanize animals to prevent suffering. The transport of animals in poor condition is prohibited under EU law, with penalties for non-compliance.

Animal welfare in the European Union is a critical component of the livestock industry, with a wide range of regulations covering all aspects of animal care, from slaughter to transport and sector-specific standards. Both the EU and national governments play an essential role in enforcing these standards, ensuring that livestock production adheres to ethical and humane practices. Ongoing research and development continue to drive improvements in animal welfare, while the EU's commitment to high welfare standards ensures compliance with both domestic and international regulations.

Animal Welfare Issues – Indian Context

Animal welfare in India's livestock industries is governed by a combination of central and state laws, aimed at ensuring the humane treatment of animals throughout their life cycle, including transportation, handling, and slaughter. These regulations seek to prevent unnecessary suffering, minimize stress, and promote ethical practices in animal care. Key areas of concern include the welfare of animals in slaughterhouses, conditions during live transport, welfare standards for calves and other young animals, and issues related to specific livestock sectors such as poultry, pigs, and cattle.

Indian Slaughterhouses

Slaughterhouses in India are regulated under the Prevention of Cruelty to Animals (Slaughter House) Rules, 2001, which are framed under the Prevention of Cruelty to Animals Act, 1960. These rules require that animals are slaughtered in a manner that minimizes pain and suffering. Indian law mandates that animals be stunned before slaughter to ensure they are unconscious and insensible to pain during the process. In religious contexts, specific exemptions exist, allowing for traditional methods of slaughter. The Food Safety and Standards Authority of India (FSSAI) monitors compliance with these regulations in both domestic and export-oriented slaughterhouses. Violations, such as improper handling or inhumane slaughter practices, are met with legal penalties.

Live Animal Transport

The transport of livestock in India is regulated by the Transport of Animals Rules, 1978, under the Prevention of Cruelty to Animals Act, 1960. These rules detail the conditions under which animals can be transported, focusing on reducing stress and suffering. Requirements include providing adequate space, ventilation, and rest periods during long journeys. Livestock transported for more than 12 hours must be given rest, water, and food to ensure their welfare. Non-compliance with these rules can result in fines and legal action. Authorities regularly monitor transport conditions to ensure adherence to these welfare standards.

Calf Welfare

In India, calf welfare is particularly relevant in the dairy industry, where male calves are often considered surplus. Although there are no specific central laws dedicated to calf welfare, the general provisions of the Prevention of Cruelty to Animals Act, 1960 apply. The focus is on ensuring humane treatment, adequate feeding, and proper housing for calves. Some states have introduced additional guidelines to prevent the unnecessary culling of male calves, and initiatives promoting ethical treatment of calves are being increasingly supported by animal welfare organizations.

Management of Wild Horses and Camels

In India, the management of wild horses and camels is largely handled through state laws and initiatives. Although not as prevalent as in the U.S., certain regions in India, particularly Rajasthan, manage camel populations under the Rajasthan Camel (Prohibition of Slaughter and Regulation of Temporary Migration or Export) Act, 2015, which aims to protect these animals from cruelty and exploitation. Welfare concerns regarding camels in traditional and tourist industries are also addressed under general animal welfare laws. Wild horses are less of a concern in India, but other feral animal populations are managed through state-driven programs that focus on humane treatment and population control.

Greyhound and Race Animal Welfare

While greyhound racing is not common in India, the welfare of racing animals such as horses is a significant concern. The Indian National Stud Book, which governs horse racing, enforces rules to ensure the welfare of thoroughbred horses. This includes monitoring training and racing conditions to prevent overexertion and injuries. Euthanasia of horses, when required, must be carried out under veterinary supervision to ensure it is done humanely. Animal welfare organizations continue to advocate for higher welfare standards in the horse racing industry.

National Animal Welfare Frameworks

India has a robust framework for animal welfare, largely shaped by the Prevention of Cruelty to Animals Act, 1960, and the work of the Animal Welfare Board of India (AWBI). These frameworks emphasize the humane treatment of animals across all sectors, including livestock farming, transportation, and slaughter. The AWBI provides guidance on best practices for livestock management and supports initiatives aimed at improving animal welfare across the country. The government also funds research into humane practices, such as the development of alternative methods for pain relief during procedures like dehorning and castration.

Pig Welfare

Pig welfare in India is regulated under the general provisions of the Prevention of Cruelty to Animals Act, 1960, but the Indian Pig Farming Standards also provide voluntary guidelines for the humane treatment of pigs. These include recommendations on housing, feeding, and handling practices that promote animal well-being. Gestation crates, which are controversial in many countries, are not commonly used in India, and the emphasis is on group housing systems that allow pigs to move freely. Pig farmers are encouraged to adopt humane euthanasia practices when necessary, and inspections are conducted to ensure compliance with welfare standards.

Poultry Welfare

Poultry farming is a major industry in India, and welfare concerns are addressed through the Prevention of Cruelty to Animals (Egg Laying Hens) Rules, 2019, and the Model Code of Practice for the Welfare of Domestic Poultry. These rules set standards for housing, feeding, and handling chickens in both cage and free-range systems. Additionally, organic poultry farms must adhere to stricter welfare guidelines, including outdoor access for birds. The Bureau of Indian Standards (BIS) also regulates poultry products to ensure compliance with welfare and safety standards.

Figure 11: Battery cage facilities in Haryana, India in 2009. Sangamithra Iyer and Wan Park, CC BY-SA 3.0, via Wikimedia Commons.

Drought and Emergency Conditions

During periods of drought and other emergency conditions, livestock producers in India are responsible for ensuring the welfare of their animals by providing adequate food and water. The National Disaster Management Authority (NDMA), along with state agricultural departments, offers support programs to help farmers manage livestock during challenging times. In extreme cases where animals cannot be adequately cared for, farmers are encouraged to sell or humanely euthanize them to prevent suffering. Transporting animals in poor condition is strictly prohibited under Indian animal welfare laws, with penalties for violations.

Animal welfare issues in India's livestock industry span a wide range of sectors and activities, from slaughterhouses and live transport to the treatment of racing animals, pigs, and poultry. Both central and state governments play crucial roles in enforcing welfare standards, while industries are increasingly adopting humane practices and technologies. These regulations and frameworks are vital for maintaining ethical practices in livestock production and ensuring compliance with both domestic and international welfare standards.

Animal Welfare Issues in Brazil's Livestock Industry

Animal welfare in Brazil's livestock industries is governed by a combination of federal and state laws, designed to ensure the humane treatment of animals throughout their entire life cycle. This includes their transportation, handling, and slaughter. These regulations aim to prevent unnecessary suffering, minimize stress, and promote ethical and legal practices. Key areas of concern in Brazil's context include the welfare of animals in slaughterhouses, live animal transport conditions, and specific welfare standards for sectors such as poultry, pigs, and cattle.

Slaughterhouses in Brazil

Brazilian slaughterhouses must comply with national animal welfare regulations, particularly those outlined by the Ministry of Agriculture, Livestock, and Supply (MAPA). Brazil has adopted the Humane Slaughter Guidelines, which stipulate that animals must be stunned before slaughter to ensure they are unconscious and do not feel pain. The Brazilian Agricultural and Livestock Inspection Service (SIF) is responsible for overseeing slaughter processes in both domestic and export-oriented slaughterhouses. Inspectors ensure that animals are handled humanely and that any violations, such as improper stunning or excessive use of prods, are addressed.

Live Animal Transport

The transport of livestock in Brazil is regulated by federal laws, which emphasize the humane treatment of animals during transit. The rules set requirements for the conditions under which animals can be transported, including adequate ventilation, access to water, and ensuring that transport times are minimized to reduce stress. The Transport of Animals Act establishes guidelines to prevent the overcrowding of vehicles and ensures that animals are not exposed to extreme conditions during transport. Violations of these laws can lead to fines and penalties for companies responsible for transport.

Welfare of Calves and Young Animals

Calves, particularly those raised for veal, are subject to specific welfare standards in Brazil. These standards govern the conditions under which young animals must be housed, fed, and transported. Many producers adhere to industry guidelines, which aim to prevent extreme confinement and ensure that young animals receive proper nutrition and health monitoring. Furthermore, state-level regulations may include additional provisions, such as banning restrictive crates for calves, in an effort to promote more humane farming practices.

Management of Wild Animals

In Brazil, wild animal management is also a significant issue. The Brazilian Institute of Environment and Renewable Natural Resources (IBAMA) plays a crucial role in overseeing the management and protection of wild animal populations. These animals, especially those living in protected areas or subject to relocation, must be treated in accordance with strict welfare standards. Techniques such as aerial captures are sometimes used to manage wild populations, though these methods are often scrutinized by animal welfare groups advocating for more humane solutions, such as fertility control.

Welfare of Racing Animals

The welfare of animals used in racing, such as horses and dogs, is a growing concern in Brazil. Regulations are in place to ensure that racing animals are kept in humane conditions and that their health is monitored. However, concerns remain, particularly in the case of greyhound racing and horse racing, where injuries and inadequate care after retirement are issues. Industry groups are working towards developing welfare guidelines that address these concerns, ensuring that animals receive proper medical care, adequate housing, and humane handling.

National Animal Welfare Frameworks

Brazil's livestock industry collaborates with government bodies, such as MAPA and Embrapa, to maintain high animal welfare standards. These standards include best practices for husbandry, feeding, transportation, and slaughter, aimed at improving animal welfare across the sector. The government also supports research and development programs that promote humane livestock management practices, such as the use of pain relief in invasive procedures and improved housing conditions for farm animals.

Pig Welfare

Pig welfare in Brazil is regulated by national laws as well as voluntary industry programs. The Brazilian Pork Quality Assurance Program sets guidelines for the humane treatment of pigs, including limiting the use of gestation crates for pregnant sows and providing standards for humane euthanasia and transportation. Producers are encouraged to adopt modern practices that align with international animal welfare standards to ensure that Brazil remains competitive in global markets, where animal welfare is a growing concern.

Poultry Welfare

In Brazil, poultry welfare is regulated through federal guidelines, including those that govern housing, feeding, and handling. The Brazilian Poultry Association (ABPA) has developed welfare standards for both conventional and free-range poultry systems. These standards ensure that poultry are raised in humane conditions, with access to outdoor spaces in free-range systems. Additionally, Brazil's growing organic market has seen the implementation of specific welfare requirements for poultry labelled as organic, which must comply with stricter standards regarding living conditions and animal treatment.

Drought and Emergency Conditions

In times of drought or other emergency conditions, livestock producers in Brazil are responsible for maintaining the welfare of their animals by providing adequate food and water. The government offers assistance programs to help farmers manage livestock during such periods, including financial aid for feed and water supplies. If producers are unable to adequately care for their animals, they are encouraged to either sell or humanely euthanize them to prevent suffering. Transporting animals in poor condition can result in legal penalties under Brazil's animal welfare laws.

Animal welfare issues in Brazil's livestock industry cover a wide range of sectors and activities, from slaughterhouses and live transport to the care of racing animals and poultry. Both the federal and state governments play a critical role in enforcing welfare standards, while the industry continues to adopt new technologies and research-based practices to improve the treatment of animals. These regulations and frameworks are essential for maintaining ethical practices in livestock production and ensuring that Brazil meets both domestic and international welfare standards, which are increasingly important for market access and consumer trust.

Animal Welfare Issues in South Africa's Livestock Industry

Animal welfare in South Africa's livestock industry is governed by a combination of national laws, provincial regulations, and industry standards aimed at ensuring the humane treatment of animals throughout their life cycle. This includes transportation, handling, and slaughter. These regulations are designed to prevent unnecessary suffering, minimize stress, and promote ethical and legal practices. Key areas of concern include the welfare of animals in abattoirs, conditions during live transport, welfare standards for young animals, and sector-specific issues such as poultry, pigs, and cattle.

Slaughterhouses in South Africa

South African abattoirs are required to comply with national animal welfare laws, particularly those outlined by the Meat Safety Act and Animal Protection Act. These laws mandate that animals be slaughtered in a manner that minimizes pain and suffering. Similar to international practices, animals must be stunned before slaughter to ensure they are unconscious and insensible to pain. The Department of Agriculture, Land Reform and Rural Development (DALRRD) is responsible for overseeing compliance with these regulations. Inspectors are tasked with ensuring humane handling and enforcing penalties for violations such as improper stunning or excessive use of electrical prods.

Live Animal Transport

The transport of livestock in South Africa is governed by national and provincial regulations under the Animal Protection Act and the Livestock Welfare Coordinating Committee (LWCC) guidelines. These rules specify the conditions under which animals must be transported, including proper ventilation, access to water, and limited travel times to reduce stress and fatigue. South Africa also follows international standards set by the World Organisation for Animal Health (OIE), ensuring that animals transported for longer periods are adequately rested, watered, and fed. Failure to comply with these regulations can result in fines and penalties for transport companies.

Welfare of Veal Calves and Young Animals

Calves, particularly those raised for veal, are subject to specific welfare standards in South Africa. National regulations address the housing, feeding, and transportation of young animals. Industry players often adhere to welfare guidelines developed in collaboration with the South African Meat Industry Company (SAMIC), which prohibit extreme confinement and require adequate nutrition and health monitoring. There is a growing move in South Africa to phase out restrictive crates for calves, promoting more humane farming practices.

Management of Wild Horses and Donkeys

In South Africa, wild horses and donkeys, especially those in rural and conservation areas, are often managed by local authorities and animal welfare organizations. These animals face challenges such as abandonment and overpopulation. Provincial conservation authorities, with the support of the National Council of Societies for the Prevention of Cruelty to Animals (NSPCA), oversee the management of wild equine populations. Humane management practices, such as fertility control and relocation, are encouraged to reduce the need for inhumane interventions.

Greyhound Welfare

The welfare of greyhounds in South Africa, particularly those used in racing, has come under scrutiny in recent years. While greyhound racing is not as widespread as in some other countries, concerns remain about the treatment of racing dogs, including injuries and the fate of retired dogs. The NSPCA plays an active role in advocating for stricter welfare regulations for greyhounds, ensuring they receive proper care, medical treatment, and humane handling throughout their lives.

Horse Welfare in Racing and Euthanasia

Horses used in racing and equestrian sports are protected under South African law, including the Animal Protection Act. These laws ensure the humane treatment of racehorses, with a focus on minimizing injury and stress during training and competition. The South African Jockey Club has developed strict welfare guidelines for racing, addressing issues such as doping, over-racing, and ensuring a humane retirement for horses. When a horse is severely injured or suffering, euthanasia is carried out under veterinary supervision, following humane procedures.

National Animal Welfare Frameworks

South Africa's livestock industry collaborates closely with government bodies, such as DALRRD and the NSPCA, to ensure high animal welfare standards. The industry adheres to both national regulations and international guidelines from the OIE, which cover best practices for husbandry, feeding, transportation, and slaughter. South Africa also invests in research and development to promote humane livestock management, including pain relief for invasive procedures and improved housing conditions for animals.

Pig Welfare

Pig welfare in South Africa is governed by both national regulations and industry standards, such as the Pork 360 Welfare Program, which provides guidelines for the humane treatment of pigs. These guidelines address issues such as the use of gestation crates for pregnant sows, which are being phased out in favour of more humane systems. The program also includes standards for humane euthanasia and transportation, ensuring that pigs are handled with care throughout their lives.

Poultry Welfare

Poultry welfare in South Africa is regulated through the South African Poultry Association (SAPA) guidelines and national laws. These regulations ensure proper housing, feeding, and handling of poultry, whether in conventional or free-range systems. South Africa's organic certification program also includes specific welfare requirements for poultry, ensuring that animals labelled as organic are raised in humane conditions with access to outdoor spaces.

Drought and Emergency Conditions

South Africa, being prone to drought, requires livestock producers to take special care of their animals during extreme weather conditions. The government offers assistance programs to help farmers manage livestock during droughts, including financial support for feed and water supplies. In cases where animals cannot be adequately cared for, farmers are encouraged to sell or humanely euthanize them to prevent suffering. Transporting animals in poor condition can result in legal penalties under South Africa's animal welfare laws.

Animal welfare issues in South Africa's livestock industry encompass a wide range of sectors and activities, from abattoirs and live transport to the care of racing animals and poultry. Both national and provincial governments play a crucial role in enforcing welfare standards, while the livestock industry continues to adopt new technologies and practices aimed at improving the treatment of animals. These regulations and frameworks are vital for maintaining ethical practices in livestock production and ensuring that South Africa complies with both domestic and international welfare standards.

CHAPTER FOUR

Signs of Welfare Issues

Basic Welfare Needs

The basic welfare needs of livestock are essential components of responsible animal husbandry and are legally mandated to ensure the well-being of animals under the care of their owners. These welfare needs are based on both ethical and legal frameworks that emphasize the humane treatment of animals, covering aspects such as water, food, shelter, and protection from environmental threats. Below is a detailed explanation of these basic welfare needs [29]:

Provision of Water, Food, Air, and Shelter

The most fundamental requirement for livestock welfare is ensuring access to basic necessities, including clean water, adequate food, fresh air, and shelter. Livestock owners must ensure that animals have a continuous supply of fresh water to maintain hydration and promote health. The animals should be fed a balanced and nutritious diet that is appropriate for their species, age, and stage of development. The availability of fresh air is vital, especially in confined or intensive housing systems, as poor ventilation can lead to respiratory issues.

Shelter is equally important, as livestock must be protected from harsh environmental conditions, such as excessive heat, cold, or rain. The shelter should be designed to provide comfort, keeping the animals dry and safe, while also ensuring proper ventilation to prevent overheating or respiratory problems.

Protection from Drought, Extreme Weather, Predators, and Disease

Livestock owners are responsible for safeguarding their animals from environmental hazards like drought and extreme weather. This includes ensuring that the animals have enough water and shade during periods of intense heat or shelter during cold, wet weather. Protective measures should be in place to guard livestock against predators, which could harm or distress them, and prevent exposure to parasites and diseases. Regular veterinary care, vaccination, and parasite control programs are critical for maintaining herd health and minimizing the risk of disease outbreaks.

Adequate Space and Freedom of Movement

Livestock must be provided with sufficient space to stand, lie down, walk, turn around, and stretch their limbs comfortably. This is essential not only for their physical health but also for reducing stress and preventing injuries that may arise from overcrowded or poorly designed living conditions. The ability to move freely and express natural behaviours is key to promoting both physical and psychological well-being in livestock.

Allowance for Natural Behaviours and Social Contact

Livestock owners must recognize the importance of allowing animals to engage in behaviours that are natural to their species. For example, animals such as pigs may root in the soil, chickens may dust bathe, and cattle may graze. Restricting these natural behaviours can lead to frustration, stress, and behavioural problems. Additionally, livestock animals are often social creatures and should have the opportunity for social interaction with others of their kind, as isolation can lead to distress and poor welfare.

Veterinary Treatment and Euthanasia When Necessary

It is a legal and ethical requirement for livestock owners to ensure that sick or injured animals receive prompt and appropriate veterinary care. If an animal's condition is incurable or if the animal is suffering severely, euthanasia may be necessary to prevent prolonged suffering. This decision must be made compassionately and in consultation with a veterinarian.

Animal Nutrition

Animal nutrition is a fundamental aspect of livestock management, and it is legally required that owners provide their animals with adequate, safe, and nutritious food to meet their daily energy needs. Proper nutrition ensures that livestock can maintain their health, produce high-quality products like meat, milk, or wool, and achieve optimal growth and reproduction rates. The nutritional needs of animals can vary greatly depending on factors such as age, reproductive status, and species, which makes understanding and fulfilling these requirements essential [29].

Nutritional Requirements for Livestock

All livestock require a balanced diet to meet their energy, protein, mineral, and vitamin needs. Energy is essential for maintaining body functions, growth, and production, while protein supports muscle development, reproduction, and lactation. Minerals and vitamins are critical for metabolic processes, bone health, and immune function. Animals must be provided with high-quality feed that is safe and free from contaminants or harmful substances.

Special Nutritional Needs for Pregnant and Lactating Animals

Certain animals, such as those that are pregnant or lactating, have higher nutritional demands. Pregnant livestock require additional energy and protein to support the growth of their developing offspring, while lactating animals need even higher levels of nutrition to produce milk. For example, dairy cows in lactation require more calories and a carefully balanced diet rich in protein and minerals to sustain milk production while maintaining their health. Failure to meet these elevated nutritional requirements can lead to poor health outcomes for both the mother and offspring, such as decreased milk production, low birth weights, or health complications [29].

Pasture Management and Carrying Capacity

For animals kept on pasture, it is critical to assess the availability and quality of the pasture to ensure sufficient feed. Pasture quality and the number of animals a paddock can support are measured by the term "carrying capacity." This is typically expressed as the number of animals that can be sustained per unit of land, based on a standard known as the Dry Sheep Equivalent (DSE). The DSE is a measure of the amount of feed (kilograms of dry matter) required to maintain a wether (castrated ram) weighing between 45 and 50 kg per day. Livestock species that require more feed have higher DSE ratings, while those that need less have lower ratings [29].

For example, a steer may have a DSE rating of 5.0, while a cow with a calf may require a DSE rating as high as 14.0. The availability of pasture must match the nutritional needs of the livestock, and if the carrying capacity is exceeded, the animals may not get enough nutrition from the pasture alone. In such cases, livestock owners must consider options like supplementary feeding with grain or hay, or agistment, where livestock are temporarily moved to another farm with adequate pasture [29].

Adjusting Feeding Based on Livestock Class

Different classes of livestock have varying nutritional needs based on their physiological state. For instance [29]:

- Sheep typically have a DSE of 1.0 to 2.0 depending on whether they are wethers or rams.

- Beef cattle range from 5.0 to 14.0 DSE, with cows with calves requiring the most feed.

- Dairy cattle can have a DSE of 3.0 to 16.0, reflecting their high-energy demands, especially for lactating cows.

- Goats have lower DSE ratings, between 0.7 and 2.0, but this varies based on age and reproductive status.

Meeting the nutritional needs of livestock is not only a legal obligation but also a crucial part of responsible animal management. Ensuring animals receive the right amount of high-quality feed allows them to grow, reproduce, and stay healthy. Regular assessment of pasture quality and carrying capacity, along with the provision of supplemental feeding when necessary, are key practices in maintaining proper livestock nutrition. Additionally, understanding the specific nutritional demands of pregnant and lactating animals ensures their well-being and the health of their offspring.

Supplementary Feeding

Supplementary feeding is a critical aspect of livestock management when natural pasture is insufficient to meet the animals' nutritional needs. This practice involves providing additional feedstuffs to ensure animals receive the necessary energy, protein, roughage, and minerals to maintain their health and productivity. Supplementary feeding becomes essential during times of drought, seasonal shortages, or when the quality of pasture is low.

Types of Feedstuffs Used in Supplementary Feeding

The specific type of supplementary feed depends on the nutritional deficiencies in the animals' diet. The most commonly used feedstuffs include:

- **Energy sources**: Grains (such as corn, barley, or wheat), molasses, and silage are frequently used to boost the energy intake of livestock. Energy is vital for body maintenance, growth, and production, especially during colder months or periods of high activity.

- **Protein sources**: Protein-rich feeds like cottonseed meal, lupins, and silage help support muscle development, milk production, and overall health. Protein is especially important for lactating animals and growing young.

- **Roughage**: Hay and silage provide the necessary fibre to maintain healthy digestion in ruminants. Roughage ensures proper gut function and helps prevent digestive disorders such as acidosis.

- **Minerals**: Mineral deficiencies can have significant impacts on livestock health. It's common to provide mineral supplements in the form of pre-prepared licks, which ensure livestock consume the appropriate amounts. These are critical for functions like bone development, reproduction, and metabolic processes.

Importance of Gradual Introduction of New Feed

Whenever supplementary feeding is introduced, it is important to do so gradually. Abrupt changes to the diet, especially the introduction of grain, can lead to serious health problems. **Grain overload**, or acidosis, can occur when livestock consume too much grain too quickly. This condition leads to digestive disturbances, dehydration, and in severe cases, death. To avoid this:

- Introduce grain slowly over time.

- Ensure all animals have equal access to the feed to prevent overconsumption by dominant individuals.

- Offer small amounts of grain more frequently, supplemented by roughage like hay to aid digestion.

Prohibited Feed Items

It is vital to adhere to strict guidelines regarding the types of feed that can be given to livestock. Certain materials are **illegal** to feed to livestock due to the risk of disease transmission. Restricted items include:

- Meat and bone meal

- Blood meal

- Poultry offal meal

- Feather meal

- Fishmeal

- Any other animal meals or manures

These restrictions are in place to prevent the spread of diseases such as Bovine Spongiform Encephalopathy (BSE), also known as mad cow disease. Care must also be taken when feeding scraps or leftovers to livestock, as even incidental contact with meat products can pose a risk. Restaurants or bakery scraps, for example, may inadvertently contain traces of meat or animal products and should be avoided.

Seeking Professional Guidance

Given the complexity of animal nutrition and the potential risks of improperly balanced diets, it is highly recommended to seek professional advice when developing a supplementary feeding program. Nutritionists can help analyse deficiencies in the diet and recommend appropriate feed supplements. This is particularly important in intensive farming systems, where the health and productivity of livestock directly impact profitability.

By carefully managing supplementary feeding, livestock owners can ensure their animals remain healthy, productive, and free from dietary-related diseases, while adhering to legal and ethical standards.

Water

Providing clean and adequate water to livestock is crucial for maintaining their health and overall well-being. Water requirements differ across species, age groups, and can significantly fluctuate depending on factors such as weather, lactation, and feed type. Ensuring consistent access to fresh water is fundamental to supporting their daily physiological processes [29].

Water Requirements for Livestock

The daily water needs of livestock vary depending on species and age. Below is an average guide for water consumption per animal per day [29]:

- **Sheep:**

    ○ Weaners: 2–4 litres

    ○ Adult dry (grassland): 2–6 litres

    ○ Adult dry (saltbush): 4–12 litres

    ○ Ewes with lambs: 4–10 litres

- **Cattle:**

    ○ Lactating cattle (grassland): 40–100 litres

    ○ Lactating cattle (saltbush): 70–140 litres

    ○ Young stock: 25–50 litres

    ○ Dry stock (400kg): 35–80 litres

- **Goats:**

    ○ Weaners: 4–6 litres

    ○ Adult dry: 5–7 litres

    ○ Doe with kid: 5–10 litres

Factors Influencing Water Consumption

- **Lactating and Pregnant Animals:** These animals require almost twice as much water compared to non-lactating or non-pregnant animals, as their bodies need more fluids to produce milk and support gestation.

- **Weather Conditions:** In extremely hot weather, livestock can consume up to 80% more water than usual. However, they avoid warm water, so it's crucial to provide shaded or deep water sources to keep the water cool and refreshing.

- **Age Differences:** Older livestock generally require more water than younger animals because of their larger body mass and slower metabolism.

Management of Water Sources

To ensure livestock always have access to adequate water, it is important to [29]:

1. **Provide Multiple Water Sources:** Having more than one water source in paddocks, sheds, and yards guarantees that livestock will always have access to water, even if one source fails or becomes contaminated.

2. **Regularly Check Water Sources:** Water systems should be routinely inspected to make sure they are functioning properly, providing sufficient water flow, and delivering clean, uncontaminated water.

3. **Maintain Clean Troughs:** Keeping water troughs clean encourages animals to drink sufficient water, as dirty or algae-filled troughs can deter animals from drinking. Additionally, unclean water sources may cause diseases, particularly if they harbor dead animals or parasites.

Water Quality Considerations

Water quality directly affects how much animals will drink. Pipes that are above ground can heat up the water, making it undrinkable in hot weather. It's essential to ensure water delivered to the animals is cool and clean to encourage adequate hydration. Dirty water can lead to health issues, such as dehydration or infections.

Environment

Providing adequate shelter is essential for livestock welfare, as it protects animals from extreme weather conditions such as cold, heat, wind, and rain. Ensuring that animals are sheltered can prevent stress and health issues, particularly during vulnerable periods, such as after birth or shearing [29].

Shelter is critical for maintaining animal well-being in adverse weather. Sheep and goats, especially when newly shorn or newborn, are particularly susceptible to cold winds or extreme heat, making protective structures essential. Without appropriate shelter, animals may suffer from hypothermia, heat stress, or exhaustion, which can affect their health and productivity [29].

Trees and shrubs can serve as natural windbreaks, protecting livestock from harsh winds and extreme temperatures. These natural shelters also encourage pasture growth by reducing wind erosion and maintaining soil moisture. In hot conditions, trees provide essential shade, allowing animals to cool down, while in cold weather, they offer protection from chilling winds.

During critical periods such as lambing, calving, or kidding, natural shelter is even more important. Animals can hide from predators in these tree-lined areas, reducing stress and increasing their chances of survival. Livestock with access to natural shelter tend to have better health and performance compared to those left exposed to the elements.

In addition to natural options, constructed shelters can offer reliable protection. Structures made from durable materials like corrugated iron, timber, or shade cloth provide shade in hot conditions and windbreaks during colder seasons. These constructed shelters should be sturdy and securely fastened to withstand strong winds or heavy rain. Ensuring shelters are well-built helps prevent accidents or injuries to animals caused by collapsing structures.

Along with weather protection, livestock must be safeguarded from chemical and physical contaminants that may pose risks to their health. Contaminants, such as residues from chemicals or heavy metals, can remain in the environment and enter livestock through exposure, potentially leading to harmful effects on the animals, humans, and the environment [29].

To minimize the risk of contamination, farmers should ensure that animals do not have access to areas where chemical residues might exist. Farm dumps, chemical storage sheds, and old orchards or machinery sites are potential hazards. Removing any leftover materials, such as old batteries, chemical drums, or discarded machinery parts from pastures, can prevent accidental exposure and contamination [29].

Shelter plays a pivotal role in animal welfare, ensuring that livestock are protected from harsh weather and potential contaminants. Providing both natural and constructed shelters, and taking measures to avoid chemical exposure, will result in healthier, more productive animals. Adequate shelter also reduces stress, enhances growth, and contributes to better overall farm management.

Body Condition Score

Body condition scoring (BCS) of livestock is a globally recognized method used to assess the nutritional status and overall health of animals, especially in relation to fat and muscle reserves. This system varies slightly between species and regions but shares the common goal of helping farmers manage the welfare, productivity, and reproductive efficiency of their animals. Below is a summary of how body condition scoring is applied to different livestock around the world:

1. Cattle (Beef and Dairy)

- **Global Overview**: Cattle body condition is often scored on a scale of 1 to 5 or 1 to 9, depending on the region and purpose.

- **North America and Europe**: Typically, a scale of 1 to 5 is used. A score of 1 indicates a severely underweight animal with prominent bones, while 5 represents an overly fat animal with no visible bone structures. In the U.S., beef cattle often use a 1 to 9 scale, providing more granularity in assessing thin and fat extremes.

- **Australia and New Zealand**: A scale of 1 to 5 is common, with 3 being the ideal for most production animals. Farmers monitor body fat distribution on the back, ribs, tailhead, and hips to adjust feeding programs.

- **South Africa**: The 1 to 5 scale is also widely used, where special attention is given to feed management during droughts and other environmental challenges to avoid animals dropping to a score of 1, which requires immediate intervention.

2. Sheep

- **Global Overview**: Sheep body condition is usually scored on a scale of 1 to 5, where 1 indicates emaciation and 5 indicates obesity. Sheep's condition is judged based on muscle and fat deposits along the spine and around the hips.

- **United Kingdom**: The BCS for sheep is often performed by feeling the animal's lumbar region, scoring from 1 (very thin) to 5 (over-fat). A score of 3 is considered ideal for breeding and production.

- **Australia and New Zealand**: BCS of sheep follows the same 1 to 5 scale, with 2.5 to 3 considered the optimal range for breeding ewes. Farmers adjust feed programs based on the seasonal availability of pasture.

3. Goats

- **Global Overview**: The body condition of goats is scored similarly to sheep, using a 1 to 5 scale.

- **United States**: Goats are scored based on the prominence of their ribs, spine, and hip bones. A score of 2.5 to 3.5 is optimal for dairy goats in milk production, while meat goats should fall between 3 and 3.5.

- **Africa**: In many countries, goat body condition scoring is an important aspect of managing livestock in environments with fluctuating feed availability due to seasonal changes or drought conditions.

4. Pigs

- **Global Overview**: Pig body condition scoring ranges from 1 to 5 or sometimes 1 to 9 in some countries.

- **Europe**: A BCS of 3 is considered ideal for sows, with scores of 1 indicating extreme thinness and scores above 4 indicating excessive fat, which can lead to reproductive issues.

- **United States**: The National Pork Board uses a 1 to 5 scale, where a score of 3 is desirable for breeding sows. Lower scores (below 2) can result in poor productivity, while higher scores (above 4) may indicate obesity.

5. Horses

- **Global Overview**: The Henneke Body Condition Scoring System is used globally for horses, on a scale from 1 to 9.

- **United States**: A score of 5 is considered optimal for working horses. A score of 1 indicates a severely emaciated horse, while a score of 9 represents obesity.

- **United Kingdom and Europe**: The same Henneke system is applied, and animal welfare regulations often reference this scale to ensure that horses are maintained in good health.

Factors Affecting Global Body Condition Scoring Practices

- **Environmental Conditions**: In drought-prone regions like Africa and Australia, farmers rely heavily on body condition scoring to manage livestock through periods of feed scarcity.

- **Species and Breeds**: Different breeds of livestock (e.g., Bos indicus vs. Bos taurus in cattle) may have different fat deposition patterns, which can affect the interpretation of BCS.

- **Production Systems**: Dairy animals may have different optimal body condition scores compared to meat-producing animals, due to the varying demands placed on their bodies for milk production versus growth.

The Body Condition Score (BCS) is a practical and widely used tool to assess the nutritional status and general health of livestock, particularly in terms of their fat and muscle reserves. It is crucial for managing livestock effectively, especially during key periods such as breeding, gestation, and lactation. A good BCS can lead to improved reproductive performance, disease resistance, and overall productivity [29].

BCS measures the fat and muscle coverage over certain areas of the animal's body, typically the ribs, loin, and backbone. The score is usually on a scale (e.g., 1 to 5 or 1 to 9) depending on the animal species, with lower scores indicating under-conditioned (thin) animals and higher scores showing over-conditioned (fat) animals. A moderate score reflects optimal condition where the animal has adequate energy reserves without being overly fat [29].

BCS is particularly important in relation to reproduction. Animals that are too thin may have reduced fertility, while over-conditioned animals may experience problems with birthing and milk production. For example, in cattle and sheep, animals with an optimal BCS at the time of breeding tend to have better conception rates. Similarly, females in good body condition during gestation are more likely to give birth to healthy offspring and produce sufficient milk.

The BCS system offers a straightforward way for farmers to assess the nutritional adequacy of their feeding programs. It provides an objective method to adjust feed intake and quality to meet the animals' energy needs at different life stages. For example, if livestock score low on the BCS scale, supplementary feeding may be necessary to improve their condition. Conversely, animals that are too fat can be put on a restricted diet to avoid complications related to obesity.

Regular monitoring of the BCS allows farmers to detect early signs of nutritional deficiency or overfeeding, making it easier to prevent long-term health issues. Adjusting feeding and management practices based on BCS can also help improve productivity and longevity, reducing the risks of disease or reproductive failure.

Body Condition Score (BCS) for Beef Cattle is a critical tool in managing the health, productivity, and well-being of livestock. It evaluates the amount of muscle and fat tissue covering the cattle's body and helps farmers make informed decisions on nutrition and care. Each score describes a particular physical condition of the animal, providing insights into its energy reserves, overall health, and readiness for breeding or market. Below is a detailed explanation of each BCS for beef cattle, ranging from 1 (very poor condition) to 5 (over-conditioned or fat) [29].

Score 1: Very Thin

At a BCS of 1, the animal exhibits a very low level of muscle and fat. The musculature is extremely minimal, and there is no evidence of fat deposits on the body. The skeletal structure, particularly the backbone, ribs, and hips, are highly pronounced and easily visible. Cattle at this condition are severely undernourished and may be at risk of health issues due to the lack of energy reserves.

Figure 12: Example of Score 1.

Score 2: Under-conditioned

A BCS of 2 indicates an animal with a slight covering of tissue over the skeletal frame. Although there is some muscle, the backbone and ribs are still clearly distinguishable. The animal has insufficient fat reserves to support health and productivity, especially during times of nutritional stress, such as pregnancy or lactation. This score suggests that the animal is in need of better nutrition to improve overall condition.

Figure 13: Example of Score 2.

Score 3: Moderate Condition

At BCS 3, the animal has a fair level of musculature but lacks significant fat. While the backbone and ribs are no longer prominent, they can still be felt beneath the skin. The hips are somewhat prominent, but the pins (bones near the tail) are filled out. This is a relatively acceptable condition, though the animal would benefit from increased nutrition to reach an optimal level for reproduction or market readiness.

Figure 14: Example of Score 3.

Score 4: Well-conditioned

A BCS of 4 represents an animal that is well-muscled with an even covering of muscle and fat. The skeletal protuberances such as the hips, ribs, and backbone are smoothly rounded, with no sharp or protruding bones. This is an ideal condition for beef cattle, as it reflects good nutritional management. Cattle at this score are typically healthy and productive, with sufficient energy reserves to handle various physiological demands.

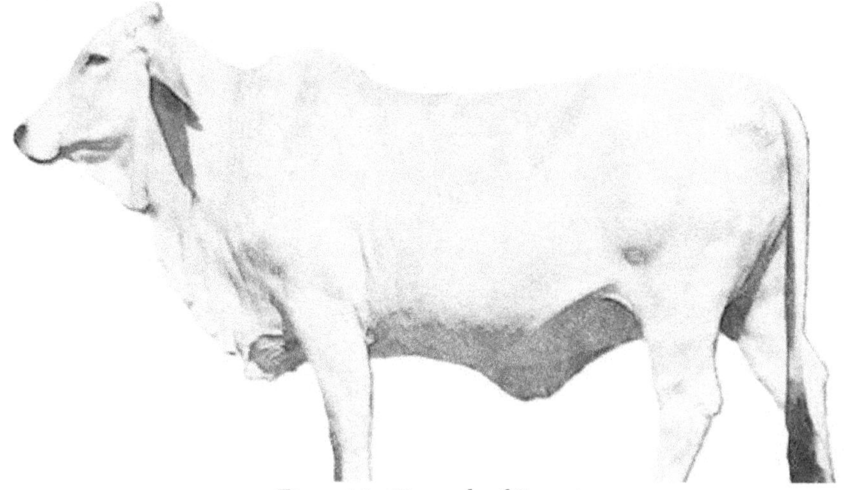

Figure 15: : Example of Score 4.

Score 5: Over-conditioned or Fat

At the highest BCS, 5, the animal has a substantial amount of fat covering the body, and in some cases, fat deposits may appear lumpy, particularly around the pins (hip bones) and flanks. Most of the skeletal structure is no longer distinguishable, as it is buried under thick layers of fat. While such animals may appear healthy, excessive fat can lead to health problems and reduced reproductive performance, making it necessary to regulate their diet.

Figure 16: Example of Score 5.

Welfare Decisions for Beef Cattle

When making welfare decisions for beef cows, it's important to assess their physical condition, particularly their body reserves, muscle wastage, and overall appearance. Beef cows can be categorized into three main risk categories based on their condition: *At risk*, *High risk 1*, and *High risk 2*. Each category requires specific actions to ensure the cows' well-being, especially during critical events such as calving, transportation, and adverse weather conditions.

At Risk Condition

Beef cows in the *At risk* category are typically lean but remain strong and healthy. They may display noticeable muscle wastage, but they still have some body reserves. This condition puts the cow at a higher risk of reduced reproductive performance, as their energy reserves are likely insufficient to support successful conception or pregnancy. The backbone and short ribs are visible, though not overly prominent. While the animal's rump muscles are concave, the cow generally maintains good mobility, appearing bright and alert.

Cows in this condition are typically suitable for transport or sale, provided they are not kept off feed for long periods. These animals must be adequately fed to prevent further weight loss and should be closely monitored, especially during calving, where some assistance may be needed.

High Risk 1 Condition

In the *High risk 1* category, cows are still healthy but exhibit significant muscle wastage, particularly around the loin and leg muscles. These cows are unlikely to conceive due to depleted energy reserves, though with proper care and feeding, they can recover. Their backbone is more visible, and the short ribs are prominent and sharp to the touch. The inside of the pin bones is noticeably sunken, and muscle loss is more evident, particularly in the rump and legs.

Cows in this condition can still move but may require moderate assistance during calving. They are unsuitable for long-distance transport or sale through saleyards but may be sold directly to a farm or abattoir with appropriate management. Immediate intervention is required in terms of providing high-quality feed to stabilize their condition and prevent further deterioration.

High Risk 2 Condition

Cows in the *High risk 2* category are in the most critical state, with very low body reserves and severe muscle wastage throughout the body. They are weak and unable to maintain normal mobility, often displaying an unsteady gait and difficulty standing up or lying down. The backbone and ribs are very prominent, and the animal's overall appearance is dull and lethargic. These cows are at high risk of death, particularly during cold, wet weather, and require intensive, high-quality care to have any chance of recovery.

In this state, cows are unfit for transport or sale and must remain on the farm. High-quality feed, water, and care are essential for these animals, and euthanasia may be considered in extreme cases where recovery is unlikely or the animal is suffering. Close supervision during calving is critical, as these cows will require significant assistance.

Action Required

Across all risk categories, the main actions focus on proper feeding and ensuring adequate care to maintain or improve the cows' condition. As the risk level increases, so does the need for immediate and more intensive interventions, such as supplementary feeding, close monitoring, and preparation for assisted calving. For cows in the highest risk category, transportation is not an option, and humane euthanasia may be necessary to prevent suffering if recovery is not feasible.

Each cow's condition should be assessed regularly to ensure appropriate actions are taken in a timely manner to protect their welfare, productivity, and overall health.

When beef cattle begin to lose fat and reach Fat Score 1, their energy reserves are critically low, and they start to mobilize muscle tissue to meet their body's energy demands. This usually occurs due to an insufficient energy density in their diet or a lack of available feed. At this stage, cattle are considered "At Risk" and require immediate intervention to restore their condition and prevent further muscle loss [30].

Fat Score 1 represents a critical threshold where cattle have lost much of their fat stores and are beginning to deplete muscle tissue to survive. The depletion of muscle is a sign that the animal's nutritional needs are not being met, which compromises their overall health and well-being. If left unaddressed, cattle will continue to deteriorate.

To further categorize the condition of animals falling below Fat Score 1, the Welfare Score system is used, which measures the degree of muscle depletion and overall body condition of beef cattle. There are three levels within this scoring system [30]:

1. High Risk 1: At this stage, muscle wastage is evident as the animal continues to use muscle tissue for energy. The animal is still mobile but shows signs of weakness and reduced muscle mass.

2. High Risk 2: Cattle in this category show significant muscle depletion and are at a high risk of death due to prolonged exposure to cold, wet weather, or additional stress. Their ability to recover depends on receiving immediate, high-quality care.

3. Downer: This is the most severe stage, where cattle are recumbent, unable to stand, and near death. The depletion of body reserves is critical, and without rapid intervention, survival is unlikely.

These Welfare Scores apply to all beef cattle breeds, including British, European, and Bos indicus breed groups and their crosses. The scoring helps farmers and livestock managers assess the urgency of the animal's condition and take appropriate action, such as improving the diet, increasing feed availability, or in extreme cases, considering humane euthanasia to prevent suffering [30].

Intervention Strategies: To improve the condition of animals at Fat Score 1 or lower, farmers must enhance the quality and quantity of feed. Providing high-energy supplements, adjusting the grazing area, and monitoring environmental factors such as weather can help improve the animal's health. In cases where recovery is unlikely due to prolonged muscle loss, euthanasia may be necessary to prevent further suffering.

Fat Score 1 – At Risk is used to describe livestock that are in a state of nutritional risk. This classification signals that the animal is beginning to mobilize its muscle tissue to meet its energy requirements due to insufficient fat reserves, placing the animal in a potentially vulnerable condition. Here's a detailed explanation of the body characteristics and behaviours associated with **Fat Score 1** cattle [30]:

Body Characteristics:

1. **Backbone Visibility**: At this fat score, the animal's backbone is easily visible, but the individual spines are not yet sharply pronounced. This indicates that the fat cover over the backbone has thinned considerably, though some muscle mass remains.

2. **Short Ribs**: These ribs are fairly sharp to the touch and easily identifiable. This is an early sign of muscle and fat loss, which can reduce the animal's insulation against cold and decrease its overall energy reserves.

3. **Tail Head**: The tail head is prominent and devoid of fat. Normally, this area would have a smooth cover of fat in healthier animals, but its prominence here suggests the animal is undernourished.

4. **Pins and Hips**: The area inside the pins (pelvic bones) is slightly sunken, and the hip bones, along with long ribs, are obvious and prominent. This indicates a noticeable reduction in muscle and fat reserves, with the body drawing on those tissues for energy.

5. **Rump and Leg Muscle Wastage**: The rump muscle appears concave (sunken), which is not typical of a healthy animal. The leg muscles are also showing signs of early wasting, where the body is breaking down muscle to compensate for inadequate dietary energy.

6. **Skin and Dewlap**: The skin remains pliable, a sign that dehydration is not yet a concern, but the dewlap (the loose skin under the neck) shows no fat deposits,

further highlighting the lack of energy reserves.

7. **Overall Muscle Condition**: Muscle wasting, particularly in the rump and leg areas, becomes increasingly evident, signifying that the animal's body is breaking down these tissues to meet its energy needs, as the fat reserves are insufficient.

Behaviour:

1. **Alertness**: Despite the physical signs of fat and muscle depletion, animals in **Fat Score 1** often maintain a bright and alert appearance. This can make it more difficult to assess their at-risk status based solely on behaviour, which is why physical scoring is so important.

2. **Mobility**: The gait of the animal remains normal and mobile, without signs of severe lameness or difficulty walking. However, without intervention, this can quickly change as the animal's condition worsens.

Implications for Management:

- **Nutritional Intervention**: Animals at **Fat Score 1** are on the threshold of greater risk and need immediate nutritional support to prevent further decline. Supplementary feeding or adjusting grazing practices is often necessary to ensure the animal does not deteriorate to a higher risk category.

- **Reproductive Health**: These animals are likely to experience reduced reproductive performance due to their compromised nutritional state. Without proper intervention, their ability to conceive and carry pregnancies to term may be negatively affected.

- **Monitoring and Recovery**: Close monitoring of these animals is crucial, and they should be given adequate high-quality feed to reverse the effects of fat and muscle depletion. Recovery can occur with timely and appropriate feeding adjustments.

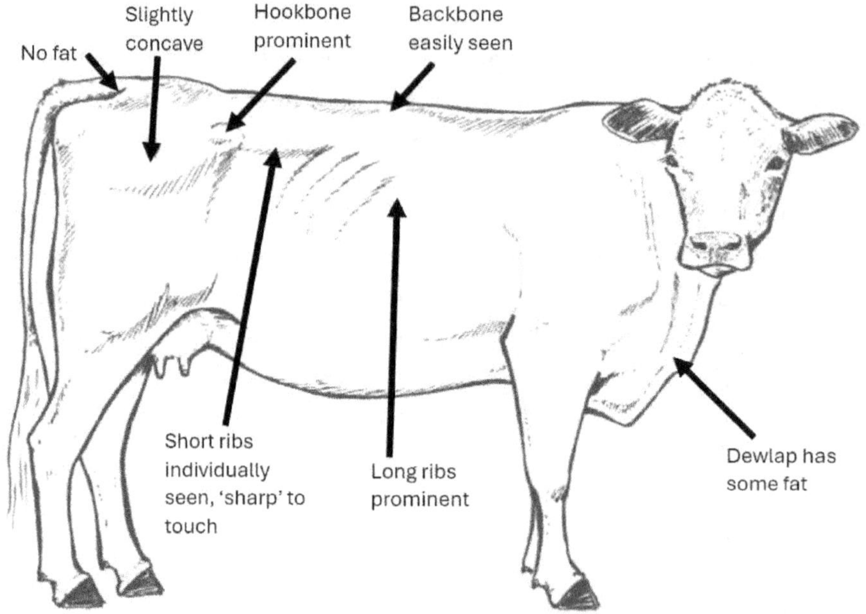

Figure 17: Fat Score 1 – At Risk.

High Risk 1 represents a significant stage of physical decline in livestock, characterized by noticeable muscle depletion and prominent skeletal features. This condition indicates that the animal's body is breaking down muscle tissue to meet its energy needs, as fat reserves have been largely depleted. Below is a detailed explanation of the body characteristics and behaviours associated with this condition [30]:

Body Characteristics:

1. **Muscle Depletion**: One of the key indicators at High Risk 1 is the evident muscle loss in the back, loin, and hind leg areas. The rump muscle becomes more concave as the animal's body mobilizes muscle tissue to compensate for inadequate nutrition.

2. **Spinal Prominence**: The spines of the backbone become easily identifiable. This sharpness along the backbone is due to the depletion of fat and muscle, making the animal's skeletal structure more visible.

3. **Sharp Ribs**: The short ribs are prominent and very sharp to the touch, a clear sign that the animal's body condition is declining. The long ribs, tailhead, and pin bones also become prominently visible, further indicating muscle wastage

and fat loss.

4. **Rump and Loin Muscle Wasting**: The rump muscle appears deeply concave, and muscle wastage in the loin and leg muscles becomes increasingly noticeable. These areas, typically well-muscled in a healthy animal, now show signs of depletion.

5. **Tail and Pin Bones**: The tail bones are just identifiable, and the inside of the pin bones are sunken. This further emphasizes the reduction in fat and muscle reserves.

6. **Stifle Joint**: In this condition, the stifle joint is not identifiable, which reflects further muscle depletion around the leg joints.

7. **Skin and Udder**: The skin, particularly over the hump in Bos indicus breeds, begins to slacken, becoming less firm due to the loss of fat and muscle. The animal's skin also becomes less pliable, a potential indication of dehydration or poor overall health. In lactating animals, the udder starts to shrink as the animal's body struggles to maintain its energy balance.

Behaviour:

1. **Mobility**: While animals in High Risk 1 can still move around, their energy levels are noticeably reduced. They may appear less energetic compared to healthier animals.

2. **Cessation of Grooming Behaviour**: At this stage, animals may stop grooming themselves. This can be a behavioural signal of declining health and well-being, as grooming is often an instinctive activity in healthy livestock.

3. **Ease in Lying Down and Rising**: Despite the muscle wastage and overall condition, these animals can still lie down and rise with ease, which may give the false impression that their condition isn't critical.

4. **Normal Dung and Cud Chewing**: The animal's dung pats remain normal, and cud chewing is still observable. These behaviours indicate that the digestive system is still functioning adequately, although the animal is in a physically depleted state.

Implications for Management:

- **Nutritional Intervention**: Animals at High Risk 1 require immediate nutritional support, including high-energy feeds and possibly protein supplementation to reverse muscle and fat loss. At this point, failure to intervene could lead to further deterioration of the animal's condition, potentially progressing to High Risk 2 or worse.

- **Monitoring for Further Decline**: Continuous monitoring is necessary for these animals, especially in harsh weather conditions, as they may be more susceptible to cold or heat stress due to their low body reserves.

- **Calving or Reproductive Challenges**: For cows, the shrinking of the udder and muscle loss may negatively impact reproductive health. Extra care is required during calving, as animals at this stage may have difficulty supporting a pregnancy or delivering calves.

High Risk 1 is a critical stage in livestock welfare that requires prompt and focused intervention. With proper care, animals at this stage can recover, but without it, they risk further muscle depletion, declining health, and potentially life-threatening conditions.

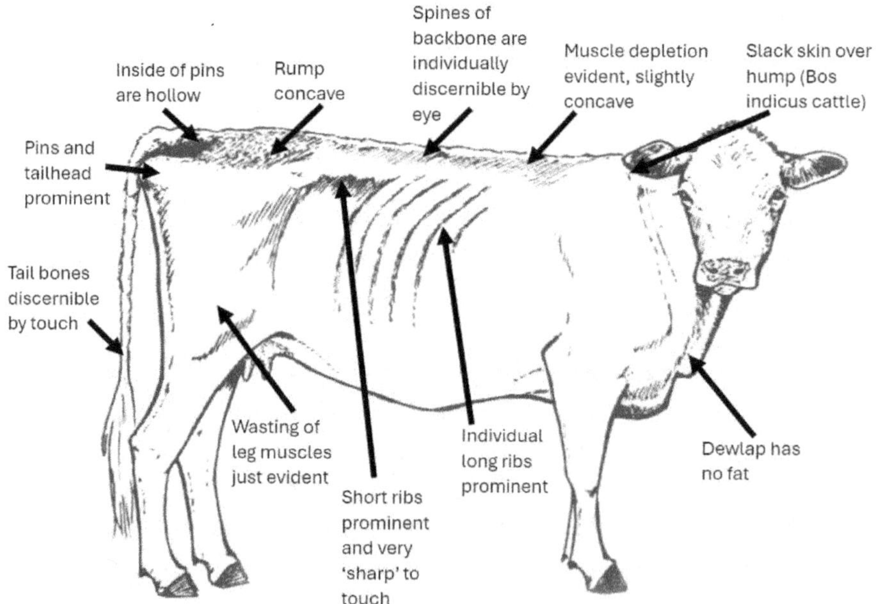

Figure 18: High Risk 1.

High Risk 2 represents an advanced stage of physical deterioration in livestock, where the animal is severely emaciated and at significant risk. Immediate intervention is necessary to prevent death. Here is a detailed breakdown of the key characteristics and behaviours associated with **High Risk 2**:

Body Characteristics:

1. **Emaciation**: At this stage, the animal is visibly emaciated. There is a significant loss of both fat and muscle tissue, leaving the animal weak and vulnerable.

2. **Prominent Spine**: The spines of the backbone are easily identifiable and sharp to the touch. This sharpness indicates severe muscle and fat depletion, as the body mobilizes remaining reserves for energy.

3. **Visible Skeletal Structure**: The hips, pins, tailhead, long ribs, and short ribs are all individually identifiable. The skeletal frame becomes increasingly pronounced due to the severe loss of muscle mass and fat, providing little cushioning around the bones.

4. **Tail Bones**: The tail bones can be easily felt, further indicating the loss of muscle and fat reserves around the tail and hip area.

5. **Sunken Pins**: The inside of the pins is deeply sunken to the bone, showcasing the depletion of soft tissue in this area. This is a critical indicator of malnutrition and muscle wasting.

6. **Muscle Wasting**: Wasting in the leg muscles is so severe that the **stifle joint** becomes identifiable, which is not typically visible in a healthy animal. The rump muscle between the hooks and pins is deeply concave, indicating significant loss of muscle mass in the hindquarters.

7. **Tight Skin**: The skin is tight due to the absence of subcutaneous fat, which would otherwise provide some elasticity and thickness. This tightness can be a sign of dehydration or poor overall health.

8. **Shrunken Udder**: For lactating animals, the udder is now shrunk and tucked up against the body, reflecting the animal's inability to support milk production due to its deteriorated state.

9. **Dewlap and Sternum**: The dewlap, which usually has some fat and muscle coverage, is now just a skinfold, and the **sternum** is identifiable, further reflecting muscle loss.

10. **Loose Skin on the Hump**: In **Bos indicus** and **Bos indicus cross cattle**, the skin over the hump becomes loose as muscle and fat reserves are depleted.

11. **Poor Rumen Function**: Signs of poor rumen function are evident, such as undigested feed in the dung, mucus, dirt, and watery stools. This is a sign of compromised digestion and poor nutrient absorption, common in severely emaciated animals.

Behaviour:

1. **Dull Appearance**: Animals in **High Risk 2** show a dull, lifeless appearance. They stop engaging in typical grooming behaviour, which is an important indicator of their deteriorating health and well-being.

2. **Slow, Unsteady Gait**: Locomotion becomes slow, and the animal displays an unsteady gait. They may begin to drag their hind feet, a sign of extreme weakness in the legs and loss of muscle mass.

3. **"Plaiting" of Hind Legs**: Animals in this condition may "plait" their hind legs, crossing them awkwardly as they walk due to reduced muscle control and coordination. This can increase the risk of stumbling or falling.

4. **Reduced Cud Chewing**: Cud chewing becomes infrequent, reflecting poor digestive function and general discomfort. Proper cud chewing is essential for ruminants to digest food effectively, and its reduction signifies a decline in health.

5. **Difficulty Standing or Lying Down**: At this stage, the animal experiences great difficulty standing up or lying down, a clear sign that it is near the end of its reserves. The ability to perform these basic movements diminishes significantly as muscle mass continues to waste away.

Management Implications:

• **Immediate Nutritional Intervention**: Immediate, high-quality feed is necessary to provide energy and prevent further deterioration. Supplementation with high-energy, easily digestible feed can aid recovery, but it must be done carefully to avoid digestive issues like acidosis.

- **Veterinary Care**: Veterinary intervention is crucial at this stage to assess the animal's overall health and provide necessary treatments, such as rehydration or medications to address infections or other health complications.

- **Close Monitoring**: Animals at **High Risk 2** should be closely monitored, especially in harsh weather conditions, as they are highly susceptible to cold, heat, and stress.

- **Consideration of Euthanasia**: In cases where recovery is unlikely or the animal is suffering significantly, humane euthanasia may be the best course of action to prevent prolonged suffering.

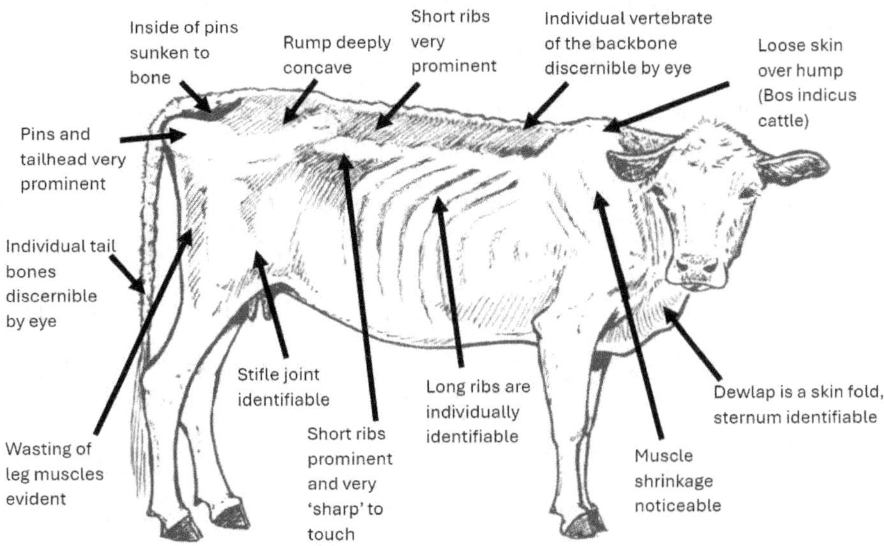

Figure 19: High Risk 2.

A **Downer** animal refers to livestock that have reached an extreme level of muscle and energy depletion, characterized by a complete inability to rise or move on their own. This condition is an extension of **Welfare Score High Risk 2** but includes additional severe physical and behavioural indicators that signify critical distress and possible organ failure.

Immediate intervention is required, as the animal is at high risk of death without prompt care or humane euthanasia [30].

Physical Characteristics:

1. **Immobility**: The animal is immobile and cannot move or stand up, even with significant external stimulation. Despite exhibiting attempted flight behaviour, it has zero flight distance, indicating that the animal has no strength left to escape or react.

2. **No Response to Stimuli**: Animals in this condition do not respond to typical external stimuli, such as sounds, touch, or other environmental factors. This lack of response suggests severe energy depletion and extreme weakness, where basic bodily functions are shutting down.

3. **Locomotion Difficulties**: The animal, if still able to move even slightly, will exhibit an unsteady gait, with a characteristic "plaiting" motion of the hind legs (where the legs cross or weave awkwardly). This is due to muscle wasting, poor coordination, and a lack of muscle strength.

4. **Inability to Stand**: When lying down, the animal is unlikely to stand without human assistance. The muscle depletion is so advanced that it cannot generate enough force to lift its body, indicating that intervention is urgently needed to prevent further suffering.

5. **Paddling Marks**: You may observe paddle marks from the animal's feet, which indicate that the animal has been attempting, unsuccessfully, to move or rise. These marks are a sign of distress and desperation as the animal struggles to regain mobility.

6. **Eye Condition**: The animal's eyes will appear tearing, sunken, and glazed, indicating dehydration, poor circulation, and systemic failure. These eye conditions are visual signs that the animal's health is severely compromised.

7. **Faecal Condition**: Brown, liquid faeces are often present, a sign that the rumen (a key part of the animal's digestive system) has limited or no function. This lack of digestive ability suggests that the animal is no longer processing food effectively, and nutrient absorption has ceased. This is another critical marker of the animal's deteriorating condition.

Behavioural Indicators:

1. **Flight Behaviour with No Response**: The animal may exhibit involuntary or automatic flight behaviour, but without any real movement. This indicates an instinct to flee from perceived threats, but the physical capability to do so is absent due to muscle failure and exhaustion.

2. **Slow or No Locomotion**: Even if some movement is still possible, it is slow and unsteady, with the hind legs often dragging due to weakness. The animal may collapse under its own weight if it attempts to stand.

3. **No Grooming**: The animal ceases to exhibit normal grooming behaviours, a sign of distress and poor health. Grooming is a basic activity in healthy animals, and its absence highlights the severity of the animal's condition.

Management and Intervention:

- **Urgent Action**: Immediate veterinary care is required to assess the possibility of recovery. For some animals, intensive care, including rehydration, high-quality feed, and close monitoring, might improve their condition.

- **Euthanasia Consideration**: In many cases, animals in the Downer condition are beyond recovery. Humane euthanasia is often the most compassionate option to prevent further suffering, particularly if the animal cannot regain the ability to stand or move.

- **Close Monitoring**: If recovery is attempted, constant supervision is essential to ensure the animal does not further deteriorate or suffer. Appropriate feeding, hydration, and shelter are necessary to provide the best chance of recovery.

The **Downer** animal condition is a serious welfare issue that requires immediate intervention. Livestock that exhibit the characteristics of this condition, such as immobility, lack of response to stimuli, and severe muscle wasting, are at the end stage of physical depletion. In most cases, the animal's chance of survival is minimal, and humane euthanasia may be the most ethical course of action. Careful monitoring of livestock to prevent them from reaching this stage is critical to maintaining animal welfare and ensuring that intervention occurs much earlier in their decline.

Figure 20: Downer.

Welfare Decisions for Dairy Cattle

In Australia, dairy cattle are body condition scored (BCS) using a numerical system ranging from 1 to 8, where 1 represents the lowest level of body fat and 8 the highest [30]. A body condition score of 3 is considered a critical threshold, marking an animal as "At Risk". At this point, intervention is necessary to prevent further deterioration and to improve the animal's condition. If a dairy cow's condition worsens beyond a score of 3, it falls into what are referred to as Dairy Welfare Scores, which identify animals that require urgent attention and, in some cases, euthanasia.

Welfare Scores Breakdown:

1. **High Risk 1:**

   - Corresponds to a **condition score of 2**.

   - At this stage, muscle and fat reserves are significantly depleted, and the cow is at serious risk of illness or death without immediate intervention.

2. **High Risk 2:**

   - Corresponds to a **condition score of 1**.

○ The cow is emaciated, with severe depletion of muscle and fat tissue. At this stage, survival is unlikely without intensive care and nutritional support.

3. **Downer**:

○ This is the most critical welfare score.

○ The animal is immobile, unable to stand or move, and typically requires euthanasia after veterinary assessment due to its inability to recover.

Scoring Method:

This scoring system is a visual assessment based on the amount of fat covering certain bony areas of the cow's body, regardless of the animal's overall size or weight. The two most common dairy breeds in Australia, Holstein Friesians and Jerseys, are significantly different in size. A mid-lactation Holstein Friesian weighs around 600 kg, while a Jersey weighs around 400 kg. However, the scoring system applies to both breeds by focusing on the proportion of body fat relative to their size.

To increase the condition score of a dairy cow, specific weight gains are required:

- A Holstein Friesian needs to gain 42 kg to improve by one BCS point.

- A Jersey needs to gain 26 kg to improve by one BCS point.

- A Holstein Friesian/Jersey cross requires about 34 kg to improve by one BCS point.

Importance of BCS in Dairy Cattle: The body condition score is vital for managing the health and productivity of dairy cows. Proper scoring helps farmers monitor their animals' nutritional well-being, and early intervention can prevent serious health problems. For instance, cows with low body condition scores are more likely to experience reduced fertility, lower milk yields, and increased susceptibility to diseases. Therefore, it is essential to monitor these scores regularly and make necessary dietary adjustments to support the health of the herd.

Exclusion of Meat Production: Dairy breeds that are managed specifically for meat production are not classified as dairy cattle under this system, as the focus is solely on animals being raised for milk production for human consumption.

Body Score 3 – At Risk describes a livestock animal, particularly cattle, that is lean and beginning to show signs of muscle and fat depletion. At this stage, intervention is necessary to prevent further deterioration and improve the animal's condition [30].

Physical Characteristics:

- **Backbone**: The backbone is visible, but individual spines are not sharply prominent. This suggests that the animal has lost some body fat but retains enough muscle mass to obscure the sharpness of the individual vertebrae.

- **Short Ribs**: The ribs are identifiable and feel fairly sharp to the touch, indicating a loss of fat tissue. However, they are not yet fully exposed, as they would be in a more emaciated animal.

- **Tailhead**: The tailhead is prominent with little to no fat around it. This lack of fat indicates that the animal is using its energy reserves, which often comes from fat stored in this area.

- **Inside of Pins**: The area between the hip bones (pins) is slightly hollow, signalling early stages of muscle and fat depletion, although it has not yet become severely concave.

- **Hip Bones and Long Ribs**: These bones are easily visible and prominent, showing that fat reserves in these areas have significantly diminished.

- **Area Between Tail and Pin Bones**: This area is concave and sunken, another clear sign of muscle and fat loss.

- **Rump Muscle**: The muscle in the rump is starting to waste, and the leg muscles are also showing signs of deterioration.

- **Tail Bones**: At this stage, the tail bones are not yet identifiable, suggesting the animal has not reached the more extreme stages of weight loss.

- **Skin**: The skin remains pliable, which indicates the animal is still hydrated and not suffering from severe dehydration.

- **Dewlap**: There is no fat in the dewlap area, reflecting the overall depletion of fat reserves.

Behaviour:

- **Bright Appearance/Alert**: Despite the physical signs of muscle and fat depletion, the animal appears bright and alert, suggesting it still retains enough energy to function relatively normally.

- **Mobile Gait**: The animal is still able to move around without difficulty, a good sign that it has not yet reached a critical level of deterioration.

Implications:

Body Score 3 signals the need for nutritional intervention. While the animal is not in immediate danger, its condition is on a downward trajectory, and without adequate nutrition, it will likely continue to lose muscle mass and fat reserves, which will lead to further health issues. Providing appropriate feed, increasing energy intake, and monitoring the animal closely are necessary steps to prevent further decline and to restore the animal's health and productivity.

Body Score 2 – High Risk 1 reflects a more serious state of muscle and fat depletion in livestock, specifically cattle. At this stage, the animal is at high risk of further decline and requires immediate nutritional intervention to prevent deterioration [30].

Physical Characteristics:

- **Muscle Depletion**: At Body Score 2, muscle depletion is notably present, particularly in the back, loin, and hind leg muscles. These areas, which usually have a healthy fat and muscle cover, now show signs of wastage, indicating that the animal is using up its muscle reserves to meet energy requirements due to inadequate nutrition.

- **Rump Muscle**: The rump muscle has become concave, signifying further muscle loss. The normally rounded rump now appears sunken, a clear indicator of poor body condition.

- **Backbone**: The spines of the backbone are identifiable at this stage, a key sign that the animal's fat reserves have diminished. As fat and muscle tissue deplete, the individual vertebrae become more prominent and can be seen clearly.

- **Short Ribs**: All the short ribs are prominent and very sharp to the touch. This sharpness is a result of the loss of the protective fat and muscle layer that usually covers these ribs.

- **Tail Bones**: The tail bones are just identifiable, which is a warning sign of significant fat loss in the hindquarters. In healthier animals, the tail bones are cushioned by fat, but in Body Score 2, they start to become distinguishable.

- **Long Ribs, Pin Bones, and Tailhead**: These areas, which should have more substantial fat and muscle cover, become prominent and stand out clearly in an animal with Body Score 2.

- **Inside of Pins**: The area inside the pin bones is hollow, showing a lack of muscle mass and fat. This further emphasizes the loss of body condition.

- **Udder Shrinkage**: The udder begins to shrink, reflecting the animal's poor nutritional status. In cows, a shrinking udder suggests that milk production is likely reduced due to the lack of sufficient nutrients.

- **Skin Pliability**: The skin is less pliable than it would be in a healthier animal, which may indicate mild dehydration or poor overall condition.

Behavioural Indicators:

- **Mobility**: The animal remains mobile, but its movement is less energetic compared to healthier livestock. This reduced energy is due to the lack of available body fat and muscle for energy production.

- **Grooming Behaviour**: Grooming behaviour, such as licking and scratching, becomes infrequent. Animals in good health tend to groom themselves regularly, but those in poor condition, like Body Score 2 animals, show a reduction in this behaviour.

- **Ability to Lie Down/Rise**: Despite the overall poor body condition, the animal can still lie down and rise with ease, suggesting that while it is in poor health, it has not yet reached a critical state of immobility.

- **Dung and Cud Chewing**: The animal's dung pats remain normal, and cud chewing is still observable. This suggests that despite the significant fat and muscle loss, the animal's digestive system is still functioning adequately, although this could change if conditions worsen.

Implications:

Body Score 2, classified as **High Risk 1**, indicates that the animal is in poor health and needs immediate intervention. If no action is taken, the animal's condition will likely deteriorate further, making recovery more difficult. Supplementary feeding, improving pasture quality, and close monitoring are necessary steps to restore the animal's health. The decrease in muscle and fat reserves means the animal is vulnerable to environmental stresses, such as cold or wet conditions, and may struggle to reproduce or maintain adequate milk production.

Body Score 1 – High Risk 2 represents a critical stage in livestock condition where the animal is severely emaciated and in urgent need of intervention. The physical and behavioural signs indicate that the animal is at a high risk of further deterioration and possibly death if immediate action is not taken.

Physical Characteristics:

- **Emaciation**: At Body Score 1, the animal is classified as emaciated. It has very little to no body fat, and muscle wastage is extensive. This is a clear sign that the animal's nutritional needs have not been met for an extended period.

- **Backbone Visibility**: The spines of the backbone are individually identifiable and pointed to the touch, with only skin covering them. This level of visibility indicates a severe lack of fat and muscle covering the skeletal structure.

- **Hips, Pins, and Ribs**: The hips, pins, tailhead, long ribs, and short ribs are all individually identifiable, with only the skin covering them. The fact that these bones are so prominent is a clear indicator of extreme weight loss and poor body condition.

- **Tail Bones**: The tail bones are easily visible, further emphasizing the animal's poor state of nutrition and muscle loss.

- **Hollow Pins**: The inside of the pin bones is sunken deeply to the bone, showing how hollow and depleted the animal's muscle reserves have become.

- **Leg Muscle Wasting**: Muscle wastage in the legs is so severe that the stifle joint, which is usually covered by muscle, has become identifiable. This suggests that the animal has used up its muscle mass to meet energy demands.

- **Concaved Rump Muscle**: The rump muscle, between the hips (hooks) and pins, is deeply concaved. This is another sign of extreme muscle depletion.

- **Tight Skin**: The skin is tight over the skeleton, indicating a lack of subcutaneous fat and general dehydration. There is no pliability, further illustrating the animal's deteriorated state.

- **Shrunken Udder**: In females, the udder has shrunk, indicating that milk production has stopped or significantly decreased due to the animal's poor nutrition and overall health.

- **Dewlap and Sternum**: The dewlap is merely a skinfold, and the sternum is easily identifiable, further emphasizing the lack of fat and muscle covering the animal's body.

- **Poor Rumen Function**: Dung will show clear evidence of poor rumen function, with signs such as undigested feed, mucous membranes, and watery consistency. This indicates that the animal's digestive system is no longer functioning properly, a sign of extreme malnutrition.

Behavioural Indicators:

- **Dull Appearance**: The animal exhibits a dull appearance, with no signs of alertness or vitality. This lack of brightness in the eyes and overall demeanour is a reflection of its poor health.

- **Locomotion**: The animal's movement is slow and unsteady, with a tendency to drag its hind feet. This unsteady gait is a result of muscle weakness and energy depletion. In some cases, the animal may "plait" its hind legs, crossing them as it walks, which is a sign of severe muscle and coordination loss.

- **Cud Chewing**: Cud chewing is reduced, indicating that the animal's digestive system is not functioning at its full capacity. This reduction in cud chewing shows that the animal's rumen is not processing food properly, which is a major concern for overall health.

- **Difficulty in Lying Down and Rising**: The animal has difficulty lying down and standing up due to muscle wastage and weakness. This is a critical sign of severe body condition decline and points to the animal's inability to function normally.

- **No Grooming Behaviour**: The animal no longer engages in grooming behaviours, a clear sign of poor welfare. Grooming is typically a sign of normal health and behaviour in cattle, and the absence of it indicates that the animal is too weak or sick to care for itself.

Implications:

An animal with **Body Score 1 – High Risk 2** is in a life-threatening condition. Immediate and high-quality intervention is needed to improve the animal's body condition and prevent further decline. This includes providing adequate, high-energy nutrition, veterinary care, and close monitoring. Without prompt action, the animal's health will continue to deteriorate, and recovery will become increasingly difficult, if not impossible. In some cases, euthanasia may be the most humane option if recovery is unlikely or if the animal is suffering significantly.

The "Downer" condition in livestock, especially cattle, is one of the most critical welfare stages, indicating that the animal is at a point of extreme physical deterioration, often near death. In this state, the animal's ability to recover is severely compromised, and humane intervention is usually required [30].

Physical and Behavioural Indicators

1. **Immobility and Zero Flight Distance**: A key feature of a Downer animal is its inability to move. Despite an instinctual attempt to escape when startled (flight behaviour), the animal remains immobile. This reflects the extreme weakness and lack of energy reserves. The animal's complete immobility means it is unable to perform normal functions, such as standing, walking, or even raising itself off the ground.

2. **No Response to External Stimuli**: In this stage, the animal shows no reaction to environmental stimuli such as sound, touch, or movement around it. This lack of responsiveness is a sign that the animal's nervous system is significantly impaired, which often accompanies severe physical exhaustion and poor overall health.

3. **Locomotion Difficulties**: When movement is attempted, locomotion is difficult or even impossible. The animal may display a "plaiting" motion with its hind legs, where the legs cross each other in an uncoordinated manner. This inability to move properly is due to muscle wastage and depleted energy reserves, resulting in severe weakness. Maintaining balance becomes nearly impossible,

contributing to the animal's inability to walk.

4. **Inability to Stand Without Assistance**: If the animal is lying down, it is unlikely to stand without help. The severe muscle depletion makes it too weak to push itself up, and it may remain in a recumbent position indefinitely. This is a sign of extreme physical debilitation, indicating that the animal's body has essentially reached its limit.

5. **'Paddle' Marks from Movement**: Despite its immobility, the animal may still attempt to move its legs or shift its body, creating "paddle" marks in the ground. This subtle movement is a sign that the animal is struggling, yet it lacks the physical strength to rise or change position. Head and body movements while sitting or lying are another indicator of this desperate attempt to move despite overwhelming weakness.

6. **Eye Condition**: The animal's eyes will appear sunken, tearing, and glazed, reflecting dehydration, malnutrition, and overall exhaustion. The glazed appearance is a sign of the body's deteriorating health, signalling that the animal's systems are beginning to shut down.

7. **Abnormal Faeces**: The animal's faeces will provide key insights into its digestive health. Brown liquid faeces indicate poor or non-functioning rumen, a critical component of the animal's digestive system. This shows that the animal is no longer properly digesting food, likely due to a lack of energy intake or metabolic disturbances. On the other hand, "hard" low-moisture-content faeces point to dehydration and the intake of high-cellulose, low-digestibility roughage, further exacerbating the animal's inability to recover.

At the Downer stage, the animal is at a critical juncture where recovery is highly unlikely without immediate, high-quality care. Given the level of muscle depletion, immobility, and overall lack of response, euthanasia is often the most humane course of action to prevent prolonged suffering. Veterinary assistance is essential to assess the animal's condition and make an appropriate decision regarding its welfare. The signs observed in a Downer animal reflect extreme neglect of the animal's nutritional, environmental, or medical needs, and prompt, humane intervention is vital.

Welfare Decisions for Sheep

When sheep experience a reduction in fat reserves and enter the Fat Score 1 category, they begin to rely on their muscle tissue to meet their energy needs. This situation arises due to insufficient feed quantity or energy density in their diet. In these cases, the body starts mobilizing muscle tissue as a secondary energy source, which signifies the animal is not receiving adequate nutrition. A sheep with a Fat Score 1 is considered to be in a critical condition and is classified as an At Risk animal, requiring immediate intervention to prevent further health deterioration.

In cases where sheep continue to lose body condition, the term Welfare Score is introduced to reflect the severity of their muscle depletion. The Welfare Score is essential for categorizing sheep that have moved beyond Fat Score 1 and need targeted management strategies. The Welfare Score system has two primary classifications: High Risk and Euthanise, based on the severity of the condition.

Manual Palpation for Assessment

Due to the wool covering that obstructs a clear view of the sheep's physical condition, the most effective way to assess body condition is through manual palpation. Visual observation alone may not give an accurate indication of the animal's condition, especially in woolly breeds. By feeling the animal's backbone, ribs, and muscle areas, stockpeople can assess the degree of muscle wastage and fat loss. Characteristics such as the prominence of bones or the concavity of the rump muscles provide important cues for the welfare assessment.

Intervention for Fat Score 1

A Fat Score 1 sheep is classified as At Risk, signalling that intervention must happen immediately to prevent further muscle depletion and to restore body condition. Without proper management, such animals can progress to more severe stages where recovery becomes increasingly difficult or impossible.

Welfare Scores: High Risk and Euthanise

- **High Risk**: This classification is for sheep that are in a critical state but may still recover with high-quality care and nutrition. Immediate steps must be taken to improve their condition by providing adequate feed, shelter, and veterinary attention.

- **Euthanise**: In cases where recovery is unlikely, or the sheep is suffering from severe muscle depletion and systemic failure, euthanasia may be the only humane option. This decision is based on the welfare assessment and aims to prevent prolonged suffering.

These classifications apply universally across all sheep breeds and their crosses. By using this structured scoring system, livestock handlers can make informed decisions about the welfare of their animals and implement appropriate interventions to manage the health of their flocks effectively. This approach is critical for ensuring that sheep maintain a healthy condition, preventing further deterioration, and addressing severe cases in a humane manner.

Fat Score 1 – At Risk is a critical classification used to identify livestock that are on the verge of nutritional failure due to inadequate fat reserves. In this stage, animals begin to show signs of severe body condition depletion, signalling that immediate intervention is necessary to prevent further health issues or death. The characteristics and behaviour of an animal in **Fat Score 1** provide essential information for livestock managers to recognize the at-risk condition.

Physical Characteristics

- **Ribs**: One of the most noticeable signs of a Fat Score 1 animal is the ease with which the individual ribs can be felt. Upon palpation, there is no tissue or muscle covering the ribs, and fingers easily fit between the ribs. This lack of covering indicates significant muscle and fat depletion.

- **Depressions Between Ribs**: The depressions between the ribs are quite pronounced, particularly in animals with short wool. This visual prominence further emphasizes the lack of tissue mass covering the ribs.

- **Pin Bones and Loin Muscle**: The pin bones are highly prominent, and the loin muscle feels concave, another sign of the muscle wastage occurring in these animals.

- **Backbone and Tail Bones**: The spines of the backbone are easily palpable, with each spine clearly distinguishable due to the lack of fat and muscle surrounding it. However, the individual tail bones are not as easy to identify.

- **Narrow Appearance**: When viewed from behind, the animal appears narrow, a result of muscle and fat depletion across the rump and hindquarters.

Behavioural Characteristics

- **Alertness and Mobility**: Despite their poor physical condition, animals in **Fat Score 1** typically remain alert and mobile. They still demonstrate a strong desire to graze and keep their heads low in search of grass.

- **Lie Down and Rise Easily**: The animal's mobility is not yet compromised, and they are able to lie down and rise with ease, which is a positive sign in terms of movement and physical capabilities.

- **Appetite**: Their appetite remains healthy, and they will actively search for food. This behaviour signifies the animal's ability to survive if provided with adequate nutrition.

- **Lamb Mortality**: For sheep that are lambing or in early lactation, the high risk of **lamb mortality** is an important factor to monitor. Animals at this score lack the energy reserves required to adequately nurse their young, leading to higher losses during these critical stages.

Immediate Intervention Required

At **Fat Score 1**, these physical and behavioural indicators mark the animal as being at a critical point where intervention is required. Failing to act may result in further deterioration to the animal's condition, potentially leading to **High Risk** classifications or even death. Providing adequate, nutritious feed and veterinary care immediately can help restore the animal's body condition and prevent further loss of health.

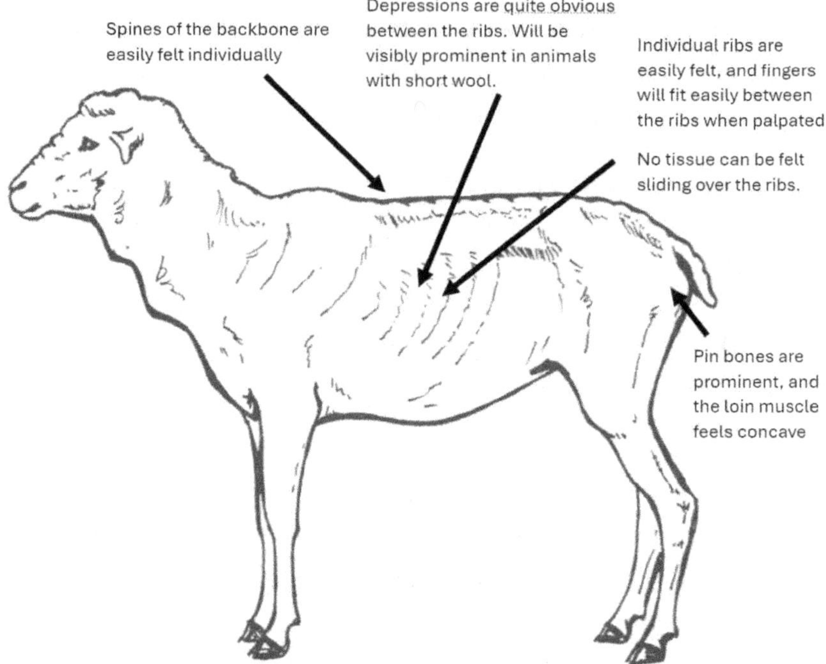

Figure 21: Fat Score 1.

High Risk describes a critical stage in the deterioration of an animal's body condition, particularly in sheep or other livestock, where significant muscle and fat depletion occurs, and immediate action is required to prevent further decline or death.

Physical Characteristics

- **Emaciation**: The animal is severely emaciated, with its overall appearance reflecting extreme weight loss. The **inside of the pins** is deeply sunken to the bone, and the **rump muscle** appears visibly concave, especially when viewed from behind. This gives the animal a distinct "tent" shape when viewed from the back.

- **Loose Skin**: The skin over the **ribs** and **backbones** is noticeably loose due to the lack of underlying fat or muscle. In sheep, the **wool** may appear dull and rough, both visually and to the touch, indicating poor nutritional health. Wool is likely to be "tender," meaning it can easily break or fall out due to the weakened state of the animal.

- **Prominent Bones**: The **spines of the backbone, hips, pins, tailhead,** and

ribs are all easily distinguishable due to the severe muscle and fat depletion. There is no soft tissue present between the skin and bones, making the bones pointed and easily palpable. The **point of the shoulder** will also be prominent in such animals, another sign of their poor condition.

- **Muscle Wastage**: There is extreme depletion of muscle in both the **loin** and **legs**, further contributing to the animal's weak physical state. The **back** may appear hunched or sunken, with the **head lowered** as the animal struggles with energy and mobility.

- **Rumen Function**: The poor nutritional status often affects the **rumen function**. Evidence of this can be seen in the animal's **dung**, which may indicate poor digestion, often reflecting infrequent **rumen contractions** and incomplete digestion of feed, further reducing the animal's ability to absorb nutrients.

Behavioural Characteristics

- **Dull Appearance**: The animal typically has a **dull** and **listless** appearance, lacking the usual alertness seen in healthy livestock. This reflects the severe energy deficiency and overall decline in health.

- **Lagging Behind**: If the animal is part of a mob being moved, it may **lag behind**, unable to keep up with the group due to weakness and fatigue.

- **Unsteady Gait**: The animal's **gait** becomes **slow and unsteady**, and it may begin to **drag its hind feet** while walking. Some animals might even **sway** when standing, showing an inability to maintain a stable posture due to their weakened muscles.

- **Difficulty in Lying Down/Standing Up**: The animal exerts **great effort** when trying to lie down or stand up, reflecting its depleted muscle strength. This is a critical indicator of the animal's frail condition.

Immediate Action Needed

Animals in this **High Risk** category require **immediate intervention**. This includes providing a high-quality, nutrient-rich diet, ensuring they have easy access to water, and possibly veterinary care. If the animal's condition continues to decline, euthanasia may be considered in the most severe cases to prevent further suffering.

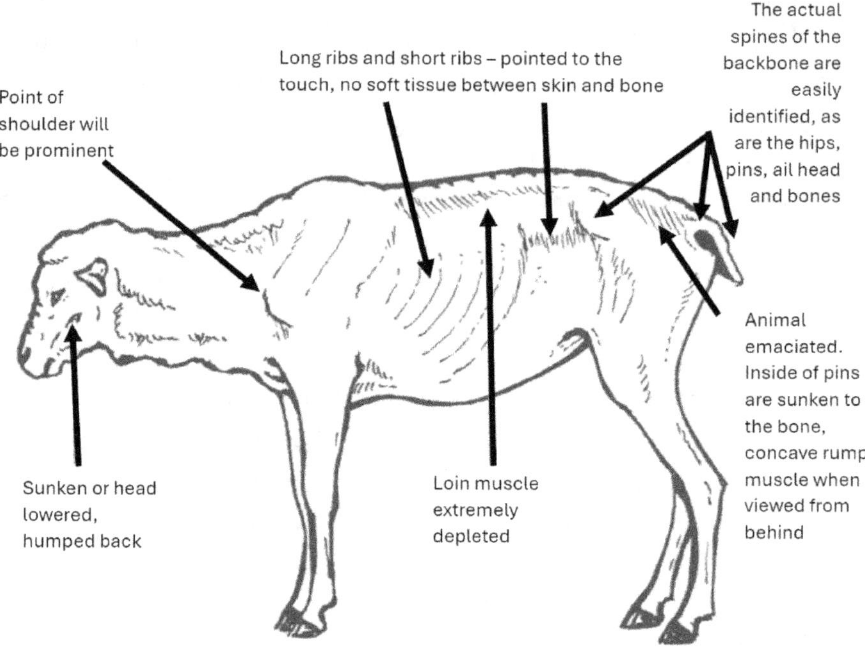

Figure 22: High Risk.

Euthanise is a category that applies to livestock, particularly in cases where the animal has deteriorated to a point where recovery is no longer viable. This decision is usually made after careful assessment and consideration of the animal's physical condition and welfare. It is a humane action taken to prevent further suffering [30].

Physical Characteristics of Euthanise Animals

Animals that fall into the euthanise category typically share characteristics with those in the High Risk classification, with some additional signs of severe deterioration [30]:

- **Severe Muscle Depletion and No Fat Cover**: These animals show extreme muscle loss, particularly around key body areas like the back, loin, and legs. In addition to this, there is **no fat** or almost no fat covering their bones. The body has used up all available energy reserves, making any further survival unlikely.

- **Immobility**: The animal is often immobile, unable to move due to the extreme weakness caused by muscle wastage. These animals may lie down for extended periods and show no ability to stand on their own, even with assistance. Their inability to move contributes to further muscle deterioration.

- **Minimal or No Response to External Stimuli**: At this stage, the animal will show little to no response to any external factors such as noise, touch, or visual stimuli. This lack of reaction indicates that the animal's cognitive and physical functions are severely impaired.

- **Inability to Walk**: Walking is no longer possible for these animals, due to both muscle weakness and the depletion of energy reserves. Any attempt at standing or moving is met with failure.

- **Lying Down and Unable to Stand**: These animals are continuously lying down and are unlikely to stand, even when given external assistance. This is a critical indicator that the animal is near the end of its life and that recovery is no longer feasible.

- **'Paddle' Marks**: Marks can be seen from the feet or head movement where the animal has been trying to move while lying down. This may result in distinct marks from the animal's attempts to shift its body position, indicating severe discomfort and struggle.

- **Sunken and Glazed Eyes**: The eyes of an animal in this condition will appear sunken and glazed, another sign of extreme dehydration and poor health. The face muscles will also be depleted, giving the animal a gaunt and hollow appearance.

- **Abnormal Faeces**: The faeces of an animal in the euthanise category will typically be abnormal, indicating either no or limited rumen function. This may manifest as very dry, low moisture-content faeces, suggesting dehydration, or poorly digested faeces, a sign of compromised rumen function.

Welfare Action

Animals in this state are past the point of recovery, and euthanasia is considered the most humane option. This process, when handled correctly, prevents further suffering and distress. Euthanasia should always be performed by a qualified individual, such as a veterinarian or a trained handler, using the most humane methods available. This decision is made out of a duty of care, ensuring that the animal does not endure unnecessary pain or discomfort in its final moments [30].

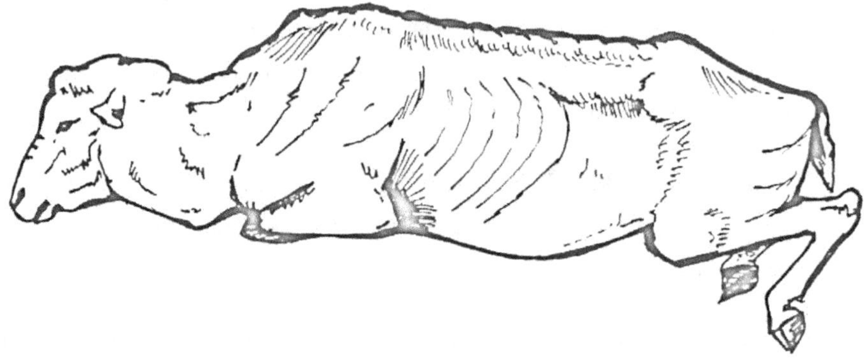

Figure 23: Euthanise.

Welfare Decisions for Horses

As horses begin to lose fat, they may fall into Condition Score 1, a point at which their bodies start using muscle tissue to meet their energy needs. This shift occurs when the horse's diet is insufficient in energy density or quantity, meaning it does not provide enough calories to support normal functions. Condition Score 1 represents an At Risk animal, and intervention is required immediately to reverse this downward trend. At this stage, horses are beginning to experience muscle depletion, which, if not addressed, will continue to worsen [30].

Condition Score 1 – At Risk

When a horse reaches Condition Score 1, their fat reserves have been depleted, and they begin to break down muscle tissue for energy. The horse's backbone, hips, and ribs become more prominent due to the loss of fat covering. The animal may still appear alert and mobile, but signs of energy deficiency are clear. At this stage, intervention is necessary to restore proper nutrition, provide a diet with higher energy content, and prevent further muscle loss.

Welfare Score System

Horses that fall below Condition Score 1 are categorized using the Welfare Score system, which is used to assess and describe the severity of muscle depletion and overall health. There are two distinct Welfare Scores for horses and ponies [30]:

- **High Risk**: Horses in this category show advanced signs of muscle depletion. Their energy reserves are critically low, and their body is severely compromised.

Without immediate intervention, these animals are at high risk of further decline and death from malnutrition or other complications.

- **Euthanise**: This score applies to horses in an even more critical state. The animal is typically immobile, unable to respond to stimuli, and has little chance of recovery. Muscle and fat reserves are almost completely depleted, and the horse can no longer sustain itself. In these cases, euthanasia is considered the most humane option to prevent further suffering.

Muscle Depletion and Energy Mobilization

As a horse's body continues to mobilize muscle tissue for energy, the depletion becomes progressively severe. The **ribs, hips, and backbone** will be even more pronounced, and the horse will likely lose the ability to stand or walk properly. Horses will also show signs of reduced energy and a lack of grooming behaviours. These physical and behavioural symptoms are clear indicators that the horse's body has been severely compromised [30].

Intervention

For horses at Condition Score 1, immediate changes to diet and care are required. A balanced and energy-rich diet, access to clean water, and veterinary supervision are essential in helping the animal regain strength and prevent further muscle depletion. If left unaddressed, the horse will progress into the High Risk and possibly the Euthanise category, where recovery becomes increasingly unlikely.

These Welfare Scores apply universally to all horses and ponies, regardless of breed, and are critical in managing and assessing the welfare of animals experiencing significant weight and muscle loss [30].

Condition Score 1 – At Risk describes an animal that is undernourished and showing early signs of muscle wastage. At this point, intervention is necessary to prevent further health decline. The specific physical characteristics and behaviour of an animal at this condition score reflect its weakened state due to insufficient nutrition.

Physical Characteristics:

- **Narrow Appearance**: When viewed from behind, the animal looks narrower than normal, indicating early signs of muscle loss, though it is still able to walk around and forage for food.

- **Neck**: The top side of the neck appears concave, and the lateral sides seem hollow. The vertebrae of the neck can be felt during palpation, showing that there is little fat or muscle coverage.

- **Withers**: The withers (the ridge between the shoulder blades) appear prominent, indicating significant muscle loss on either side.

- **Shoulders**: The bones of the shoulder are accentuated, and the scapula (shoulder blade) is visible due to muscle wastage.

- **Ribs**: The ribs are visible and easily felt, as there is very little fat or muscle covering them.

- **Spine**: The spine is highly visible and can be easily felt. The spinous processes (the bony projections along the topline) are discernible, showing a lack of muscle or fat in this area.

- **Tail Head**: The tail head is prominent, and there is a noticeable cavity under the tail due to the lack of fat and muscle.

- **Rump and Pelvis**: The rump appears sunken, and the angle of the pelvis is visible, indicating a lack of tissue covering this part of the body.

Behaviour:
- **Alertness**: The animal remains alert but may appear lethargic due to its weakened state.

- **Mobility**: Despite its condition, the animal is able to lie down and rise easily, showing that it still has some strength left.

- **Movement**: The animal is able to move around voluntarily, but when driven or herded, it may lag behind the group due to its reduced energy levels.

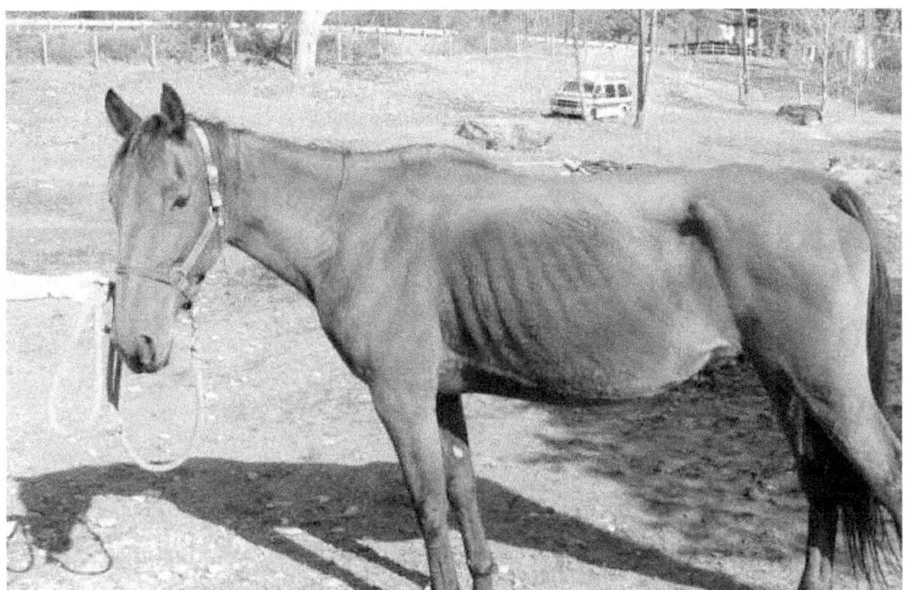

Figure 24: Horse Condition Score 1 – At Risk and scoring a 2 on the body condition scoring scale (Henneke Horse Body Condition Scoring System). eXtensionHorses, CC BY-SA 2.0, via Wikimedia Commons.

At this stage, although the animal is still able to move and interact with its environment, the clear signs of muscle wastage and fat depletion indicate that immediate nutritional intervention is needed to prevent further decline. This might involve increasing feed quality, providing supplements, and ensuring access to nutritious food sources to allow the animal to regain strength and recover from its deteriorated state [30].

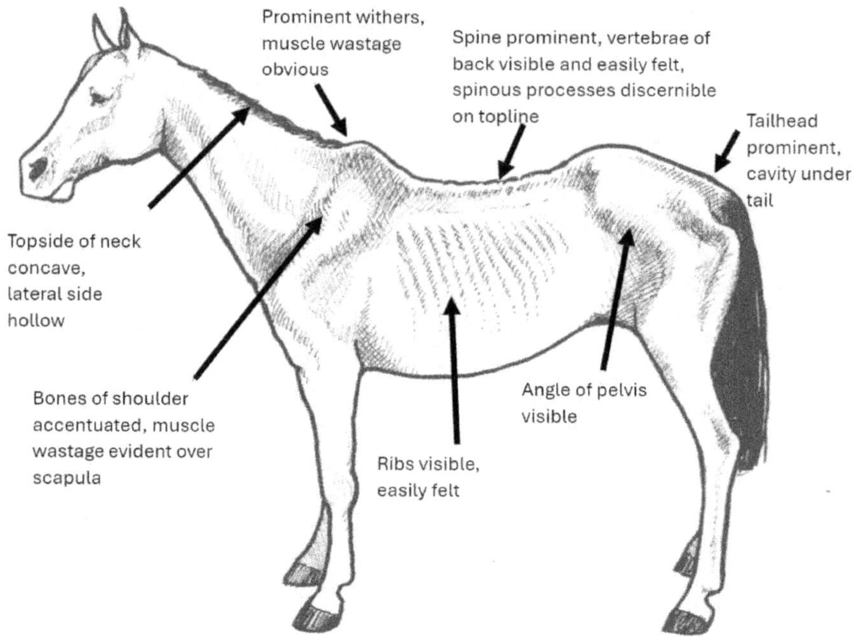

Figure 25: Condition Score 1 – At Risk.

High Risk refers to a severe stage of malnutrition or physical decline in an animal, where significant muscle wastage and body deterioration are visible. Immediate action is required to prevent further deterioration or death. Here's a detailed breakdown of the physical characteristics and behavioural changes in animals classified as "High Risk" [30]:

Physical Characteristics:

- **Significant Muscle Wastage**: The animal is experiencing severe muscle depletion, which is visible across different parts of the body, indicating that the body is now mobilizing muscle tissue for energy due to insufficient fat reserves.

- **Neck**: The side of the neck appears deeply concave, with the vertebrae clearly visible. This is a sign of significant muscle loss in the neck region.

- **Withers**: The withers (the highest point between the shoulder blades) are highly prominent due to a lack of surrounding muscle and fat tissue.

- **Shoulder and Scapula**: The bones of the scapula (shoulder blades) are very prominent, almost protruding from the body. This accentuated visibility of the bones highlights extreme muscle wastage over the shoulders.

- **Ribs:** The ribs are clearly visible and appear very prominent, with deep depressions between each rib. This indicates a critical loss of muscle and fat tissue, leaving the ribs exposed and skeletal in appearance.

- **Spine:** The spine is highly visible, with the tissue on either side of the backbone sunken. This gives the animal's back a concave or hollow appearance, further evidence of the extreme depletion of body reserves.

- **Tail Head:** The tail head is extremely prominent, with a deep cavity visible underneath. The lack of fat and muscle around the tail area makes it appear skeletal.

- **Rump and Pelvis:** The rump is deeply concave, and the angle of the pelvis is highly visible with tight skin stretched over it. This is a sign that the animal's fat and muscle reserves in the hindquarters are severely depleted.

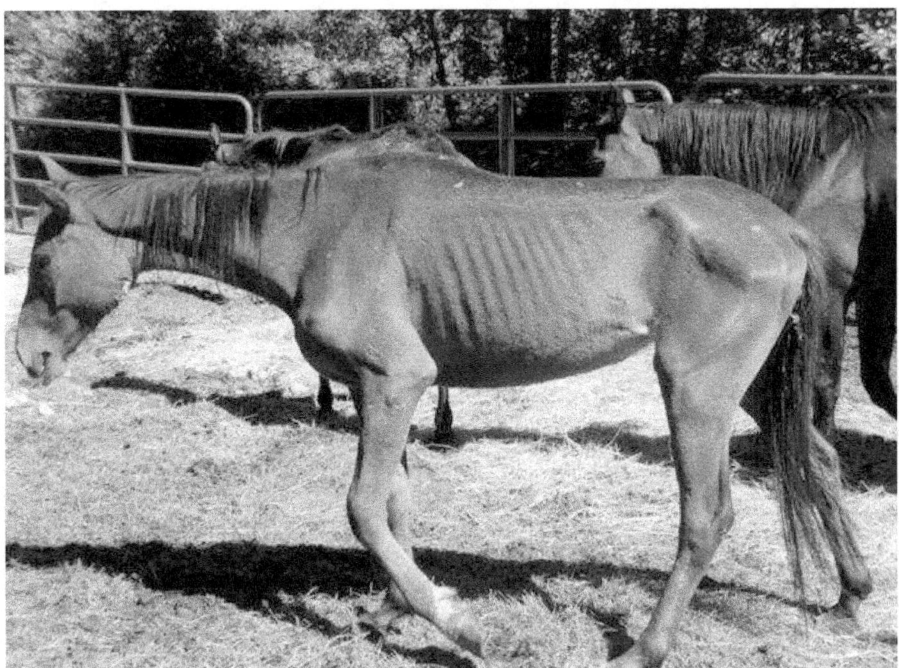

Figure 26: Horse Condition High Risk. This horse is a 1 on the Henneke Body Condition Scoring Scale. eXtensionHorses, CC BY-SA 2.0, via Wikimedia Commons.

Behavioural Changes:

- **Alertness and Lethargy**: While the animal remains aware of its surroundings, it exhibits extreme lethargy, lacking the energy to engage in normal activities.

- **Response to Stimuli**: The animal shows a very limited response to external stimuli, indicating severe physical and mental fatigue.

- **Head Position**: The head is held in a low position, a sign of weakness and lack of vitality.

- **Locomotion**: Movement is slow and unsteady, and the animal may drag its feet while walking. This unsteady gait is a sign that the animal is struggling with coordination and energy.

- **Difficulty Rising**: The animal may find it difficult to rise from the ground due to muscle weakness, reduced energy, and physical deterioration.

In this "High Risk" state, animals are severely compromised and require urgent intervention. Immediate nutritional support, medical care, and an improved living environment are necessary to stabilize and restore the animal's health. If these conditions are left untreated, the animal's condition may worsen, potentially leading to death or the need for humane euthanasia [30].

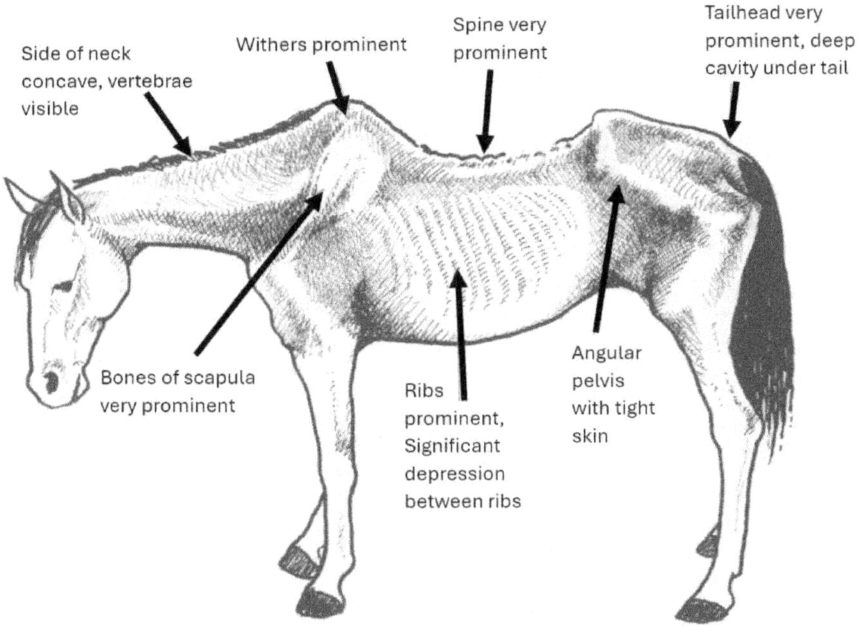

Figure 27: High Risk.

When an animal reaches the stage where euthanasia is necessary, it means that its condition has deteriorated beyond recovery, and humane intervention is required to prevent further suffering. This stage is often described as the animal being in **Condition Score 0**, which is the most severe classification and signals critical body deterioration. Below is a detailed explanation of the physical and behavioural signs observed in an animal that needs to be euthanised [30]:

Physical Characteristics:

- **Weak with Low Body Reserves**: At this stage, the animal is extremely weak, with almost no body reserves left. It is unable to sustain itself due to severe muscle wastage and lack of fat.

- **Skin Stretched Over Skeleton**: The skin appears tightly stretched over the skeleton because there is no fat or muscle left to cushion the bones. The animal's body essentially consists of just skin and bones, with very little tissue remaining to provide support.

- **Rub Marks and Hair Loss**: Due to extended periods of lying down and lack of mobility, the animal develops rub marks and hair loss over bony prominences. This is caused by friction between the skin and hard surfaces, as the animal can no longer move to alleviate pressure on certain parts of its body.

Behavioural Signs:

- **Little or No Awareness of Stimuli**: The animal shows minimal or no response to external stimuli. This includes not reacting to sounds, movement, or touch, indicating that it is in a state of severe physical and mental shutdown.

- **Low Head Position**: If the animal is still able to stand, it holds its head very low, a sign of extreme fatigue, weakness, and a lack of energy to hold its head upright.

- **Walking Difficulties**: The animal has great difficulty walking. It may stumble, fall, or exhibit an unsteady gait due to the depletion of muscle mass and energy, making it hard to maintain balance and coordination.

- **Recumbency and Inability to Rise**: In many cases, the animal will be recumbent (lying down) and may be unable or unwilling to rise. This is a critical sign that its strength has been entirely depleted, and it no longer has the capability to stand or move.

- **No Interest in Eating**: The animal has lost interest in food, which is a clear sign that its bodily functions are shutting down. Inability or refusal to eat further indicates that the animal has little energy or will left to sustain itself.

When an animal exhibits these signs, particularly the lack of response to stimuli, inability to stand, and severe emaciation, euthanasia is the most humane option to prevent prolonged suffering. At this stage, the animal's quality of life is severely compromised, and recovery is no longer feasible. In these cases, consultation with a veterinarian is essential to assess the situation and ensure that euthanasia is performed as humanely and painlessly as possible. This decision is guided by the principles of animal welfare, prioritizing the prevention of unnecessary suffering and promoting the well-being of the animal even in its final moments [30].

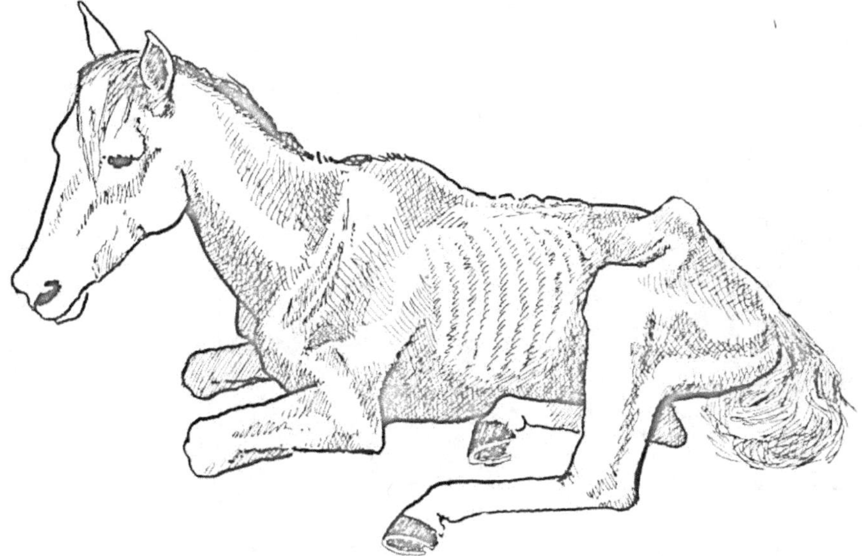

Figure 28: Condition Score 0 - Euthanise.

Henneke Horse Body Condition Scoring System

The Henneke Horse Body Condition Scoring System is a widely accepted method for evaluating the body condition of horses based on fat coverage in key areas of their body. Developed by Dr. Don Henneke in the 1980s, this scoring system provides an objective way to assess a horse's overall condition, ranging from emaciation to obesity [31].

The Henneke Body Condition Score (BCS) is based on a scale of 1 to 9, with 1 being extremely thin and 9 representing obesity. This system assesses fat deposits in six main areas of the horse's body:

1. Along the neck

2. Withers

3. Behind the shoulder

4. Ribs

5. Topline (backbone)

6. Tailhead

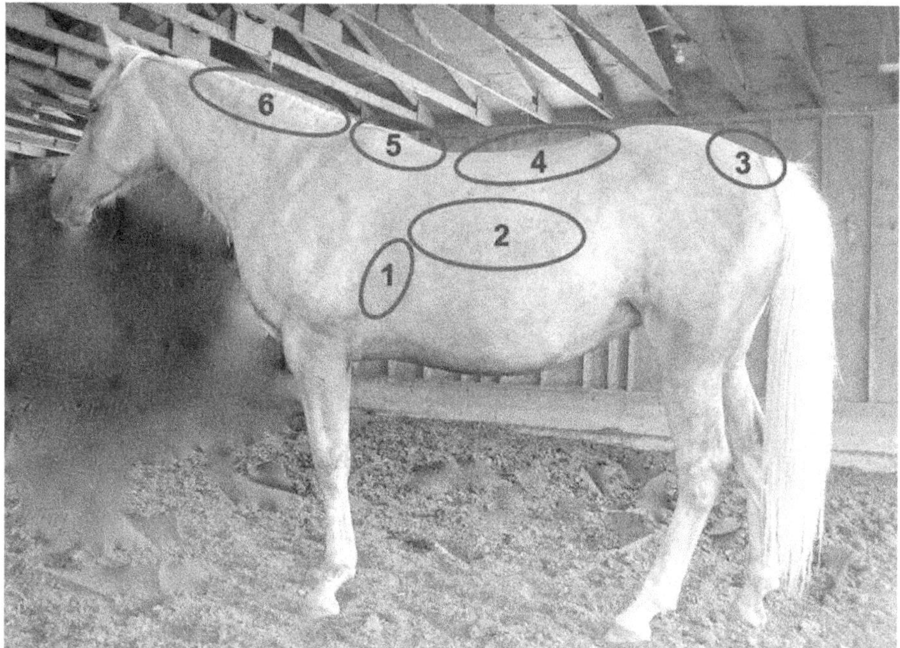

Figure 29: Hennecke fat deposits. Montanabw, CC BY-SA 3.0, via Wikimedia Commons.

Scoring Breakdown:

- **Score 1 (Emaciated):** The horse has no discernible fat and is extremely thin. The bones are very prominent, including the ribs, hips, and spine. See Figure 26.

- **Score 2-3 (Thin):** There is very little fat coverage. The ribs are easily visible, and the tailhead and backbone are prominent. Muscle wastage may be present. See Figure 24.

- **Score 4-6 (Moderate/Ideal):** This range is considered healthy for most horses. The ribs may or may not be visible but can be easily felt. There is some fat over the ribs, with smooth contours around the withers, shoulders, and tailhead.

- **Score 7-9 (Overweight/Obese):** Horses in this range show obvious fat deposits in the neck, withers, and shoulder areas. Fat may form around the ribs and tailhead, and the horse may develop a "cresty" neck. At the highest end, fat patches or lumps can be present.

In the Henneke Horse Body Condition Scoring System, fat deposits accumulate in a predictable order as a horse gains weight. This system is used to visually assess the body condition of a horse by evaluating fat deposits across key areas of the body. Fat storage tends to follow a certain progression, with internal fat accumulating around the major organs first. Once that internal fat storage is saturated, fat begins to be deposited in external, more visible areas in the following order:

1. Behind the Elbow (Scapula)

- Fat is first noticeable behind the elbow, particularly along the scapula. This area can feel more padded as fat accumulates and indicates the beginning of noticeable external fat storage. In leaner horses, this region is flat or slightly indented, while in horses with higher fat stores, this area becomes rounded.

2. Between the Ribs

- As fat storage progresses, it begins to accumulate between the ribs. At lower condition scores, the ribs are clearly visible or easily felt with little to no fat between them. As the condition score increases, the fat between the ribs makes the ribs less distinct and harder to feel.

3. Beside the Tailhead

- Fat next accumulates at the tailhead, where fat pads begin to form on either side of the base of the tail. In horses with low body condition scores, this area is bony and prominent. As fat increases, the area becomes smoother, and the tailhead appears to sink into a rounded mass of fat.

4. Along the Back

- As the body condition score continues to increase, fat begins to accumulate along the spine and the top of the back. Horses with a low body condition have a sharp, easily felt backbone. However, as the horse gains weight, fat covers the spine, smoothing its appearance and sometimes forming a "gutter" along the topline in obese horses.

5. Over the Withers

- Fat accumulation over the withers follows next. In lean horses, the withers appear sharp and pronounced. As fat increases, the withers become more rounded and less distinct. Fat deposited here can smooth the area significantly.

6. In the Neck

- Finally, fat deposits accumulate in the neck, causing it to thicken. In higher body condition scores, the neck can develop a pronounced "cresty" appearance due to fat buildup. In extreme cases, the crest becomes rigid and hard to the touch.

Example: Horse with a Body Condition Score of 7

A horse with a body condition score of 7 (as shown in Figure 29), although still fit, will show substantial fat accumulation in the aforementioned areas. At this stage:

- The ribs are barely discernible.

- There will be noticeable fat behind the elbow and around the tailhead.

- The back will be smooth, and the withers may be rounded.

- A "cresty" neck is likely to develop, and fat deposits will be visible along these key areas.

By following this predictable pattern of fat deposition, horse owners, veterinarians, and equine professionals can assess and manage a horse's nutritional status and overall health using the Henneke Body Condition Scoring System. The system helps ensure horses maintain an optimal balance between muscle and fat for both health and performance.

Welfare Decisions for Pigs

Sow Condition Scoring: An Overview

Condition scoring is an essential practice used to evaluate the health and nutritional status of sows throughout their reproductive cycle. Originally developed for cattle and sheep, this technique has been adapted for pigs, particularly sows and boars, to guide feeding policies, ensuring the optimal body condition of these animals at different life stages. Condition scoring also aligns with animal welfare standards, such as the Model Code of Practice for the Welfare of Animals - Pigs, which requires that minimum body condition levels be maintained for the welfare of all pigs [32].

Sow condition scoring helps farmers assess and adjust feeding strategies, ensuring that each animal maintains optimal body condition for reproduction and productivity. By maintaining proper body condition, farmers can optimize reproduction rates, minimize health risks, and increase the overall profitability of pig farming operations. The condition scoring system for sows ranges from 0 (emaciated) to 5 (grossly fat), combining both visual appraisal and physical palpation to accurately determine the sow's fat cover.

Manual Condition Scoring

The manual scoring technique is easy to learn and involves pressing specific areas of the sow's body to assess fat cover. The primary areas evaluated include the pin bones, tail setting, loin, backbone, and ribs. A score is given based on how much pressure is needed to feel the bones and how much fat cover is present. Visual inspection alone is insufficient because it may overlook certain conditions, making physical palpation crucial.

For instance:

- A **score of 0** reflects an emaciated sow with prominent pin bones, sharp vertebrae, and visible ribs.

- A **score of 3** reflects a sow in ideal condition, with a full flank and backbone, and ribs that are difficult to feel.

- A **score of 5** reflects a grossly fat sow, with deep fat deposits and no visible or palpable bones.

Global Applications and Adaptations

Body condition scoring is widely adopted in different regions around the world. For example:

- **In the United States**, condition scoring is integral to the swine industry, particularly for ensuring the welfare of breeding sows. Many U.S. producers use condition scoring as part of their breeding and farrowing strategies to optimize reproductive performance and maintain animal health.

- **In the European Union**, condition scoring has been aligned with welfare regulations to ensure that sows are not overfed or underfed, especially during gestation and lactation periods. Producers in countries like Denmark and Germany incorporate condition scoring into routine management practices to enhance productivity and meet the strict EU animal welfare guidelines.

- **In Australia**, where the Model Code of Practice for the Welfare of Animals - Pigs applies, condition scoring is used to prevent extreme body conditions in sows, thus ensuring both productivity and compliance with national welfare standards.

Benefits of Condition Scoring for Sows

1. **Optimized Feeding Programs**: By assessing the body condition at different stages of the reproductive cycle (e.g., before mating, during gestation, and after farrowing), farmers can tailor their feeding programs to meet the specific nutritional needs of each sow. For instance, sows with a score of 2 or below may require increased feed intake, while those scoring 4 or above might need a more controlled diet to prevent reproductive and health issues.

2. **Improved Reproductive Success**: Sows that are too thin or too fat are less likely to conceive or may have difficulty farrowing. Maintaining a score of around 2.5 to 3 ensures that sows are in the best condition for breeding and farrowing.

3. **Increased Productivity and Longevity**: Regular condition scoring helps maintain sows' overall health and longevity in the breeding herd. Sows that are kept in optimal condition tend to have better longevity and higher lifetime productivity, which is economically beneficial for producers.

CHAPTER FIVE

Taking Action to Address Welfare Concerns

Taking action to address livestock welfare concerns involves several critical steps to ensure animals are treated humanely and their health and wellbeing are prioritized. Addressing welfare concerns promptly and effectively helps to prevent suffering and ensures compliance with legal and ethical standards in livestock management.

1. Observation and Identification of Welfare Issues

The first step in addressing livestock welfare concerns is recognizing the signs of distress, illness, or inadequate living conditions. Livestock handlers and owners must be vigilant and regularly monitor animals for common signs of poor welfare, which can include:

- **Behavioural signs**: Isolation from the herd, reluctance to move, aggression, or lethargy.

- **Physical signs**: Visible injuries, poor body condition, respiratory distress, and abnormal posture.

- **Environmental signs**: Lack of access to clean water, inadequate shelter, overcrowding, or unsanitary living conditions.

Early detection of issues allows for quicker interventions, minimizing suffering and preventing further complications.

2. Documentation and Reporting of Concerns

When welfare concerns are identified, it is important to document them in detail. Records should include:

- **The specific concern**: What behaviour or condition was observed?

- **Date and time**: When was the issue noticed?

- **Location**: Where is the livestock being held?

- **Action taken**: What immediate actions were taken to address the issue (if any)?

Documentation is essential for tracking the progression of the animal's condition and can be used in reports to authorities or veterinarians if needed. If you are part of a larger operation, it is important to report the issue to a supervisor or animal welfare officer promptly.

3. Immediate Action and Intervention

Once a welfare issue is identified, immediate action should be taken. Interventions may include:

- **Veterinary care**: Sick or injured animals should receive prompt treatment from a veterinarian. For minor concerns, the livestock handler may be able to address the issue, but severe cases require professional intervention.

- **Euthanasia**: In cases where an animal is suffering with no chance of recovery, humane euthanasia must be considered. This must be done promptly and humanely, ensuring minimal suffering.

- **Environmental changes**: If the concern is related to poor living conditions, changes to the environment, such as providing clean water, food, shelter, and bedding, must be made immediately.

4. Engaging Experts and Veterinary Support

If the concern cannot be easily remedied or requires professional knowledge, it is important to contact a veterinarian or animal welfare expert. Experts can assess the situation, diagnose any underlying health issues, and prescribe the necessary treatment or management changes.

Veterinary support may also be necessary for:

- Administering vaccines, antibiotics, or other medical treatments.

- Performing procedures such as hoof trimming, dehorning, or castration in a humane manner.

- Advising on nutrition and feeding practices, especially for sick, pregnant, or lactating animals.

5. Monitoring and Follow-Up

Once corrective actions have been taken, continuous monitoring is required to ensure the issue is fully resolved. If treatment was administered, monitor the animal's recovery and make adjustments as advised by the veterinarian. In cases of environmental or management changes, continue observing the animals to ensure they remain healthy and comfortable.

It is also important to keep detailed records of:

- The progress of the animal's health.

- Ongoing treatment or interventions.

- Environmental improvements and the resulting changes in animal behaviour and condition.

6. Training and Education

Livestock handlers and owners must be properly trained in identifying and addressing welfare concerns. Regular training on animal handling, behaviour, and the signs of pain or distress is essential for early intervention. Additionally, having up-to-date knowledge of the legal requirements and welfare standards in your region ensures compliance with regulations.

7. Legal Compliance and Ethical Responsibility

Every region has specific legal frameworks governing the welfare of livestock. Compliance with these regulations is not just a legal responsibility but an ethical one. Failing to meet welfare standards can result in fines, loss of business licenses, and damage to reputation.

Farmers and livestock handlers should regularly review local, national, and international animal welfare laws, including:

- **Veterinary medicines regulations**: Understanding the legal use of medicines and treatments for animals, including record-keeping of treatments and observing withdrawal periods.

- **Euthanasia regulations**: Ensuring humane methods of euthanasia are followed and that the person performing the task is appropriately trained.

- **Transport and housing standards**: Ensuring that animals have adequate space, food, water, and environmental protection during transportation and in their living quarters.

Low-Stress Stock Handling

Understanding animal behaviour and practicing low-stress stock handling are crucial for improving livestock welfare and productivity. Research shows that improper handling of animals can negatively impact their behaviour, causing fear and chronic stress, which in turn can result in up to a 20% variation in productivity, reproduction, and the quality of products like milk, meat, or wool. This reduction in efficiency comes as a result of stress-induced hormonal changes that affect the animals' health, growth, and reproductive abilities.

Inappropriate handling methods—such as hitting, shouting, or rough physical contact—cause animals to develop a fear of humans. The frequency and regularity of such negative interactions can lead to chronic stress. Chronic fear is detrimental not only to the animals' well-being but also makes them difficult to handle, leading to potentially dangerous situations for handlers and animals alike. Fear and stress responses manifest in behavioural signs like increased aggression, avoidance, or difficulty in herding and milking.

The key to reducing fear and stress in animals lies in understanding their natural behaviours and responses. Animals perceive humans based on the nature of the interactions they experience. For example, if animals frequently experience rough treatment, they will associate human presence with fear and become harder to handle over time. Livestock handlers who practice calm, deliberate movements, avoid shouting, and allow animals time to move without force are more likely to see lower stress levels and better productivity in their herds.

To further reduce stress in livestock, low-stress stock handling methods have been developed and widely promoted. These methods focus on using the animals' natural instincts and behaviour to move them efficiently and humanely without inducing fear. Proper handling reduces incidents of injuries, lowers the risk of production losses, and improves the overall welfare of animals.

Additionally, professional stockperson training programs, such as those developed by the Animal Welfare Science Centre, are aimed at livestock industries, transporters, and abattoirs. These programs teach the best practices for handling livestock in ways that promote calmness and reduce the likelihood of fear-induced stress. By implementing low-stress stock handling techniques, farmers and livestock handlers not only improve animal welfare but also enhance the operational efficiency of farms and feedlots.

On the farm, biosecurity measures also play a crucial role in maintaining herd and flock health. These include providing adequate access to feed and water, offering nutritional supplements when needed, and protecting animals from pests and diseases. If any welfare concerns arise, it is critical that they are promptly reported and addressed to maintain the health and productivity of the livestock. Proper care, in combination with low-stress handling techniques, ensures a higher quality of life for animals and a more efficient, productive farming operation.

Animal Sentience

Animal sentience refers to an animal's capacity to experience a wide range of emotions, including both positive and negative feelings such as joy, contentment, pain, fear, and frustration. Sentient animals are capable of perceiving their environment, processing information, learning from their experiences, and making decisions based on those experiences. Sentience implies that animals can experience subjective feelings, and this ability influences their well-being and behaviour [33].

The concept of animal sentience is crucial for understanding animal welfare because it recognizes that animals, like humans, can feel pleasure and suffering. This understanding shifts the focus of animal care from simply meeting basic physical needs, such as food and shelter, to ensuring the emotional and psychological well-being of animals. Historically, humans recognized their own sentience but often overlooked or ignored the emotional capabilities of other animals. However, modern science has provided compelling evidence that many animals, particularly vertebrates, are sentient. There is even growing research that suggests some invertebrates, such as octopuses and certain insects, may also possess sentient capacities [33].

The formal recognition of animal sentience has had significant implications for animal welfare policies. For example, in 2008, the Treaty of Lisbon recognized animals as sentient beings, obligating the European Union to consider animal welfare in its policies. Similar recognition has been adopted in various countries, including New Zealand, Canada, and specific Australian regions, where laws now acknowledge animal sentience as a key component of animal welfare. This recognition has led to the development of welfare frameworks that account for animals' mental states alongside their physical conditions.

One such framework, known as the Five Domains, emphasizes the need to address not only the physical health and nutrition of animals but also their mental and emotional well-being. The Five Domains—nutrition, environment, health, behaviour, and mental state—are used to assess animal welfare comprehensively, ensuring that animals live with minimal suffering and experience positive states of being.

A crucial aspect of assessing sentience is observing an animal's behaviour. Positive emotions in animals are often indicated by behaviours such as playing, exploring, and engaging in social interactions with other animals. Negative emotions, on the other hand, can be observed through behaviours such as avoidance, freezing, or showing fear and aggression. Technologies that assess brain function are also being developed to further understand how animals experience emotions [33].

Another method to assess an animal's emotional state is by offering them choices in their environment. This method acknowledges that sentient animals seek pleasure and avoid pain, which can be measured by observing their preferences in tests designed to gauge the importance of various stimuli, such as food, shelter, or social contact. By understanding these choices, animal caretakers can adjust environments and care practices to better meet the animals' emotional needs [33].

Recognizing animal sentience allows for the development of policies and procedures that ensure the mental and physical well-being of animals, promoting a "life worth living." It also holds important ethical implications, as it calls for changes in how humans interact with animals, whether in farming, research, or as companions. By observing and responding to signs of sentience, those responsible for animal care can ensure that animals experience minimal suffering and greater welfare overall.

Minimising Pain

Minimising pain in livestock is a crucial aspect of good animal husbandry, aimed at ensuring animal welfare and improving productivity. The responsibility for managing pain lies with the owner and involves a combination of proper training, timely intervention, and, when necessary, the use of veterinary-prescribed pain relief options [29].

Pain Relief and Veterinary Prescriptions

Pain relief products for animals are available but must be prescribed by a veterinarian. These products range from non-steroidal anti-inflammatory drugs (NSAIDs) to local anaesthetics, which can effectively reduce pain during procedures or illness. It's essential to refer to the Australian Pesticides and Veterinary Medicines Authority for a full list of approved pain relief products in Australia for example [29]. Proper training is needed to administer these medications safely, and it's important to follow the withholding periods—the time between the last treatment and when the animal or its products (like meat or milk) can be safely used for human consumption. This is to ensure that no residual chemicals are present in the final product, protecting consumer health.

Different countries have their own regulatory bodies that oversee the approval and use of veterinary medicines, including pain relief products. Below are the equivalents of the Australian Pesticides and Veterinary Medicines Authority (APVMA) in various parts of the world:

- **United States:**

  - The Food and Drug Administration (FDA) through the Center for Veterinary Medicine (CVM) regulates veterinary drugs. The CVM ensures that all veterinary products, including pain relief medications, are safe for animals and humans. Veterinary drugs approved by the FDA can be found in their Green Book.

- **European Union:**

  - The European Medicines Agency (EMA) is responsible for the scientific evaluation, supervision, and safety monitoring of veterinary medicines in the EU. The Committee for Medicinal Products for Veterinary Use (CVMP) within the EMA oversees these matters.

- **United Kingdom**:

  - The Veterinary Medicines Directorate (VMD) is the regulatory body responsible for the authorization of veterinary medicines in the UK, ensuring the safety and efficacy of products used in animals.

- **Canada**:

  - In Canada, the Veterinary Drugs Directorate (VDD), a branch of Health Canada, is responsible for approving veterinary drugs. They ensure that veterinary drugs meet the required standards for safety, quality, and efficacy before being marketed.

- **New Zealand**:

  - The New Zealand Food Safety Authority (NZFSA) regulates veterinary drugs, including pain relief products, through the Agricultural Compounds and Veterinary Medicines (ACVM) group.

- **South Africa**:

  - In South Africa, the Department of Agriculture, Forestry and Fisheries (DAFF), through the Registrar of Act 36, manages veterinary products. The DAFF's Fertilizers, Farm Feeds, Agricultural Remedies, and Stock Remedies Act governs the approval of veterinary medicines.

These agencies ensure that the products available in their countries are safe for animals and meet the appropriate standards for use.

Recognizing Pain in Livestock

Recognizing pain in livestock can be challenging because many prey species have evolved to hide signs of pain to avoid attracting predators. Despite this, livestock may still display behavioural signs that indicate discomfort or pain. Obvious signs of pain include [29]:

- Lameness or reluctance to bear weight on a limb

- Wounds with discharge or inflammation

- Self-protective behaviour, such as favouring a part of the body

However, **subtle signs** that might also indicate pain include:

- **Vocalization**: Animals may grunt, moan, or otherwise vocalize in response to pain.

- **Teeth grinding**: A common sign of discomfort, especially in cattle and sheep.

- **Reluctance to move**: Animals in pain may avoid movement to prevent worsening discomfort.

- **Rapid or shallow breathing**: Pain can alter breathing patterns.

- **Isolation**: Animals may separate themselves from the herd or flock.

- **Foot stamping**: Seen particularly in cattle and sheep as an expression of irritation or discomfort.

- **Posture changes**: Shifting weight or adopting unusual postures can indicate discomfort.

- **Dullness or depression**: A sudden lack of interest in food or social interactions is a significant sign of pain or illness.

Recognizing signs of pain in livestock is crucial for their well-being and can indicate a need for intervention, such as treatment or pain relief. Each species of livestock exhibits different signs of pain, and understanding these signs helps farmers and caregivers respond effectively. Below is an explanation of general signs of pain in various livestock species based on the observed behaviours in a scientific context.

Vocalization: Livestock often vocalize in response to pain, which serves as a clear sign of discomfort. This behaviour is especially noticeable in cattle, pigs, sheep, and goats. For instance, cattle may make low-pitched moaning sounds, while pigs may squeal when in pain. This vocalization often indicates acute distress, whether from injury, illness, or routine procedures like dehorning or castration.

Grinding Teeth: Teeth grinding is another common sign of pain, particularly in cattle, pigs, sheep, and goats. It is often an involuntary response to both physical discomfort and stress. In pigs and cattle, for example, grinding may suggest internal pain, such as digestive issues or musculoskeletal discomfort.

Reluctance to Move: Animals in pain may become unwilling or hesitant to move due to discomfort or injury. This is a prevalent sign in cattle, pigs, and goats, and it indicates musculoskeletal issues or pain from injuries such as lameness. For example, a cow suffering from hoof issues may exhibit reluctance to walk, while pigs may refuse to move due to abdominal pain.

Rapid/Shallow Breathing: Rapid or shallow breathing can be an indicator of distress in livestock, particularly in goats and pigs. This response is often associated with acute pain, such as during illness or after invasive procedures. It may also indicate respiratory distress or systemic infections that cause pain.

Isolation from Group Mates/Unresponsive to Social Contact: Livestock, particularly social animals like pigs, sheep, and goats, tend to isolate themselves from their herd when in pain. This behaviour is a protective mechanism, as injured or sick animals may distance themselves from others to avoid conflict or further harm. Isolation is an important red flag indicating that the animal is experiencing discomfort.

Foot Stamping: This behaviour is most commonly seen in goats and is often a response to irritation or discomfort, particularly from conditions like parasitic infections or hoof issues. Foot stamping can indicate pain or irritation in the feet, legs, or lower body, and it requires investigation into potential causes, such as injury, infection, or environmental irritants.

Abnormal Posture or Frequent Postural Changes: Livestock in pain may adopt unusual postures or frequently change their posture in an attempt to alleviate discomfort. For example, cattle, pigs, goats, and poultry may hunch their back or appear to crouch as a way of protecting sore areas or avoiding further pain. This is often observed when there are abdominal, musculoskeletal, or internal injuries.

Head Tucked/Eyes Closed: Animals in pain, such as pigs and poultry, may exhibit behaviours like tucking their head or closing their eyes. This lethargic appearance is often a response to both pain and discomfort. It suggests the animal is experiencing severe discomfort or illness, and it warrants immediate attention.

Decreased Production: Pain in livestock often correlates with a drop in productivity. For example, goats experiencing pain may produce less milk, while hens in pain may lay fewer eggs. This decrease in production is the body's way of redirecting energy from growth and reproduction to dealing with pain and maintaining essential functions.

Recognizing the signs of pain in livestock is vital for ensuring animal welfare and taking prompt corrective measures. While some signs, such as vocalization and reluctance to move, are easy to identify, others, like grinding teeth or abnormal posture, may require close observation. Understanding these behaviours allows for early intervention, whether that means providing medical treatment, improving living conditions, or addressing nutritional needs. Keeping a close eye on these indicators helps maintain the overall health and well-being of livestock.

Routine Husbandry Practices

Many routine animal husbandry procedures, if not done correctly or at the proper time, can cause significant pain. These include [29]:

- **Tail docking**

- **Castration**

- **Mulesing** (removing strips of wool-bearing skin to prevent flystrike in sheep)

- **Horn and hoof trimming**

- **Ear tagging**

- **Disbudding and dehorning**

Each of these procedures requires careful handling and, when possible, the use of pain relief. In most cases, the younger the animal is, the better it responds to these interventions with less pain and faster recovery.

Proper Training and Welfare Standards

For livestock handlers and farmers, being properly trained to handle these procedures, assess pain, and use pain relief effectively is critical. In addition, keeping detailed records of when procedures are performed and any medications administered is necessary to ensure compliance with welfare standards and facilitate proper veterinary care.

Euthanasia of Livestock

If you are responsible for livestock, it is essential to have the ability and resources to humanely euthanize an animal or quickly contact a trained professional, such as a veterinarian, to carry out the task without delay. Euthanasia should be performed promptly, safely, and humanely, ensuring minimal distress to the animal [29].

Key Considerations for Humane Euthanasia:

1. **Promptness**: Delaying euthanasia for an animal that is suffering can lead to unnecessary pain and distress. Livestock owners must act swiftly once it becomes clear that euthanasia is necessary, either by performing the procedure themselves if trained, or by contacting a professional to do so.

2. **Safety**: Whether euthanasia is carried out by the livestock owner or a veterinarian, it must be performed in a way that is safe for the person doing it and for others around. This includes taking precautions with firearms, captive bolt guns, or other euthanasia tools to prevent accidents.

3. **Humaneness**: Euthanasia methods must result in the rapid loss of consciousness, followed by death while the animal remains unconscious. This is critical to prevent the animal from experiencing prolonged suffering. Tools such as firearms and penetrating captive bolt guns are often used, but they must be employed correctly to ensure effectiveness.

Legal and Ethical Responsibilities: It is crucial to comply with regional regulations concerning euthanasia tools, especially when it comes to firearms or captive bolt guns. Different areas have specific laws governing the use of such equipment, and operators must be appropriately trained to use them. In many cases, formal training or certification is required to ensure that the person performing the euthanasia is capable of doing so humanely and in line with the law.

Additionally, in some regions, especially in residential or semi-rural areas, regulations may govern not only the euthanasia of animals but also the keeping of livestock for personal use. For example, selling or gifting meat from an animal killed at home without the appropriate license may be prohibited. Public health regulations are often in place to ensure that meat sold or distributed is safe for consumption, and these laws must be followed to avoid legal penalties.

When it comes to the humane euthanasia of livestock, countries around the world have put in place specific laws and regulations governing the use of tools such as firearms and captive bolt guns. The main purpose of these regulations is to ensure that animals are treated ethically and humanely, especially at the end of their life, while ensuring safety for both the operator and the public. Compliance with these laws is a critical part of ethical animal management.

1. European Union (EU)

The EU has comprehensive animal welfare regulations that cover the humane slaughter and euthanasia of animals. Regulation (EC) No 1099/2009 on the protection of animals at the time of killing mandates that animals must be spared avoidable pain, distress, or suffering during slaughter or euthanasia. It emphasizes the need for competent handling and the use of proper tools, such as captive bolt guns, which must be properly maintained and calibrated. Anyone carrying out euthanasia must be appropriately trained and certified.

Key Points:

- Certification is required for operators.

- Specific rules govern the use of tools like captive bolt guns.

- Compliance with welfare standards is required, with severe penalties for non-compliance.

2. United States

In the U.S., regulations governing animal euthanasia differ from state to state, but the general framework is provided by the Humane Methods of Slaughter Act (HMSA) and guidelines issued by organizations like the American Veterinary Medical Association (AVMA). Euthanasia must result in the rapid loss of consciousness, followed by death without pain or distress. The use of firearms or captive bolt guns must be performed by trained personnel, and there are detailed recommendations on which methods are appropriate for different species.

Key Points:

- Federal and state laws may vary, but the AVMA guidelines offer standard procedures.

- The use of euthanasia tools requires training and certification.

- Strict regulations govern the use of animals for personal use, including meat consumption and sales.

3. Australia

In Australia, the Australian Animal Welfare Standards and Guidelines govern the humane killing of livestock. The regulations require that those euthanizing animals must use methods that minimize pain and suffering, which include the use of firearms or captive bolt guns for larger animals. Livestock owners must ensure they have the competence to euthanize animals or that they can call a veterinarian or another competent person. Each state and territory has its own specific regulations, but national guidelines set the minimum standard.

Key Points:

- Euthanasia methods must be approved and performed by trained individuals.

- Operators must be licensed to use firearms or captive bolt guns.

- Detailed record-keeping is required to ensure compliance with animal welfare laws.

4. Canada

Canada follows the National Farm Animal Care Council (NFACC) Codes of Practice, which require the humane euthanasia of animals. Similar to the U.S. and Australia, euthanasia tools such as firearms and captive bolt guns must be used by competent and trained individuals. Canadian regulations ensure that animal owners and handlers have access to pain relief and must minimize stress during euthanasia.

Key Points:

- Euthanasia must result in immediate unconsciousness.

- Operators need to be trained and licensed to use firearms or captive bolt guns.

- Laws exist to ensure compliance with humane treatment, including restrictions on home-killed meat sales.

5. New Zealand

In New Zealand, the Animal Welfare Act 1999 governs the humane slaughter and euthanasia of animals. Livestock owners are required to minimize animal suffering, and anyone performing euthanasia must be trained and competent. The use of captive bolt guns and firearms is regulated, and operators must follow guidelines to ensure the process is humane. There are also restrictions on the home slaughter of livestock, particularly regarding the sale of meat.

Key Points:

- Competency in euthanasia methods is required by law.

- Licensed professionals or trained individuals must carry out euthanasia using approved tools.

- Home-killed meat is regulated to prevent the sale of uninspected meat.

Public Health and Meat Sale Regulations

Around the world, public health regulations also influence euthanasia practices. In many regions, the sale or distribution of meat from animals killed at home is prohibited without the appropriate licenses. This is because the meat has not been inspected by government authorities for safety and hygiene standards, which could pose risks to public health.

Common Regulations:

- Animals slaughtered for personal use must not be sold or gifted unless they comply with local health regulations.

- Meat sales typically require a license, and animals must be slaughtered in licensed facilities under government inspection.

Training and Competency: Anyone responsible for livestock should seek proper training in euthanasia techniques. Understanding the correct application of equipment like captive bolt guns or firearms is vital. This training ensures that the euthanasia process is carried out quickly and painlessly, minimizing the stress for both the animal and the handler. Being familiar with the signs of effective euthanasia, such as immediate unconsciousness, is also crucial to ensuring the process has been successful.

Cattle

When it comes to the humane killing of cattle, the primary methods recommended are the use of a firearm or a penetrating captive bolt aimed at the brain. These methods ensure that the animal experiences a rapid loss of consciousness, minimizing suffering. The distance between the end of the firearm barrel and the animal should be between 10 and 100 cm, with the only approved target being the brain. Captive bolts are placed directly against the skull, while firearms should be used at close range but without direct contact with the head to avoid misdirection [29].

High Frontal Shot: The high frontal shot is the best practice for adult cattle. The target area is the brainstem, located deep under the front of the skull, about halfway between the base of the ears. This shot ensures the most direct impact on the brainstem, resulting in immediate unconsciousness if applied correctly.

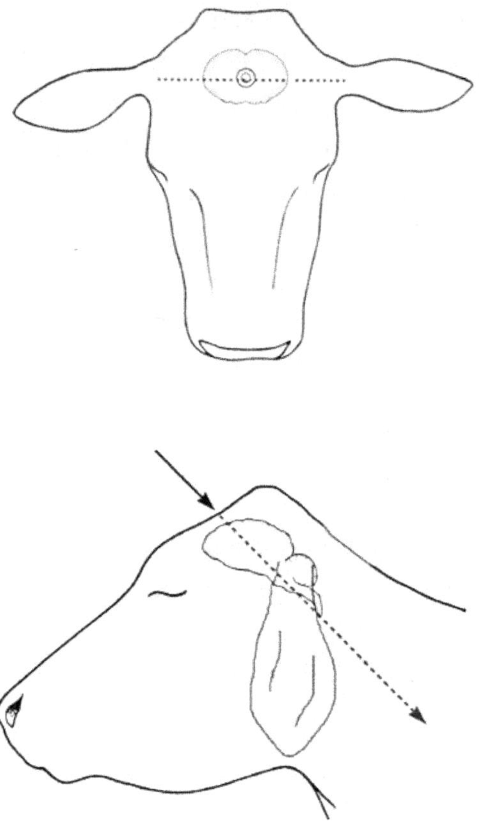

*Figure 30: The optimal method for euthanizing cattle is to aim
for the high frontal shot, targeting the brainstem, which is positioned deep beneath the front of the skull, roughly halfway
between the base of the ears.*

Once the cattle are shot, the procedure should be followed by bleeding out (exsanguination) to ensure the animal is dead. This additional step is essential when using a
captive bolt, as it ensures no possibility of recovery.

Signs of Death: To confirm that the animal has been humanely euthanized, at least
three signs of death should be observed:

    1. **Loss of consciousness** and movement, including eye movement.

    2. **Absence of corneal blink reflex** when the eyeball is touched.

3. **Absence of rhythmic breathing** for at least five minutes.

If normal breathing returns or if any doubt arises, the animal should be re-shot to ensure it does not regain consciousness.

Firearm and Captive Bolt Specifications: For adult cattle, a firearm delivering at least the muzzle energy of a 0.22 magnum calibre cartridge is recommended. Larger animals, such as bulls, may require a 0.30 calibre high-power cartridge. For younger or smaller livestock, such as calves, sheep, and goats, a 0.22-calibre long rifle cartridge is sufficient.

In all cases, the firearm should be used at close range to ensure maximum impact, though it should not be placed directly on the animal's head. When using a penetrating captive bolt, the tool must be pressed firmly against the skull to be effective.

Handling and Restraint: Before administering the euthanasia, cattle must be restrained to avoid unnecessary stress. This restraint ensures that the animal's head is still, providing a clear and safe shot. Proper handling reduces distress and ensures the procedure is carried out effectively and humanely.

Sheep and Goats

The recommended methods for humanely killing sheep, lambs, goats, and kids focus on ensuring that the process is quick, effective, and minimizes suffering. These methods include using a close-range firearm to target the brain, the discharge of a penetrating captive bolt to the brain, or in some cases, a veterinarian-administered lethal injection. After using a captive bolt, it is crucial to follow up with a secondary procedure, such as bleeding out (exsanguination), to ensure the animal is completely dead [29].

To confirm the effectiveness of the method used for humane killing, it is advised to observe at least three signs of death. These signs include [29]:

1. **Loss of Consciousness**: The animal should show no signs of awareness or deliberate movement, including any eye movement.

2. **Absence of Corneal Reflex**: When the animal's eyeball is touched, there should be no blinking response, indicating that the brain is no longer functioning.

3. **Pupil Dilation**: The pupils should be maximally dilated.

4. **Respiratory Movements**: There should be no rhythmic breathing for at least five minutes, signifying the cessation of respiratory function.

For sheep and goats, the best practice is to aim for the low poll shot. The target area is the brainstem, which is located just beneath the poll (the highest part of the head between the ears), roughly midway between the base of the ears. This shot is designed to ensure the most efficient and humane result.

Overall, these guidelines emphasize the importance of ensuring a swift, humane death, while also ensuring that operators are trained to recognize when additional measures, like re-shooting or exsanguination, are needed to guarantee the animal's death and minimize suffering.

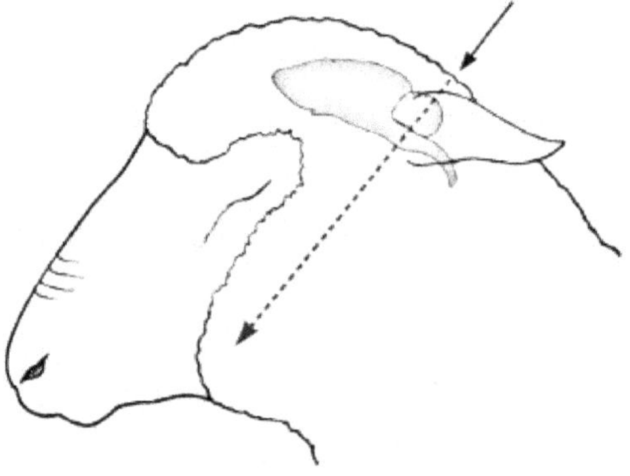

Figure 31: For both sheep and goats, the recommended method is to aim for the low poll shot. This approach applies equally to both species and is considered the most effective.

Dealing with Animal Welfare Emergencies

Proper planning for emergencies is essential for the welfare of all animals, including livestock, horses, companion animals, and wildlife. Emergencies such as bushfires, floods, and other natural disasters can severely impact the safety and survival of animals. As a result, owners and caretakers bear the responsibility of ensuring their animals' welfare by being prepared and having a plan in place [34].

Figure 32: Cattle affected by Hurricane Harvey flooding in Port Arthur, Texas, August 30, 2017. Defense Visual Information Distribution Service, Public Domain, via Picryl.

Emergency Preparedness: Emergency preparedness for animals begins with the development of an emergency plan that addresses potential hazards. The plan should include information on evacuation routes, transport options, and identification of safe areas where animals can be sheltered. For livestock and other animals that cannot be evacuated, it is crucial to have safer areas on the property where they can be moved in times of danger (e.g., paddocks without vegetation during fire risk or elevated areas during floods). The plan should also include a list of emergency contacts such as veterinarians and animal shelters. Keeping updated records of animals, including identification details, is essential for ensuring their return in case they become separated from their owners.

Emergency Kit: Owners and caretakers should prepare an emergency kit containing essential supplies such as food, water, medications, and identification documents for their animals. The kit should be easily accessible and ready to use in case an emergency arises. Having items like halters, leads, and first-aid supplies for animals ensures that caretakers can handle injuries or illnesses promptly.

Property Preparation: In addition to preparing a plan and emergency kit, property owners should take steps to mitigate hazards that could affect the animals during emergencies. This might include installing gates on internal fences to avoid moving animals along public roads, securing emergency fodder and water supplies, and marking gate locations on a property map. Property owners should also consider installing fire extinguishers, sprinkler systems, and smoke detectors to protect both animals and property. Ensuring access to a reliable water supply is also crucial, especially in the event of fires or floods.

Evacuation and Relocation: When there is an imminent risk, evacuation or relocation to safer areas is often necessary. While it is ideal to relocate animals to safer areas well before the danger escalates, it may not always be feasible to transport them off the property. In such cases, moving animals to a safer area within the property, such as an elevated paddock during floods, can help protect them. Owners should also practice moving animals to reduce stress during actual evacuations. Animals unfamiliar with containment methods such as crates or floats may become stressed or difficult to manage during emergencies.

Animal Identification: Identification is a critical aspect of ensuring animals can be returned to their owners if they become separated during an emergency. Effective identification methods include brands, National Livestock Identification Service (NLIS) devices, microchips, and name tags. Keeping animal records up to date with the vet, local government, and insurance provider is vital. Owners should also keep photos of their animals as they may assist in confirming identification.

Sample Emergency Plan for Animal Welfare

1. Introduction

This emergency plan outlines steps to ensure the safety and well-being of all animals under care, including livestock, horses, companion animals, and wildlife. The plan addresses potential hazards such as bushfires, floods, or other natural disasters and focuses on evacuation, relocation, and providing continuous care during emergencies.

2. Emergency Preparedness

2.1 Hazards

Potential hazards identified for the area:

- **Bushfires:** High risk due to dry vegetation.

- **Flooding:** Risk during heavy rain, with low-lying areas prone to waterlogging.

- **High Winds:** Potential damage to shelters and trees falling onto animal enclosures.

2.2 Evacuation Routes

- **Primary Route:** Main road to nearby safe shelter (Animal Boarding Facility 5km away).

- **Secondary Route:** Access via the north field, leading to a higher ground shelter on the property.

2.3 Safe Areas on the Property

- **Livestock:** Move to paddocks cleared of vegetation for fire risks or higher ground paddocks during floods.

- **Horses:** Designated shelter with secure fencing and shaded areas in the south paddock.

- **Companion Animals:** Small animals should be moved to indoor safe zones if relocation off-property is not possible.

2.4 Emergency Contacts

- **Local Vet:** Dr. Jane Smith (Phone: 123-456-7890) – 24/7 availability.

- **Animal Shelter:** Safe Haven Shelter (Phone: 987-654-3210) – Ready for emergency boarding.

- **Animal Transport Services:** Reliable Livestock Movers (Phone: 111-222-3333).

3. Emergency Kit

3.1 Contents

- **Feed & Water:** At least three days' supply of feed for all animals and 50L of clean drinking water.

- **Leads and Halters:** For all livestock and horses.

- **Medications:** First aid kit with essentials like bandages, antiseptic, pain relief, and prescription medications.

- **Documentation:** Identification, vaccination records, and contact details.

- **Other Items:** Blankets, rugs, and basic grooming supplies for comfort.

4. Property Preparation

4.1 Safety Measures

- **Internal Gates:** Installed to avoid moving livestock along public roads.

- **Water Supply:** Ensure a backup water supply system for five days.

- **Fodder Storage:** Emergency fodder stored in safe, flood-proof locations.

- **Fire Safety Equipment:** Installed fire extinguishers and smoke detectors in animal shelters.

- **Flammable Materials:** Remove rugs, halters, and flammable items from near animals' areas during fire risk.

5. Evacuation and Relocation

- **Companion Animals:** Relocate to the shelter or safe indoor space.

- **Livestock and Horses:** Move to designated safe paddocks if evacuation is not possible, or transport to external facilities (Animal Boarding Facility).

- **Timing:** Evacuate animals before conditions worsen, triggered by weather alerts or visible fire/smoke.

6. Animal Identification

- **Livestock:** NLIS tags applied to all cattle and sheep, updated in the local government registry.

- **Companion Animals:** All pets microchipped, and collars with ID tags attached.

- **Photographs:** Recent photos of all animals kept in both digital and hard copies for identification in case of loss.

7. Emergency Response

- **Primary Evacuation:** Triggered by weather alerts for imminent fire or flood danger. All animals will be moved to safe areas as outlined.

- **Secondary Action:** If evacuation is not possible, ensure all animals are in safe zones within the property and have adequate food, water, and protection from the elements.

8. Post-Emergency Care
- **Immediate Checks:** All animals will be inspected for injuries, stress, or signs of illness.

- **Food & Water:** Ensure animals have continuous access to clean water and nutritious food after the emergency.

- **Veterinary Care:** Engage the vet for animals showing signs of stress, dehydration, or injury.

Conclusion

Regularly updating this emergency plan and practicing evacuation drills is essential for the safety of all animals. Proper identification, emergency kits, and property preparation will ensure swift action during emergencies, reducing the risk of injury or loss.

In situations where animals cannot be evacuated during an emergency, it becomes the responsibility of their owner or caretaker to ensure they are appropriately sheltered or relocated on the property and provided with sufficient food and water for an extended period. This planning is critical because, during emergencies like bushfires, floods, or storms, the animals' welfare and survival could depend on the preparation and resources left behind.

Acting Early to Minimize Risk

One of the most important actions during an emergency is acting promptly. Evacuating animals early when danger is first anticipated gives enough time for safe relocation, reducing risks for both the animals and the people involved. If early evacuation isn't possible, owners should have an emergency plan in place that outlines steps for securing animals within the property to ensure their safety.

Animals' Response to Emergency Situations

Animals often become frightened or stressed in emergencies, which can make them harder to manage. Some animals may hide, resist handling, or exhibit abnormal behaviours. When it is impossible to evacuate them, they should only be left behind in cases where it poses a safety risk to move them. This decision should only be made after all other options are exhausted.

Precautionary Preparations for On-Property Shelter

For animals that must remain on the property, various precautions should be taken:

- **Water and Food:** Ensure animals have access to enough water and food for at least five days, especially if you might not return for several days.

- **Remove Hazards:** In the event of a fire, remove any synthetic materials like rugs or blankets from animals, as these can melt and cause burns. For floods, move feed and supplies to higher ground.

- **Shelter and Security:** Animals should be moved to secure areas protected from hazards such as fire or floods. These areas should be free from dangerous debris or flammable materials.

- **Communication:** It is important to inform a friend or relative where your animals are located and leave a note for emergency responders indicating the types of animals present and your contact information.

Relocation to Safer Areas

If there is an emergency warning, consider relocating animals to safer areas on the property, such as paddocks with low fire fuel loads or areas of higher ground during a flood. For example:

- **Horses:** They should be moved to open areas where they can move freely, which helps them avoid fire and escape extreme heat.

- **Livestock:** Move to areas with less vegetation or cleared paddocks during fire threats. Ensure access to sufficient drinking water.

- **Companion Animals:** If left indoors, they should be secured in rooms with minimal hazards, supplied with enough food and water, and protected from dangers like sharp objects or chemicals.

Evacuating Animals

Evacuating animals is the safest option whenever possible. Large animals like horses and cattle require careful coordination to avoid endangering owners and responders. It's important to plan routes, ensure trailers or floats are ready, and animals are trained to enter transport equipment calmly. Smaller animals should be confined in crates or cages to minimize stress during transportation.

Identification and Tracking

Ensuring animals are identifiable during emergencies is vital for their safe return if they become lost. Methods like microchips, National Livestock Identification Service (NLIS) tags for livestock, or name tags for companion animals can help authorities or rescuers return the animals to their owners.

Post-Emergency Care

Once the emergency has passed, animals should be checked for injuries, stress, and illnesses. It's important to provide continuous access to water, food, and safe shelter, and to seek veterinary care for animals showing signs of distress or trauma.

Monitoring and Maintaining Accurate Records

Monitoring and maintaining accurate records is essential for animal welfare management across the world, whether for regulatory compliance, breeding programs, or research. This practice is integral to ensuring ethical animal care and facilitating the health and wellbeing of animals in various settings, from farms to research facilities.

In many countries, record-keeping is mandated by animal welfare laws and guidelines. For example, in Australia, institutions must comply with the Australian Code for the Care and Use of Animals for Scientific Purposes . This Code requires thorough monitoring and documentation to safeguard animal welfare in both breeding and research contexts. Similarly, European Union regulations (such as Directive 2010/63/EU) and U.S. regulations (like the Animal Welfare Act) mandate stringent monitoring and record-keeping protocols for research and agricultural animals.

The general principles behind these systems worldwide include:

- **Animal Health Monitoring**: This involves tracking the health of animals on a daily basis, noting any changes in condition, signs of illness, distress, or adverse effects. Daily records of animals, particularly those undergoing procedures or recently introduced to a facility, are crucial.

- **Environmental Monitoring**: It's common to monitor environmental factors such as room temperature, humidity, and other conditions that may impact animal health. These environmental metrics should be logged regularly, ensuring they are within species-appropriate ranges.

- **Breeding Records**: Maintaining detailed records of animal breeding is essential for managing colonies. This includes tracking births, mortality, and any complications during reproduction.

- **Adverse Event Reporting**: Monitoring unexpected events, such as illness, injury, or procedure-related issues, must be reported and documented. In research settings, this can help identify potential concerns related to animal welfare or genetic phenotypes.

- **Data Accessibility**: Records need to be maintained for several years and made accessible for inspection by authorities like animal ethics committees (AECs), ensuring compliance with welfare standards.

- **Use of Technology**: Increasingly, facilities around the world are using databases and software systems to centralize record-keeping. These systems allow for easier data access, trend analysis, and more efficient monitoring.

Measures to Prevent Recurrence and Minimise Risk of Animal Welfare Emergencies

It is essential for animal owners and caregivers to have a well-prepared emergency plan in place to ensure the safety and welfare of their animals, including livestock, horses, companion animals, and wildlife. Emergencies such as bushfires, floods, or severe storms can occur unexpectedly, and having a pre-established plan reduces the likelihood of making last-minute, risky decisions that could compromise the safety of both the animals and their caretakers.

Importance of Acting Early

During emergencies, swift and early action is key to minimizing risks to both lives and property. Acting as soon as high-risk conditions or incidents arise allows for more time to focus on securing animals, ensuring their safety, and addressing immediate concerns. Animals can often behave unpredictably during emergencies, becoming frightened, stressed, or difficult to manage. A pre-prepared plan that outlines clear actions helps to streamline the process and ensures that all steps are taken calmly and efficiently.

Owners and carers are legally and ethically responsible for ensuring the welfare of their animals during an emergency. This includes providing adequate food, water, shelter, and veterinary care. The primary preference, where safe and feasible, is to evacuate animals to ensure their safety. However, when evacuation isn't possible, either due to time constraints or because the situation poses a risk, alternative solutions like relocating animals to safer areas within the property must be considered. These areas should offer maximum protection against the imminent threat, such as a paddock with minimal vegetation during fire risk or higher ground in case of floods.

Being aware of the potential risks to animals and property during high-risk periods is crucial for all animal owners globally. To stay informed, utilizing reliable resources that provide real-time, map-based updates on various hazards like fires, floods, storms, or other emergencies is essential. In different parts of the world, these resources may vary. For example, platforms like FEMA's Ready.gov in the United States, Met Office Weather Warnings in the UK, or Geoscience Australia provide hazard alerts specific to their regions. Additionally, international platforms such as ReliefWeb and the Global Disaster Alert and Coordination System (GDACS) offer global disaster and emergency alerts.

In conjunction with these resources, maintaining open communication with neighbours, family, and local communities is critical. Being aware of local conditions through multiple communication channels, such as social media, radio broadcasts, and emergency alert systems, helps to ensure that you are always up to date. Social media platforms like Twitter and Facebook are widely used by emergency services globally to provide real-time updates, while local and national radio stations often broadcast important warnings and instructions during emergencies. Always make sure to stay tuned to official sources in your region, and rely on credible emergency apps or broadcasts to make informed decisions.

As soon as risk levels escalate or an incident occurs, it's time to implement the emergency plan. A well-prepared plan helps identify the right moments (or "triggers") to take action, such as when to evacuate animals or when it becomes necessary to make alternative arrangements, like relocating them to safer areas within the property. It is crucial to review this plan regularly to ensure that all necessary supplies and logistics are up to date and functional.

The approach to ensuring the welfare of animals during an emergency depends on the circumstances:

- **Evacuation**: This involves moving animals entirely outside of the danger zone. Evacuation requires prior preparation, including identifying transport options for large animals and planning where to take them.

- **Relocation within Property**: When animals can't be evacuated, relocating them to a safer area within the property (e.g., paddocks with less vegetation or barns with solid shelter) can minimize the risks. It's important to have pre-designated areas that can protect animals from fire, flood, or other threats.

In summary, preparedness is essential to animal welfare in emergencies. By having a clear plan, staying informed, and acting early, owners and carers can significantly improve the safety of their animals and themselves during high-risk situations.

Staff Induction

Identifying animal welfare induction and training needs is an essential component of maintaining a safe, ethical, and productive environment in industries involving livestock and other animals. Proper training ensures that everyone involved is equipped with the knowledge to meet legal obligations and uphold animal welfare standards. Here's a detailed explanation of the key elements involved in this process [35, 36]:

Duty of Care to Livestock

The duty of care is a legal obligation placed on all individuals involved in managing livestock. It ensures that animals are provided with proper food, water, and living conditions, including protection from predators and environmental hazards. People responsible for livestock include owners, managers, transport drivers, contractors, agistors, and abattoir workers. Each of these individuals must understand their duty to provide for the animals' basic welfare needs, as their actions directly impact the well-being of the animals.

Livestock Codes, Standards, and Guidelines

Across the globe, livestock industries are governed by national and international standards and guidelines that outline best practices for animal welfare. These include specific rules for the transport, handling, and housing of livestock, as well as procedures for slaughterhouses. These guidelines aim to ensure the humane treatment of livestock in both commercial and hobby farming environments. National animal welfare guidelines often extend to peri-urban settings, where livestock may be kept on smaller properties.

Instructing Staff

Training staff on how their actions impact animal welfare is crucial. This process often involves both theoretical and practical instruction. The training should cover animal handling techniques, health and safety requirements, hygiene practices, and quality assurance protocols. In addition, record-keeping of training sessions is important, as it provides evidence of compliance with industry regulations. Trainees are usually supervised by an experienced handler until they are deemed competent. Regular in-service training ensures that skills are updated, and new employees are always informed of the latest best practices.

Providing Fresh, Palatable, and Unspoiled Food

A key part of animal welfare is ensuring livestock receive adequate and nutritious food. For those using commercially formulated feeds, feed mills must adhere to quality standards to ensure the wholesomeness of the product. On-farm, however, issues such as feed palatability and spoilage need to be monitored closely, as they can affect the voluntary intake of food by animals. For those who mix their feed on-site, stringent standard operating procedures (SOPs) are necessary. These should cover the selection of ingredients, storage conditions, and regular testing for quality. Keeping silos clean and free from contamination is critical to prevent food spoilage.

On and Off-the-Job Animal Welfare Training

Both on-the-job and off-the-job training are essential for workers handling livestock. Effective induction programs for new workers should cover a variety of important topics, including animal handling techniques, relevant legal and regulatory requirements, workplace health and safety, and compliance with the National Livestock Identification Systems (NLIS) and other animal welfare standards. Communication skills are vital during induction, as clear instructions ensure that workers understand their responsibilities and the consequences of not following best practices.

Benefits of a Structured Induction Program

A structured induction program can increase workplace safety, employee retention, productivity, and overall morale. It also ensures that new employees understand their role and responsibilities from the start. When working with animals, this is particularly important because poor training can lead to serious welfare violations, including harm to the animals. In addition to improving employee performance, proper induction programs can reduce the risk of legal penalties and ensure that employees understand how to handle animals safely and ethically.

Continuous Monitoring and Updates

Finally, an effective animal welfare program is not a one-time effort. Ongoing training, periodic skill assessments, and the continuous review of procedures are necessary to ensure that welfare standards are maintained. As new laws and guidelines emerge, employees must be kept up to date through training refreshers. Monitoring systems should also be in place to assess the health and well-being of animals regularly, identifying any areas where care needs to be improved.

Proper induction and training for livestock welfare are critical in ensuring that all individuals involved in the care and management of animals meet their legal and ethical responsibilities. This training helps prevent mistakes that could lead to animal suffering and ensures that animals are treated humanely in every aspect of their care.

CHAPTER SIX

Transporting Livestock

Handling Livestock

Handling livestock properly is critical not only for animal welfare but also for the safety of those managing the animals. Proper handling reduces stress on the animals, minimizes injury risks, and can even improve overall productivity. Here's a detailed breakdown of the key aspects of handling livestock [29]:

Low-Stress Handling Techniques

Low-stress livestock handling is essential for both the well-being of the animals and the handlers. Animals, like humans, can experience stress during handling, and this can lead to injuries, both for the animals and for the people working with them. Low-stress handling methods aim to use the animals' natural behaviours to guide them without causing fear or anxiety. For example, animals tend to move together in groups (a natural herd behaviour), and good handlers take advantage of this by calmly guiding them instead of forcing them into stressful situations. Reducing stress also helps maintain the animals' immune systems and can lead to better meat or milk production.

Importance of Proper Handling Facilities

Ensuring that your facilities are "fit for purpose" is a critical component of low-stress handling. The design of livestock facilities should take into account the type and breed of the animals. For cattle, for instance, a well-designed cattle crush or chute is essential for drenching, vaccinations, or veterinary care. Similarly, sheep or goats require specialized areas like shearing sheds for crutching (removing wool from the rear end) or shearing. These facilities should allow the animals to move comfortably without being crowded or panicked, which can cause injury or stress.

Properly maintained facilities reduce handling times, prevent injuries (to both animals and handlers), and improve animal health outcomes by ensuring that necessary treatments can be delivered in a calm and controlled manner.

Impact of Stress on Livestock

When livestock experience stress during handling, their overall health and productivity can be negatively affected. Stress can suppress immune function, making animals more susceptible to disease. In dairy cattle, stress can reduce milk production, while in beef cattle, stress before slaughter can affect meat quality, leading to undesirable effects like "dark cutters," where the meat becomes dark and dry due to elevated stress hormones. Additionally, high levels of stress in sheep and goats can lead to weight loss, reduced reproductive success, and even death in extreme cases.

Respectful Treatment of Animals

Respecting livestock and handling them with care is an ethical obligation for farmers and animal handlers. Treating animals with dignity and compassion is not only the right thing to do but also leads to better long-term outcomes for the animals. Stock handlers should be trained in proper techniques and should continuously update their skills, as improper handling methods can cause significant harm. For example, striking animals or shouting at them only increases fear and anxiety, leading to greater difficulties in controlling the herd.

The Benefits of Low-Stress Handling

Incorporating low-stress handling techniques has several benefits. Animals handled in a calm and controlled manner are easier to manage, reducing the physical strain on handlers. Low-stress techniques can also lead to improved reproductive rates, better weight gain, and overall higher productivity, whether in meat, milk, or wool production. Furthermore, by reducing the risk of injury to both animals and humans, low-stress handling can help cut down on veterinary bills and lost workdays due to injury.

Figure 33: Low stress handling and loading for transport. K-State Research and Extension, CC BY 2.0, via Flickr.

Transport of Livestock

Transporting livestock involves several legal and ethical responsibilities to ensure the welfare and safety of the animals throughout the journey. Improper handling during transport can lead to unnecessary suffering, stress, injury, and in extreme cases, death. Many countries have strict regulations in place to protect livestock during transit, and failure to comply with these regulations is considered a serious offense. Below is a detailed explanation of key considerations that must be taken into account when transporting livestock, along with a broader context of global practices [29].

It is illegal in many countries to transport animals that are considered unfit for loading. An animal is deemed unfit if it is sick, injured, or incapable of standing or moving without pain. Transporting such animals can cause extreme suffering, and strict penalties are imposed for violations. Globally, countries like Australia, the United States, the European Union, and Canada have laws that govern the fitness of animals for transport. In the European Union, the Animal Welfare in Transport Regulation (Council Regulation (EC)

No 1/2005) provides specific guidelines on animal welfare during transit. In Australia, the Land Transport Standards and Guidelines dictate the conditions under which livestock can be transported.

Figure 34: Sheep in B Double truck, Moree, NSW. Cgoodwin, CC BY-SA 3.0, via Wikimedia Commons.

Providing adequate feed and water is one of the most critical aspects of animal transport. Animals must have access to water and food before loading and may require it during long journeys. Different species have specific needs, and these must be addressed to avoid dehydration and malnutrition. For example, regulations in the United States (such as the 28-Hour Law) mandate that livestock being transported for more than 28 hours must be unloaded, rested, and provided with food and water. In Australia, guidelines specify water requirements before and during transport, with curfew times based on animal type and journey duration.

During transportation, animals must be shielded from environmental extremes such as excessive heat, cold, or rain. Overexposure to heat can lead to heat stress, while cold and wet conditions can cause hypothermia. Transportation trucks and containers need to be ventilated to prevent overheating, and shelters should be used when loading or unloading animals to protect them from direct sun, wind, and rain. In some regions, including Canada and the EU, strict rules on thermal environments and ventilation are enforced to maintain appropriate conditions for the animals.

Transporting livestock in a manner that could lead to injury is illegal in many countries. Overcrowding, poor handling, or inappropriate transport equipment can cause bruises, broken bones, or other injuries. The transport vehicle and the handling equipment must be appropriate for the size and type of animal. Additionally, hygiene during transport is important for preventing the spread of diseases. Vehicles should be cleaned and disinfected between trips to limit cross-contamination of infectious diseases, a practice reinforced in Europe and in North America.

Figure 35: A ship carrying sheep from Australia docked in Oman. Tom Jervis from UK, CC BY 2.0, via Wikimedia Commons.

Animal welfare during transport depends heavily on providing enough space for the animals. Overcrowding can lead to increased stress, injury, and even trampling. Conversely, too much space may cause animals to be thrown around during transit, particularly in rough terrain. Regulations usually specify the minimum and maximum stocking densities based on the type of animal, their size, and the duration of the journey. For example, the European Union regulates space requirements to prevent overcrowding, mandating that animals must be able to stand and lie down comfortably.

Mixing unfamiliar animals during transport can lead to fighting, bullying, and injuries, particularly among species like pigs and cattle that establish social hierarchies. Aggressive animals should be separated from others to avoid harm. Globally, animal transport regulations encourage the separation of aggressive animals from more docile ones. For instance, in Australia, the National Livestock Transport Standards advise against mixing unfamiliar groups of livestock to minimize the risk of aggression and ensure safer transit.

Transporting livestock requires careful planning, adherence to regulations, and consideration for the welfare of the animals involved. Every stage of the journey, from preparation to unloading, should prioritize the animals' health and safety, ensuring they are fed, hydrated, sheltered, and handled with care. By following best practices and legal standards, the risks of stress, injury, and suffering can be significantly reduced. Global frameworks and national regulations provide guidance to ensure humane treatment during transit, helping to protect the animals and avoid legal consequences for improper handling.

Handling Livestock During Transport

Handling livestock during transport is a globally significant aspect of animal welfare, and improper handling can lead to stress, injury, or death. Animal welfare regulations and guidelines emphasize the importance of minimizing pain and distress to ensure the safety and well-being of animals throughout their journey. The following points outline key considerations for handling livestock during transport within a worldwide context [29]:

Appropriate Handling Techniques

Individuals responsible for handling livestock must use methods that align with the species and class of the animals to avoid injury or unnecessary stress. Global standards, such as those established by the OIE (World Organisation for Animal Health), specify that animals should never be lifted by a single body part, such as the leg, ear, neck, or wool, as this can cause pain or injury. Instead, animals must be supported during any mechanical lifting process, such as hoisting onto a vehicle, to ensure they remain secure and unharmed.

Avoiding Harmful Practices

The mishandling of livestock can lead to serious welfare concerns. Under worldwide guidelines, livestock must never be thrown, dropped, or handled in an excessively rough manner. Striking, punching, or kicking animals is strictly prohibited across most regions, including the European Union, North America, and Australia, to prevent physical harm and distress. Electric prods, often used to move animals, are restricted, and their use is heavily regulated. Electric prods should not be used excessively, on sensitive areas like the face or genitals, or on young or weakened animals that cannot move away.

Handling Non-Ambulatory Animals

Special care is required for livestock that are unable to stand. Dragging animals, except in emergencies where movement is required to provide treatment or humane destruction, is prohibited. Global best practices highlight the importance of using appropriate equipment to move non-ambulatory animals, ensuring that the handling does not exacerbate their condition. In situations where an animal cannot be moved safely, the handler should consult a veterinarian or consider humane euthanasia as necessary.

Special Considerations for Young Livestock

Particular care is required when transporting young animals such as calves, kids, and lambs. Unweaned calves under five days old should not be transported without their mothers due to their vulnerability and lack of physical resilience. Calves younger than 30 days require gentle handling, as they may not instinctively follow handlers and may become easily fatigued. In these cases, handlers should prioritize calm, controlled movement to minimize stress and avoid injuring these animals.

Control of Dogs

Dogs used during the loading, transporting, and unloading of livestock must be under control at all times to prevent them from causing harm or distress to the animals. In several regions, including Australia and the EU, muzzles are recommended, particularly when working with young animals such as calves, to prevent accidental bites or injuries.

In a worldwide context, transporting animals, including dogs, with livestock requires careful consideration of safety and animal welfare standards. Various international guidelines and regional regulations aim to ensure that both livestock and companion animals are transported in a humane and safe manner, minimizing risk and distress.

In most regions, it is advised that dogs should not be transported in the same pen as livestock. The only exception to this rule applies to bonded guardian dogs, such as Maremma Sheepdogs, which are specifically trained to live and work closely with livestock, providing protection from predators. These dogs are integrated with the livestock and can cohabitate safely without causing harm or distress to the animals they protect.

However, for non-bonded dogs or working dogs, mixing them with livestock in the same pen during transport is not advisable. Transporting them separately ensures the safety of both the dog and the livestock, preventing injuries or stress that could occur from conflicts or the natural instincts of some dogs to herd or chase.

Floor Area Considerations

The floor area allocated for each animal during transport is a critical factor in maintaining animal welfare. International animal welfare guidelines, such as those from the World Organisation for Animal Health (OIE), emphasize that as the wool or fibre length of animals such as sheep or goats increases, the space provided to each animal must also increase. This is because long wool or fibre reduces the ability of animals to dissipate heat and move easily, increasing their vulnerability to overcrowding. Conversely, newly shorn sheep or goats may require less space due to their lighter wool cover, but they still need adequate room to lie down and rise without assistance.

In addition, animals with horns should also be given more space. Horns can become entangled or cause injuries to other animals if the space is too confined. This consideration is particularly important when transporting species such as goats or cattle that may have long or sharp horns.

Movement and Loading Considerations

Animals must have sufficient space to move during transport and should be able to stand up unassisted from a lying position. If they are too tightly packed, there is a risk of smothering, trampling, or even death due to the inability to shift positions or move. Proper spacing reduces the risk of physical harm and ensures that the animals' stress levels are minimized during long journeys.

When loading and unloading, animals should be handled carefully. Forcing animals to jump from vehicles is dangerous and can result in severe injuries, such as broken legs or hips. Purpose-built ramps, designed with non-slip surfaces and gentle slopes, are essential for safely moving animals on and off transport vehicles.

Prohibition on Transport in Car Boots and Tying Legs

Across most jurisdictions, transporting animals in car boots is prohibited. This practice endangers the animal's well-being due to the lack of ventilation, potential for overheating, and restricted movement, which could lead to injuries or suffocation. Animals must be transported in properly ventilated areas where they have enough space to stand or lie down comfortably.

Additionally, it is universally recognized that animals should not have their legs tied during transport. Tying legs can cause severe pain, restrict circulation, and lead to long-term injuries. It may also prevent the animal from standing or shifting positions, increasing the risk of injury during the journey. Ensuring that animals can move freely, without restraints on their limbs, is a basic requirement for humane transport.

Global Welfare Standards

The focus on humane handling during livestock transport is enshrined in international animal welfare frameworks, such as those established by the OIE, the European Union's Animal Welfare During Transport regulation, and other national bodies, such as Australia's Livestock Welfare Standards and Guidelines. These regulations aim to ensure that handling methods across the world meet humane standards, protect the animals, and promote safer transport processes. Compliance with these guidelines is not only a legal responsibility but also a moral one for those involved in the transport and care of livestock.

Transport Vehicles

Transporting livestock safely requires careful attention to the vehicle being used, ensuring that it meets the specific needs of the animals and adheres to international animal welfare standards. Proper transport vehicles are essential to minimize stress, injury, and suffering in livestock during transit, particularly over long distances [29].

Appropriateness of Transport Vehicles

Vehicles used for livestock transport must be designed to safely contain the species being transported. Different livestock species, such as cattle, sheep, goats, and pigs, have varying needs regarding space, temperature regulation, and ventilation. It is crucial to ensure that the vehicle is suitable for the type of livestock being transported to prevent stress, injuries, or even death.

Global guidelines, such as those from the World Organisation for Animal Health (OIE), recommend species-specific adjustments in transport to account for the behaviour and needs of different animals. For example, sheep may require increased ventilation due to their wool covering, while cattle need ample space to stand and move.

Ventilation and Airflow

Effective airflow is critical to maintain a comfortable environment for livestock during transport. Poor ventilation can lead to heat stress, especially in warmer climates or during summer months. Animals such as pigs and poultry are particularly vulnerable to overheating due to their lower tolerance for high temperatures. Therefore, vehicles must be designed to provide sufficient air circulation, particularly during long-distance transport.

Globally, livestock transportation is subject to strict regulations regarding ventilation. For instance, in the European Union, vehicles used for transporting animals over long distances must be equipped with ventilation systems that can maintain an appropriate temperature regardless of external conditions.

Flooring and Injury Prevention

The flooring of transport vehicles should minimize the risk of livestock slipping or falling. Non-slip surfaces are essential for all species, as wet or smooth surfaces can lead to serious injuries, including broken limbs or bruises, which are not only harmful to the animals but can also result in economic losses for producers.

In addition, flooring should be designed to allow proper drainage of urine and manure to maintain hygiene and reduce the likelihood of animals contracting diseases during transport. The European Commission, for instance, requires that transport vehicles have flooring that provides grip and supports cleanliness during transit.

Protection from Internal Protrusions and Sharp Edges

Vehicles should be free from internal protrusions, sharp edges, or other objects that could cause injury to livestock. This is particularly important for horned animals like goats or cattle, which are more likely to be injured in confined spaces. Livestock crates or partitions within the vehicle should have smooth surfaces to prevent bruising, cuts, or other forms of physical trauma.

Sufficient Vertical Clearance

Sufficient vertical clearance in the vehicle is necessary to minimize the risk of injury, especially for larger animals like cattle and horses. Ensuring enough headroom allows animals to stand comfortably without hitting their heads on the roof of the transport vehicle. This is an essential welfare consideration, as cramped conditions can cause stress and discomfort.

In regions like Australia and New Zealand, specific regulations govern the vertical clearance for livestock transport to ensure that animals can move freely and are not subject to physical harm during the journey.

Fixed Partitions and Segregation of Livestock

Fixed partitions are critical when transporting livestock in hilly areas, high-traffic zones, or when carrying smaller numbers of animals. These partitions help prevent livestock from being thrown around inside the vehicle, reducing the likelihood of injury or distress. Segregating aggressive or unfamiliar animals is also essential to avoid fighting or other conflicts during transport.

For example, in North America, guidelines for cattle transport specify the need for partitions to prevent injuries during transport, especially over uneven terrain. This ensures that animals remain stable and protected throughout the journey.

Solid Ramp Extensions and Loading Procedures

Loading and unloading livestock should be done using solid ramps that are securely fastened to the vehicle. Any gaps between the loading ramp and the floor of the vehicle must be covered to prevent animals from falling or getting their limbs stuck. Proper ramps reduce the risk of injury during loading and unloading, which is a critical part of the transport process.

Internationally, the OIE provides standards for ramp inclines, recommending that they be as gentle as possible to prevent stress or injury during loading. In countries like the UK and the EU, regulations mandate the use of solid ramps with non-slip surfaces for livestock transport.

Cold Stress and Wind Chill Protection

Livestock, particularly young animals or recently shorn sheep and goats, are vulnerable to cold stress during transport. Vehicles should have enclosed fronts or the ability to cover the roof or sides to prevent wind chill and ensure animals remain warm during colder months or in regions with harsh weather. Transporters must be equipped to modify vehicle structures according to weather conditions to protect livestock from extreme cold.

In regions like Canada or Northern Europe, where temperatures can drop significantly, transport vehicles are often fitted with covers or windbreaks to ensure that livestock are not exposed to freezing conditions during long journeys.

Properly designed and maintained transport vehicles are vital to ensuring the welfare of livestock during transport, reducing stress and injury risks. Global animal welfare standards emphasize the need for vehicles to be species-appropriate, well-ventilated, and free from hazards that could harm the animals. These measures are critical not only for animal welfare but also for complying with international regulations aimed at humane livestock transport.

tines to stop the spread. Some diseases, like rabies or brucellosis, pose serious
health risks, underscoring the need for prompt reporting.

Reporting Protocol

Across the world, producers and livestock owners have a legal duty to report suspected
of notifiable diseases. In Australia, for instance, any farmer or producer who suspects
mal is showing signs of a notifiable disease must immediately contact a local veteri-
or report the case through the Emergency Animal Disease Watch Hotline (1800
8). Other countries have similar reporting mechanisms, such as the Animal and
Health Agency (APHA) in the United Kingdom or the United States Department
culture (USDA) in the U.S.

necessity for rapid reporting ensures that authorities can quickly enact measures
ain and manage the disease outbreak. Delays in reporting or failure to do so can
-reaching consequences, including the rapid spread of disease, disruption of trade,
n legal penalties for those who fail to comply.

Examples of Notifiable Diseases

Notifiable diseases vary by region but often include:

- **Foot-and-Mouth Disease (FMD)**: A highly contagious viral disease that affects
  cloven-hoofed animals such as cattle, pigs, and sheep. FMD can severely impact
  livestock production and international trade.

- **African Swine Fever (ASF)**: A viral disease that affects domestic and wild pigs,
  with high mortality rates. It poses a significant threat to pig populations and the
  pork industry globally.

- **Bovine Spongiform Encephalopathy (BSE)**: Commonly known as mad cow
  disease, this disease has serious public health implications as it can be transmitted
  to humans in the form of variant Creutzfeldt-Jakob disease (vCJD).

- **Newcastle Disease**: A contagious viral disease affecting birds, particularly
  poultry, with severe effects on the poultry industry.

- **Anthrax**: A bacterial disease that can infect livestock and wildlife and is trans-
  missible to humans, making it a serious zoonotic concern.

International Cooperation

CHAPTER SEVEN

Animal Health Assessment

Assessing Animal Health Status and Identifying Potential Health Issues for Different Mobs and Classes of Livestock

Diseases affecting cattle, sheep, and goats can arise from various sources, including infections, parasite infestations, nutritional imbalances, and metabolic disorders. Each of these
factors can compromise the health, welfare, and productivity of livestock, even when
animals may not show obvious signs of illness [37].

Causes of Diseases in Livestock

1. **Infections**: Livestock are vulnerable to infections caused by bacteria, viruses,
   and fungi. These pathogens can lead to conditions such as foot-and-mouth
   disease (caused by a virus), tuberculosis (caused by bacteria), or fungal infections
   like ringworm. Infections often spread quickly through herds or flocks, especially under poor hygiene or crowded conditions, leading to widespread health
   issues.

2. **Parasite Infestations**: Internal parasites (such as liver flukes, gastrointestinal
   worms) and external parasites (such as lice or mites) can significantly affect
   livestock. These parasites can weaken the animals, causing issues like anaemia,
   weight loss, poor growth, and decreased milk production. In extreme cases,

severe infestations can lead to death.

3. **Nutritional Deficiencies and Imbalances**: Livestock rely on balanced diets for optimal health. Deficiencies in essential nutrients like vitamins and minerals can lead to health problems. For instance, a deficiency in calcium or phosphorus may cause weak bones (osteoporosis), while an excess of certain nutrients can also be harmful. Metabolic disorders such as ketosis in dairy cows or pregnancy toxaemia in sheep and goats are often linked to dietary imbalances during specific stages, like pregnancy or lactation.

4. **Metabolic Disorders**: These disorders disrupt normal bodily functions, usually as a result of dietary problems. For instance, conditions like ketosis in cows or acidosis in sheep can occur due to imbalances in energy or carbohydrate intake. These conditions may lead to decreased production, reproductive issues, and in severe cases, can be life-threatening.

Impact on Productivity

Diseases can have a profound impact on livestock productivity:

- **Reduced growth rates**: Ill animals may grow more slowly, delaying the time to market or reducing overall size and weight.

- **Reduced reproductive rates**: Diseases can cause fertility issues, leading to lower birth rates, failed pregnancies, or smaller offspring.

- **Carcass condemnation**: Some diseases lead to infections or abscesses that result in parts of or entire carcasses being condemned at slaughterhouses.

- **Reduced milk production**: Diseases affecting dairy cattle, such as mastitis, can drastically reduce milk yield and quality.

- **Fleece quality**: In sheep and goats, diseases can affect the quality of wool or fleece, resulting in lower fibre strength, thinner fibres, and reduced staple length, diminishing their commercial value.

- **Hide and fleece damage**: Parasites and infections can also damage the skin, hides, or fleece of livestock, reducing their market value.

Impact on Welfare

Diseases also negatively affect animal welfare, particularly whe[…] present. Diseased animals may:

- **Become weak and lethargic**: This can hinder their a[…] drink, or seek shelter.

- **Become vulnerable to predators**: Weak animals are […] attacked, particularly in environments where predators […]

- **Suffer from pain and discomfort**: Many diseases, s[…] cause significant discomfort and suffering, requiring […] ment.

Zoonotic Risks

Some livestock diseases can be zoonotic, meaning they can i[…] such as Q-fever (caused by *Coxiella burnetii*), leptospirosis, an[…] gious ecthyma) pose risks not only to the animals but also to the p[…] These diseases highlight the importance of proper hygiene, vacc[…] measures in preventing transmission between livestock and hum[…]

The health of cattle, sheep, and goats is essential not only […] also for maintaining productivity and minimizing economic lo[…] causes of diseases and implementing preventative measures—such[…] nutrition, and parasite control—are critical to safeguarding b[…] public health, particularly given the zoonotic potential of some […]

Notifiable Diseases

Notifiable diseases in livestock are those that, by law, must […] thorities if suspected or confirmed in animals. These diseases […] public health, animal welfare, and economic stability, as many of […] and have severe consequences, including the potential to cross i[…] (zoonoses). Globally, governments maintain lists of notifiable […] immediate action to prevent further transmission.

Importance of Notifiable Diseases

Notifiable diseases are highly contagious and can lead to se[…] context of livestock, diseases such as foot-and-mouth disease, avi[…] swine fever are common examples that, if not controlled, can […] flocks, causing massive economic losses. These diseases can also a[…] with restrictions placed on the movement of animals, culling […]

Notifiable diseases require international cooperation to manage their spread. Organizations like the World Organisation for Animal Health (OIE) work with countries to standardize responses to outbreaks and share information to limit the global spread of diseases. Many countries are required to notify the OIE when outbreaks of certain diseases occur, helping to create a global surveillance network.

Principles of Herd or Flock Health

Managing herd or flock health is a fundamental responsibility of livestock producers worldwide. The primary goal is to prevent disease, promote animal welfare, and ensure high productivity. By understanding local disease risks, improving nutrition, and employing preventative measures, producers can maintain the overall health of their livestock.

One of the key steps for producers is to be informed about the common diseases that affect cattle, sheep, or goats in their region. These can vary significantly depending on local climate, geography, and farm practices. Diseases may be caused by infections from bacteria, viruses, or fungi, as well as nutritional deficiencies and metabolic disorders. For example, foot-and-mouth disease or Johne's disease in cattle is common in many parts of the world, while scrapie is a concern in sheep.

Globally, the principle of "prevention rather than treatment" is widely adopted. Disease prevention can be achieved through vaccination, regular health monitoring, and biosecurity measures. These practices not only improve livestock welfare but also reduce the need for costly treatments later on. For instance, vaccination against diseases such as clostridial diseases in cattle or blue tongue in sheep is standard practice in many countries.

A herd or flock disease management plan is essential to monitor the health of animals and prevent the introduction of new diseases. Producers should quarantine new animals to prevent the spread of disease, as outlined in livestock management standards in regions like Australia and the EU. Quarantining and proper record-keeping ensure that no new infections are introduced when bringing new stock to a farm.

Good nutrition is fundamental to supporting the immune system of livestock. Providing a balanced diet helps animals resist infections and recover faster if they do fall ill. In regions like New Zealand or the UK, where grazing plays a crucial role, ensuring animals receive proper supplements and access to good-quality pasture is part of holistic livestock management.

Where vaccines are available and cost-effective, they should be utilized to protect the herd or flock from preventable diseases. Vaccination programs are widely adopted across various countries for diseases like brucellosis or bovine viral diarrhea (BVD). Additionally, implementing biosecurity measures, such as disinfection and controlling farm access, is important to limit disease spread.

Monitoring the health of livestock is a crucial part of maintaining their welfare. Globally, this includes checking body condition, reproductive health, and behavioural signs that may indicate illness. Any diseased animals should be treated promptly, and in severe cases, euthanasia may be necessary to prevent suffering, as outlined in welfare codes of practice in countries like the U.S. and Australia.

In cases of sudden or unexplained deaths, carcasses must be quarantined and handled according to local health regulations. This reduces the risk of spreading diseases to other animals or humans, especially in zoonotic diseases such as Q-fever or leptospirosis.

Animal welfare is a top priority in herd or flock management, with clear guidelines established in many countries. Farmers must maintain standards of care by ensuring proper nutrition, disease prevention, and low-stress handling practices.

Maintaining appropriate condition scores for livestock is critical to ensure productivity and welfare. In many countries, cattle are expected to maintain condition scores of 2.5 or above to meet reproductive and welfare goals. For example, gastrointestinal parasites and bloat can severely impact herd health if not managed properly, with an integrated approach needed to prevent and treat these issues.

Adhering to national and regional welfare codes ensures that animal welfare standards are maintained. For example, Animal Welfare Standards in Australia or EU Animal Welfare Directives provide frameworks to guide livestock producers in ensuring welfare needs are met in terms of housing, handling, and disease management.

Low-stress livestock handling techniques, such as those promoted in New Zealand and Australia, improve both animal welfare and productivity. Understanding animal behaviour is key to handling them safely and efficiently, reducing the risks of stress and injury.

In transport, global standards such as those found in the OIE (World Organisation for Animal Health) guidelines must be followed to reduce stress on animals and prevent harm during transportation. These include provisions for space, ventilation, and access to food and water, ensuring animals are not overcrowded or exposed to extreme conditions.

Producers globally are encouraged to develop disaster management plans that address issues like droughts or floods, which can significantly impact livestock welfare. In countries prone to natural disasters, such as Australia or parts of the United States, having a disaster preparedness plan ensures livestock can be protected or evacuated quickly, minimizing stress and loss.

The principles of herd or flock health are globally consistent in focusing on prevention, proactive monitoring, and adherence to welfare standards. Producers must continually assess the health, welfare, and productivity of their livestock while ensuring they meet national and international guidelines for animal care.

Cattle Assessment

Cattle assessment is a vital tool for beef producers to determine the readiness and suitability of their animals for specific markets. The ability to assess cattle accurately ensures producers meet market requirements for traits like weight, fatness, age, and more, ultimately influencing profitability [38].

Key Aspects of Cattle Assessment: Cattle assessment focuses on a variety of factors, each of which plays a role in determining the market value of the animal. These factors include [38]:

- **Fatness**: The fat cover on an animal impacts the market suitability, as different markets require specific fat ranges. Excessive fat or inadequate fat can result in price penalties.

- **Muscle Score**: This refers to the muscle-to-fat ratio in cattle. Markets often reward cattle with higher muscle scores due to the greater yield of saleable meat, as well as improved efficiency for processors.

- **Frame Score**: Frame size indicates how the animal will mature and its growth pattern. Larger frame animals tend to grow faster and are leaner, while smaller-framed animals fatten earlier, influencing both market preference and breeding decisions.

- **Dentition**: The age of cattle is often measured by their dentition (the state of their teeth), particularly in regions where specific markets have age restrictions.

- **Liveweight**: Liveweight is an important indicator for meeting weight specifications in different markets and can be used to estimate dressing percentage, which influences the carcase weight.

- **Dressing Percentage & Carcase Weight**: Dressing percentage refers to the proportion of the liveweight that remains after slaughter, impacting the actual marketable yield of meat.

- **Other Criteria**: Additional criteria like sex, breed, temperament, and structural soundness also play a role in assessing cattle for specific markets.

Practical Uses of Live Cattle Assessment

Cattle assessment is not only important for market readiness but is also used in daily farm management [38]:

- **Breeding Stock Selection**: Proper assessment allows farmers to select breeding animals that will improve the herd's overall productivity, whether through better muscle scores, improved structural soundness, or appropriate frame size.

- **Market Specifications**: By assessing the traits of cattle, farmers can ensure their livestock meets market specifications, minimizing price deductions and improving profitability.

- **Growth and Maturity Patterns**: Producers can gauge the growth potential and maturity of cattle, allowing them to manage feeding programs and market timing for optimal returns.

- **Nutritional Status**: Regular assessment of cattle, particularly in relation to their body condition score, provides valuable insights into their nutritional needs and helps guide feeding strategies.

Structural Assessment

Structural assessment is crucial, especially when selecting breeding stock. Structural soundness ensures the longevity and productivity of animals [38]:

- **Feet and Legs**: Poor structural integrity in legs can lead to lameness, impacting an animal's mobility and resulting in production losses or early culling.

- **Pelvic Structure**: For females, good pelvic structure reduces the risk of difficult births (dystocia), improving calf survival rates.

- **Sheath Structure in Bulls**: Bos indicus bulls, in particular, benefit from a sound sheath structure to prevent injuries or infections that can result in infertility.

Body Condition Score: The body condition score (BCS) is a widely used system to evaluate the fat and muscle cover on cattle. The BCS provides a useful metric to assess an animal's energy reserves, which is important for managing their health, reproductive performance, and overall welfare. By monitoring BCS, producers can adjust feeding programs to ensure cattle maintain optimal energy levels, particularly during key stages of the production cycle such as breeding and calving.

Frame Score: Frame scoring helps in understanding the maturity pattern of an animal and when it will start to fatten. Larger-framed animals grow faster but remain leaner for longer, while smaller-framed cattle mature earlier and accumulate fat sooner. Choosing cattle with the right frame size helps producers meet market requirements more effectively.

Muscle and Fat Score: Fat and muscle scores are key indicators of how well an animal will perform in the market. Higher muscle scores generally mean a higher yield of saleable meat, while fat scores help ensure cattle meet the fat cover specifications for different markets. An accurate visual assessment of muscle and fat scores can improve returns and avoid penalties from under- or over-fat animals.

Estimated Breeding Values (EBVs) and Nutrition: Estimated Breeding Values (EBVs) offer an additional layer of assessment by providing a genetic estimate of an animal's traits. EBVs can be used alongside visual assessments to evaluate an animal's growth potential, fat deposition, and muscle development. Additionally, a well-balanced nutrition program ensures that animals reach their genetic potential in terms of muscularity and growth.

Cattle assessment is an essential skill for beef producers worldwide, influencing everything from breeding decisions to market readiness. Through regular assessment of fatness, muscle score, frame size, dentition, and liveweight, producers can better meet market demands, improve productivity, and increase profitability. Incorporating structural assessment, body condition scoring, and genetic data like EBVs ensures a comprehensive approach to cattle management.

Copper Deficiency in Sheep and Cattle

Copper is an essential trace element required for many biological functions in animals, including body, bone, and wool growth, pigmentation, nerve fibre health, and immune response. Inadequate copper levels can lead to copper deficiency, which is common in livestock such as sheep and cattle.

Copper deficiency can arise from two primary causes:

1. **Low Copper Levels in Soil and Plants**: Some soils naturally lack sufficient copper, leading to lower copper levels in the plants consumed by livestock. In areas where copper fertilisers are not applied, the deficiency in copper can affect grazing animals.

2. **Induced Deficiency**: This occurs when livestock consume excessive levels of molybdenum and sulphur. These elements form compounds that reduce the bioavailability of copper in the body, mimicking the symptoms of primary copper deficiency.

Copper deficiency in sheep and cattle typically arises during periods of rapid pasture growth, particularly after winter rains when the concentration of copper in grasses declines. Late winter and spring are critical periods for copper deficiency, especially in pastures that contain more grasses than legumes, as legumes tend to absorb more copper. Liming of soil, which increases pH levels, can also induce copper deficiency by increasing the release of molybdenum, which binds copper in the digestive tract, making it unavailable for absorption.

The symptoms of copper deficiency vary between species:

- **Cattle**: Cattle may exhibit depigmentation, particularly around the eyes, leading to a "bespectacled" appearance. Other symptoms include sudden death due to heart failure ("falling disease") and lameness.

- **Sheep**: Swayback or enzootic ataxia is common in lambs, causing an inability to coordinate movement. Lambs may be born with this condition or develop it within six months after birth. Other signs include depigmentation in black-woolled sheep and an increased incidence of bone fractures.

- **Goats**: Goats can show signs such as rough coats, scouring, and poor fertility. Swayback is also observed in kids, with paralysis developing in the hindlimbs.

Copper deficiency is diagnosed through a combination of symptoms and laboratory tests:

- **Liver copper concentration**: Liver biopsies, particularly after post-mortem examinations, provide the best indicators of copper status. However, individual variability exists in how animals manifest copper deficiency.

- **Blood tests**: Blood copper concentrations are unreliable in cases of molybdenum-induced deficiency, so specific plasma tests are often used to assess copper availability.

- **Pasture tests**: Testing pastures for copper, molybdenum, and sulphur content can provide insights into whether the pasture is contributing to the deficiency.

Animals at the highest risk for copper deficiency include newborn lambs, pregnant or lactating sheep and cattle, and animals that have been overstocked or grazed on poor pastures. Cattle, particularly those with heavy worm infestations, are more susceptible to copper deficiency than sheep.

There are several methods to prevent copper deficiency in livestock:

- **Fertilisation**: Applying copper fertilisers to copper-deficient soils can prevent deficiency for up to eight years. However, in cases of induced deficiency due to excess molybdenum, direct supplementation to animals may be necessary.

- **Supplements**: Copper can be provided to livestock through slow-release ruminal boluses, injections, or multi-mineral supplements. Injections must be administered every few months, while boluses can provide long-term supplementation.

- **Pasture Management**: Limiting the use of molybdenum fertilisers or applying them cautiously can reduce the risk of induced copper deficiency.

Treatment involves copper supplementation, which can be delivered via injections, oral drenches, or boluses. However, copper supplementation must be approached cautiously, particularly in sheep, as they are highly susceptible to copper toxicity. In areas where liver damage has occurred due to other causes, copper supplementation can lead to secondary copper toxicity.

Copper deficiency in livestock can have severe consequences, including reduced productivity, increased susceptibility to disease, and even death. Proactive management through soil and pasture treatment, as well as direct supplementation, is essential to ensure the health and welfare of sheep, cattle, and goats. Monitoring copper levels through diagnostic tests and maintaining proper fertilisation and supplementation practices can help prevent deficiency and ensure optimal livestock performance.

Cobalt Deficiency in Sheep and Cattle

Cobalt deficiency in livestock, particularly in sheep and cattle, is a critical issue globally, as it impairs the synthesis of vitamin B12, which is essential for energy metabolism and the production of red blood cells. While all ruminants require cobalt, its deficiency in soil can lead to a lack of vitamin B12, causing various health issues.

Sheep are more susceptible to cobalt deficiency than cattle, with young and weaning animals being the most affected due to their high energy requirements for growth. Common signs of cobalt deficiency include:

- **In both sheep and cattle**: Reduced appetite, ill-thrift, and anaemia.

- **Sheep-specific**: Weepy eyes, wool break, and small lambs born to affected ewes.

- **Cattle-specific**: Rough coats, reduced milk production, scours in calves, and a depraved appetite (pica), where cattle may eat unusual items such as bark or dirt.

Figure 36: Cobalt deficient ewes produced by feeding cobalt-deficient wheaten hay chaff. CSIRO, CC BY 3.0, via Wikimedia Commons.

Factors Affecting Cobalt Availability

- **Soil Type**: Cobalt deficiency mainly occurs in deep sandy soils and leached soils, which are commonly found in regions like Western Australia. On the other hand, poorly drained soils tend to have higher cobalt content, with waterlogged conditions increasing its availability to plants.

- **Pasture Species**: Grasses typically contain less cobalt than legumes, so animals grazing on predominantly grassy pastures are more at risk of deficiency.

- **Climate**: Cobalt deficiency is more common in regions with high winter rainfall. The leaching of cobalt from the soil combined with rapid pasture growth can dilute cobalt levels in the plants.

Animal Factors Influencing Cobalt Requirement

- **Age**: Newborn and young animals are at higher risk of deficiency because they have low reserves of cobalt and rely on colostrum for vitamin B12. Milk contains very low levels of vitamin B12, so supplementation is often necessary.

- **Species Susceptibility**: Sheep and goats are more prone to cobalt deficiency than cattle, and young animals with high worm burdens are especially vulnerable.

- **Worm Burdens**: Parasitic infections, especially from Barber's pole worm and scour worms, exacerbate cobalt deficiency by causing protein loss and reduced feed intake, making young animals particularly susceptible.

Diagnosis and Testing

- **Blood Tests**: Blood tests can measure methylmalonic acid (MMA) levels, which increase when vitamin B12 is deficient.

- **Post-Mortem Assay**: Testing the liver for vitamin B12 content at post-mortem is a reliable indicator of cobalt deficiency.

- **Pasture Tests**: While plants don't require cobalt, testing for cobalt and molybdenum in pasture can provide insights into potential deficiency risks. Levels of less than 0.04 ppm cobalt in pasture indicate a potential deficiency.

Preventing Cobalt Deficiency

Several methods can prevent and manage cobalt deficiency in livestock:

- **Vitamin B12 Injections**: This is the fastest method to prevent deficiency, with effects lasting for 6-8 weeks. It is especially useful for lambs and calves.

- **Intraruminal Cobalt Pellets**: These slow-release pellets provide long-term protection (up to 3 years for sheep and 1 year for cattle).

- **Cobalt Licks and Drenches**: While effective, these methods have variable intake rates among animals and must be given frequently, making them less reliable than other options.

- **Cobalt Fertiliser or Pasture Sprays**: Cobalt can be applied to pastures through fertilisers or sprays. Rotating grazing animals through treated paddocks can help prevent deficiency without needing to treat the entire property.

Globally, the prevalence and severity of cobalt deficiency vary depending on regional soil types, climate conditions, and pasture management practices. For example, in the UK, New Zealand, and parts of Australia, cobalt deficiency is a known issue due to the local soil conditions, while it is less common in regions where cobalt-rich soils are present. Farmers and livestock managers in areas with a history of cobalt deficiency must remain vigilant, regularly test soil and pastures, and ensure proper supplementation to avoid production losses and health issues in their herds and flocks.

Ovine Brucellosis

Overview of Ovine Brucellosis Ovine brucellosis is a reproductive disease primarily affecting rams, caused by the bacterium *Brucella ovis* (B. ovis). The disease is prevalent across most sheep-producing regions of the world and is especially concerning for breeders due to its impact on fertility, lambing percentages, and ram productivity. Ovine brucellosis can lead to significant economic losses in sheep flocks due to reduced reproductive efficiency, the necessity for extended lambing seasons, and the premature culling of rams. If established in a flock, eradicating the disease is labour-intensive, requiring repeated blood testing and culling of infected rams.

Global Occurrence and Impact: Ovine brucellosis is a global issue, though its prevalence varies by region. It naturally occurs only in sheep, with rams being the most severely affected. In many countries, voluntary control schemes exist to assist breeders in managing and eradicating the disease from their flocks. For instance, in Western Australia, breeders can participate in the Ovine Brucellosis Accreditation Scheme, which helps to maintain disease-free flocks through regular testing and monitoring.

Pathology and Transmission: The bacterium *Brucella ovis* primarily affects the reproductive organs, particularly in rams. It causes inflammation of the epididymis, which is the structure that transports semen from the testes. The condition leads to sperm blockages, swelling, and hardening of the epididymis, which can result in infertility. Infected rams may show abnormalities in the size and texture of their testes. In rare cases, infected rams may recover physically but remain carriers of the disease, continuing to pose a risk of transmission to other animals.

Transmission of ovine brucellosis occurs through contact with infected semen, vaginal discharges, or aborted foetal material. Rams are often infected by exposure to vaginal discharges from infected ewes or through homosexual behaviour, which is common in young rams. Additionally, the disease can spread when infected rams are introduced into a clean flock, often through the purchase or borrowing of rams.

Clinical Signs in Rams and Ewes: In rams, the initial infection can result in an elevated temperature and reduced appetite, although these signs may go unnoticed. The primary signs of infection are related to the reproductive organs, with rams showing swelling or hardness in the scrotum. Chronic infections may result in a grossly enlarged epididymis and shrunken testes. In ewes, infection with *B. ovis* is typically less severe and short-lived, as ewes usually clear the infection after one or two reproductive cycles. However, it can cause early embryonic death, abortions, or inflammation of the placenta in pregnant ewes.

Identifying ovine brucellosis, a reproductive disease caused by *Brucella ovis* in sheep, requires a combination of clinical signs observation, physical examinations, and laboratory tests. The disease primarily affects rams, causing inflammation in the epididymis, leading to fertility issues, but ewes can also be transient carriers. Below are the key methods for identifying ovine brucellosis:

1. Clinical Signs Observation

- **In Rams:**

  - Affected rams often show inflammation of the epididymis, which is the vessel that transports semen from the testes. This may cause swelling, hardening, or changes in the size of the scrotum.

  - In severe cases, rams may have an enlarged epididymis and shrunken testes, particularly in chronic infections.

  - Rams may initially display mild signs like an elevated temperature and a reduced appetite, but these signs usually pass unnoticed.

- **In Ewes:**

  - Ewes typically do not show outward clinical signs, but the disease can lead to early embryonic death or inflammation of the placenta during pregnancy. Ewes rarely remain infected for more than two oestrus cycles.

2. Palpation of the Scrotum

- Palpating the testes and epididymides is a common diagnostic technique. Rams are examined by physically feeling their scrotum for asymmetry, lumps, or swelling. Both hands are used to feel the testes from top to bottom, ensuring there are no abnormalities.

- Infected rams may exhibit a hard or enlarged epididymis, and the swelling is often more noticeable in the lower part of the scrotum.

- It's important to note that not all infected rams show detectable abnormalities through palpation.

3. Semen Examination

- Semen analysis can help identify infection by examining the quality and volume of sperm. Brucella infection often leads to poor sperm motility, reduced semen volume, and abnormal sperm structure.

- B. ovis can also be cultured directly from semen samples, which is a strong indicator of infection.

4. Blood Testing

- Blood testing for *Brucella ovis* antibodies is one of the most reliable diagnostic methods. The presence of these antibodies indicates exposure to the bacteria.

- Regular blood tests are often necessary to monitor the health of a flock, particularly during an eradication program.

5. Ultrasound and Imaging

- In some cases, ultrasound imaging may be used to get a clearer view of the reproductive organs, identifying abnormalities not detectable by palpation alone.

6. Epidemiology and Risk Factors

- The disease spreads primarily through infected rams via semen or vaginal discharges. Homosexual activity among young rams can also spread the disease. Identifying the presence of infection in rams is critical as they are the primary carriers of the disease.

- Introduction of new rams into a flock without proper screening is one of the main ways the disease spreads, so testing newly purchased rams is essential for preventing brucellosis in previously clean flocks.

7. Laboratory Testing for Confirmation

- In addition to semen culture, laboratory tests of blood or tissue samples can confirm the presence of *Brucella ovis*. A combination of serology and bacteriology testing helps veterinarians accurately diagnose the disease.

Early detection and management are essential to prevent the spread of ovine brucellosis within a flock. Regular palpation, blood testing, and semen analysis are the primary methods of identifying the disease. Producers should be vigilant about screening new rams and culling infected animals to prevent economic losses and maintain flock health. Where possible, consult a veterinarian to confirm diagnosis and establish an eradication program if necessary.

Effect on Flock Fertility: The use of infected rams can lead to reduced lambing percentages and an extended lambing season. As conception rates decline, more ewes return to oestrus, resulting in a longer, more drawn-out lambing season. Over time, continued use of infected rams can lead to a significant decline in flock fertility. The high culling rates of rams due to brucellosis-related reproductive issues also contribute to economic losses.

Diagnosis and Control: Three primary methods are used to diagnose ovine brucellosis: palpation of the scrotum to detect abnormalities, semen examination for poor quality or volume, and blood testing for *B. ovis* antibodies. Palpation involves physically examining the ram's scrotum for swelling or abnormalities in the epididymis. However, this method is not foolproof, as some rams may be infected without showing detectable abnormalities.

Blood testing, which detects the presence of *B. ovis* antibodies, is one of the most reliable methods for diagnosing infection. Once diagnosed, infected rams must be removed from the flock to prevent further transmission.

Eradication Efforts: Eradicating ovine brucellosis from a flock requires a combination of testing, culling, and strict management practices. Manual examination, blood testing, and the removal of infected rams are essential steps in controlling the disease. Infected rams should be culled immediately or sold for slaughter, as prolonged treatment with antibiotics is generally not economical or effective in rams with severe damage to the reproductive organs.

Prevention and Global Management Practices: Preventing the introduction of ovine brucellosis into a flock is crucial for long-term control. Flock owners should ensure that any rams introduced into their flock come from disease-free sources. In some regions, such as Western Australia, certification schemes exist that allow breeders to verify the health status of the rams they are purchasing. A single negative blood test may not be sufficient, as antibodies may take up to seven weeks to appear after initial infection. Therefore, ongoing monitoring and testing of rams is necessary to maintain flock health.

Globally, different countries have varying levels of control over ovine brucellosis. In New Zealand, for example, stringent testing and monitoring programs have been successful in reducing the prevalence of the disease. In contrast, in countries where sheep farming is more extensive and less regulated, the disease may be more widespread.

Ovine brucellosis is a significant reproductive disease in sheep that requires proactive management to prevent economic losses. The disease affects ram fertility, resulting in reduced lambing percentages and longer lambing seasons, which ultimately reduce the profitability of sheep farming. Diagnosis through physical examination, semen analysis, and blood testing is essential, and once the disease is identified, infected rams must be removed from the flock. Effective prevention strategies include purchasing disease-free rams, regular testing, and participation in voluntary accreditation schemes. Through diligent management, the global sheep farming community can mitigate the impact of this costly disease.

Diarrhoea in Adult Cattle

Diarrhea in adult cattle can arise from multiple causes, both nutritional and infectious, with significant implications for herd health and productivity. Understanding the underlying causes and taking appropriate measures can prevent major losses in production.

Diarrhea, or scours, in cattle results when fluids in the intestines are not properly reabsorbed, causing loose, watery stools. Normally, the intestines absorb most of the water, but any disruption in this function leads to excessive fluid loss through faeces. This can be due to several factors, broadly categorized as nutritional or infectious.

Causes and Risk Factors

- **Nutritional Causes:**

  - **Acidosis (Grain Overload):** When cattle consume excessive amounts of grain, it can lead to acidosis, upsetting the stomach's normal digestive processes and causing diarrhea.

- **Lush Green Feed:** Rapid intake of young, fast-growing pastures can also lead to diarrhea due to the high water content in the plants, which can overwhelm the digestive system.

- **Mineral Deficiencies:** Nutritional imbalances such as deficiencies in cobalt, selenium, or copper can disrupt normal bodily functions and result in diarrhea.

- **Infectious Causes:**

  - **Parasites:** Worms can irritate the digestive tract, leading to inflammation and fluid loss.

  - **Viruses and Bacteria:** Common viral causes include bovine viral diarrhea virus (BVDV), while bacteria like *Salmonella* and *Yersinia* are known culprits. These infections can spread within herds and, in some cases, pose zoonotic risks to humans.

- **Other Diseases:** Conditions like liver disease, heart failure, and chemical or plant poisonings (e.g., lead or nitrate poisoning) can present diarrhea as a secondary symptom, making it harder to diagnose without veterinary help.

Diagnosis

Diagnosing diarrhea in cattle requires early intervention to minimize stock and production losses. Veterinarians typically collect faeces, blood, or tissue samples for laboratory analysis to pinpoint the cause. Diagnostic tests may also be subsidized in regions where there are risks to market access, such as in Australia.

Treatment depends on the underlying cause, and a veterinarian will usually recommend specific strategies. Common treatments include:

- **Rehydration:** As diarrhea leads to dehydration, rehydration fluids are critical for restoring electrolyte balance.

- **Antibiotics:** These may be necessary for bacterial infections like *Salmonella* but should be used judiciously to avoid contributing to antimicrobial resistance.

- **Anti-Inflammatories and Worm Treatments:** These may be prescribed for parasitic or inflammatory causes.

- **Nutritional Support:** If the diarrhea is linked to nutritional causes like mineral deficiencies, mineral supplementation may be required.

Preventing diarrhea involves addressing the root cause. Measures include:

- **Grazing Management:** Managing grazing practices, such as controlling access to lush pastures, can prevent nutritional causes of diarrhea.

- **Biosecurity and Quarantine:** Introducing biosecurity measures to prevent the introduction of infectious diseases and quarantining new or sick animals can protect herd health.

- **Vaccination and Deworming:** Regular vaccination programs and deworming schedules can help prevent infectious causes of diarrhea.

Bacterial infections like *Salmonella* and *Yersinia* are common causes of diarrhea in adult cattle. These bacteria cause damage to the intestines, leading to symptoms like foul-smelling diarrhea, fever, and loss of appetite. *Salmonella* often spreads through faecal contamination of feed and water, while *Yersinia* tends to affect animals after harsh winter conditions. Both are zoonotic, meaning they can be transmitted to humans, so high levels of hygiene are required when managing infected animals.

This reportable bacterial infection, caused by *Mycobacterium avium* subspecies *paratuberculosis*, is another significant cause of diarrhea in adult cattle. Johne's disease progresses slowly, and infected cattle may not show signs until years after infection. It results in severe weight loss and chronic diarrhea, leading to major economic losses in infected herds.

Good biosecurity practices are the most effective way to prevent bacterial diarrhea. These practices include regular health checks, isolating new or sick animals, and ensuring that facilities are clean and well-maintained to avoid faecal contamination.

In summary, diarrhea in adult cattle is a multifaceted issue with a variety of causes, ranging from nutritional imbalances to infectious agents. Proper diagnosis and timely intervention are essential for minimizing the impact on herd health and productivity. Implementing preventive measures such as good grazing management, biosecurity practices, and vaccination can help safeguard herds from the detrimental effects of diarrhea.

African Swine Fever (ASF)

African Swine Fever (ASF) is a highly contagious viral disease affecting pigs and wild boars [39]. Although it does not impact humans, ASF is a critical threat to global pig production due to its devastating effects on pigs, leading to high mortality rates and significant economic losses in affected regions. The virus is different from *Classical Swine Fever* but can present similar symptoms in pigs. Both diseases are exotic to Australia and many other regions where biosecurity measures are critical to prevent their introduction.

Historically, ASF was confined to Africa and Sardinia, but it has spread significantly since 2007. Wild boars are believed to have accessed contaminated food waste at an international airport, allowing the virus to spread to the Caucasus and Eastern Europe. Since then, ASF has rapidly spread across Eastern Europe, China, and other parts of Asia. China, home to the world's largest pig population, has been especially hard hit by ASF, leading to severe impacts on its pig industry.

ASF now poses a major threat to pig-producing countries worldwide. For example, testing in Australia has shown that contaminated pork products brought into the country through airports pose a significant risk. The virus can survive for months in pork products that have been smoked, dried, or frozen, making it difficult to control its spread.

ASF can be challenging to distinguish from classical swine fever based on clinical signs alone, necessitating laboratory testing for accurate diagnosis. The severity of the symptoms depends on the virus strain and the immunity of the affected pigs. Since pigs in many countries, including Australia, have no immunity to ASF, the impact of an outbreak could be catastrophic. Common signs of ASF include:

- Increased death rate in the herd

- High fever and loss of appetite

- Skin reddening and blueness of extremities (such as the ears)

- Coughing and respiratory distress

- Diarrhea and vomiting

- Abortions in pregnant sows

ASF primarily spreads through direct contact between pigs or through contaminated feed, equipment, vehicles, or even clothing. One of the most likely ways ASF could enter a country like Australia is through illegally imported pig meat or products. ASF can survive for extended periods in processed pork products, making the risk of contaminated meat a significant concern. In some regions, ticks are also vectors for ASF, though in places like Australia, the tick species capable of carrying ASF do not commonly feed on pigs.

Given the severity of ASF, prevention is crucial. There is no vaccine or treatment for ASF, and its mortality rate can reach 100%. The primary ways to reduce the risk of ASF on farms include:

- **Do not feed pigs meat or products that have come into contact with meat.** This practice, known as swill feeding, is illegal in many countries due to its potential to introduce ASF and other diseases.

- **Maintain strict biosecurity measures** on farms. All visitors, workers, and vehicles entering a farm should follow strict protocols to prevent contamination.

- **Monitor the health of pigs closely** and consult veterinarians immediately if any unusual symptoms or behaviours appear in the herd.

Globally, ASF remains one of the most dangerous threats to pig production, with the potential to devastate pig industries across continents. Countries must continue strict biosecurity and import regulations to prevent its spread and ensure that pig herds are kept safe from this fatal disease.

Foot-and-Mouth Disease (FMD)

Foot-and-Mouth Disease (FMD) is a highly contagious viral infection that primarily affects cloven-hoofed animals such as cattle, sheep, pigs, goats, buffalo, deer, and other similar species. The disease is particularly damaging because of its ability to spread rapidly, leading to severe economic consequences in the livestock industry. It does not affect non-cloven hoofed animals such as horses, dogs, cats, or birds, and is distinct from hand, foot, and mouth disease in humans.

The virus primarily manifests through the development of blisters (also known as vesicles) in sensitive areas like the mouth, nostrils, teats, and feet. These blisters often rupture, leading to symptoms like:

- Slobbering and drooling due to mouth lesions

- Lameness as a result of sores on the hooves

- Reluctance to move, general weakness, and depression

- Significant drops in milk production in dairy cattle

- Abortion in infected pigs

- Sudden death in young animals

In some species, particularly sheep, the signs may be mild and hard to detect, with lameness being the only outward symptom. The incubation period ranges from 3 to 14 days, and animals can spread the virus even before showing visible symptoms.

FMD poses one of the greatest threats to the global livestock industry. The disease's ability to spread rapidly and its debilitating effect on productivity means that affected regions face severe economic losses. In countries like Australia, which currently remains FMD-free, an outbreak would have catastrophic effects. A multi-state outbreak could cost the Australian economy as much as $52 billion over 10 years, according to estimates by the Australian Bureau of Agricultural and Resource Economics and Sciences (ABARES). This figure includes lost export markets, disruption to domestic industries, and the cost of disease management.

Globally, FMD outbreaks have caused significant damage in various regions, especially in Asia, Africa, and parts of Europe. The disease is endemic in some countries, which results in continuous losses and strict biosecurity measures to prevent its spread.

The FMD virus spreads through direct contact between infected animals or via contaminated materials such as feed, water, vehicles, clothing, and even air over short distances. Pigs are particularly susceptible to the disease and can excrete large amounts of the virus, further spreading the infection to other livestock. Contaminated meat and dairy products, especially those illegally imported, are major sources of transmission. Human activity, particularly travellers returning from infected countries, can also introduce the virus via contaminated clothing or equipment.

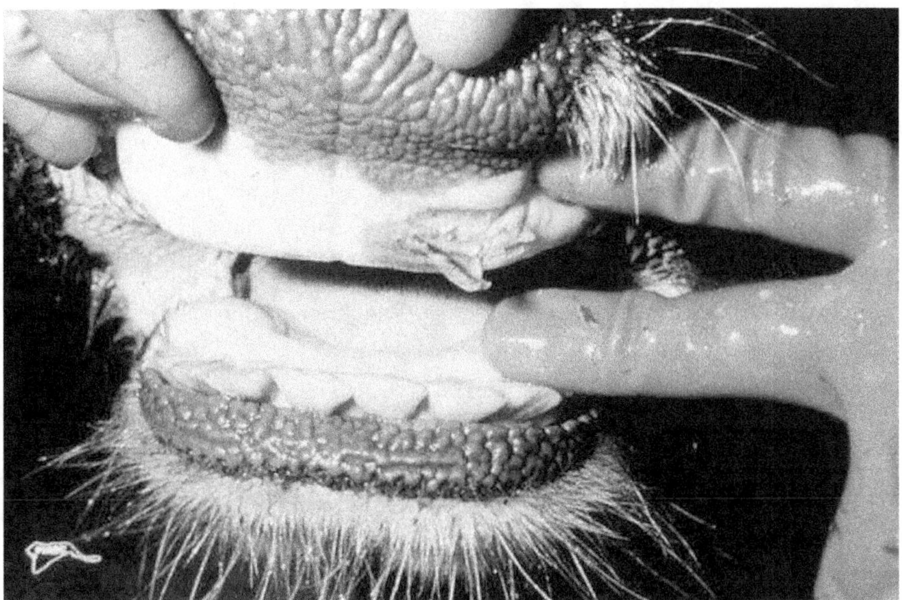

Figure 37: Ruptured oral vesicle in a cow with Foot-and-mouth disease. U.S. Department of Agriculture, Public Domain, via Picryl.

Countries that are free from FMD, such as Australia, New Zealand, and parts of North America and Europe, maintain strict biosecurity measures to prevent the disease's entry. These measures include:

- Strict controls on the importation of meat, dairy, and livestock products.

- Prohibition of feeding pigs any products containing or in contact with meat, which could be a source of the virus.

- Rigorous disinfection and quarantine protocols for travellers returning from regions with known FMD outbreaks.

Strong biosecurity is essential for preventing outbreaks, and governments worldwide have implemented measures such as quarantining affected farms, culling infected animals, and establishing zones of control to prevent the spread of the disease.

FMD is endemic in many parts of Africa, Asia, and parts of South America. Its ability to spread across borders makes it a serious concern for international trade and food security. Countries with active FMD control programs often require strict veterinary oversight and extensive vaccination campaigns to manage outbreaks.

Reporting the disease immediately upon detecting any signs is crucial for mitigating the spread. Authorities recommend that any unusual signs, such as blisters, lameness, or sudden death in livestock, be reported to a veterinarian or local agricultural department. Early detection allows for rapid response measures, helping to contain the disease and minimize economic damage.

Foot-and-mouth disease is a serious viral infection that can cause significant harm to the livestock industry worldwide. While it is not a threat to human health, its potential to cripple agricultural economies means that strict biosecurity measures and early detection are critical in preventing and controlling outbreaks.

Barber's Pole Worm (Haemonchus contortus) in Sheep

Barber's pole worm (*Haemonchus contortus*) is a highly pathogenic roundworm that infects sheep and goats, leading to the disease known as haemonchosis. It poses significant risks to livestock due to its blood-sucking nature, which can result in severe anemia and even sudden death in infected animals. This parasitic worm is a global problem, particularly prevalent in areas with warm and moist climates, such as parts of Africa, Australia, Europe, Asia, and the Americas.

The barber's pole worm resides in the abomasum, or the fourth stomach of sheep, where it attaches to the stomach lining and feeds on blood. Female worms are prolific egg layers, with thousands of eggs passing through the animal's feces and hatching into larvae on pasture. These larvae can be ingested by grazing animals, completing the life cycle. Within three weeks, they mature into adult worms, starting the cycle again.

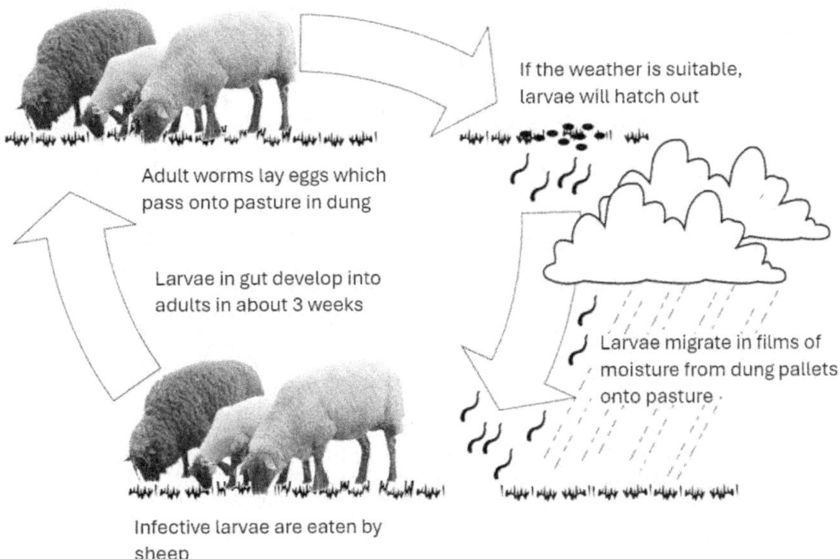

Figure 38: Life cycle of sheep worms.

Due to the high egg production and quick development, large worm burdens can accumulate rapidly, particularly under favourable environmental conditions. Warm, humid climates and high rainfall support the survival and spread of larvae on pasture, making haemonchosis more prevalent in such regions.

Barber's pole worm is predominantly found in regions with warm and wet climates, such as coastal and high-rainfall areas. For example, it is a major problem in many areas of Australia, particularly in Western Australia's coastal regions, and has a widespread presence in tropical and subtropical areas worldwide. In regions where rainfall exceeds 600 millimetres annually, the parasite is more likely to thrive. In dry areas, outbreaks can still occur following unseasonal heavy rainfall, but these tend to be less common.

Sheep and goats are particularly susceptible to barber's pole worm due to the worm's blood-feeding behaviour. The parasite drains blood from the host, leading to severe anaemia, loss of condition, weakness, and in acute cases, sudden death. The disease is most often seen in young or immunocompromised animals, such as lambs, weaners, and ewes post-lambing, as their immune systems are less capable of controlling the worm burden.

Signs of haemonchosis include:

- **Anaemia**: Pale mucous membranes in the mouth and eyes.

- **Weakness and Collapse**: Affected animals may collapse if driven.

- **Bottle Jaw**: Swelling under the jaw caused by fluid retention.

- **Good Body Condition**: Paradoxically, affected sheep may still appear well-nourished, as the disease affects the blood rather than the body fat.

The mortality rate can be significant in heavily infested flocks, especially if left untreated. The infection also reduces productivity, as affected animals experience weight loss, poor wool quality, and decreased reproductive performance.

Management and Control

Preventative Strategies:

- **Worm Control Programs**: Regular monitoring of worm burdens through fecal egg counts can help identify high-risk periods and inform drenching decisions. Rotational grazing and the use of safe pastures can also reduce the risk of infection by minimizing the exposure to larvae.

- **Drenching**: Anthelmintic drenches (such as benzimidazole, levamisole, and macrocyclic lactones) are commonly used to control barber's pole worm. However, resistance to these drugs is an increasing problem globally, necessitating careful management. In regions with significant drug resistance, targeted drenching using narrow-spectrum drugs like closantel or naphthalophos may be necessary. Persistent-action drenches like moxidectin provide extended protection by killing larvae during grazing.

- **Vaccination**: The **Barbervax** vaccine, developed in Australia, offers an additional method of control, especially in areas where drug resistance limits treatment options. The vaccine must be administered multiple times throughout the risk season to maintain efficacy.

High-Risk Situations:

- **Lambing and Weaning**: Ewes and lambs are at high risk due to a temporary suppression of immunity around lambing. Pre-lambing drenching or vaccination can help mitigate this risk.

- **Environmental Factors**: In areas with persistent green pasture, such as those near water sources or irrigated fields, the risk of infection remains high year-round, and regular monitoring and treatment are essential.

Anthelmintic resistance in barber's pole worms is a significant concern worldwide. In many regions, overuse of broad-spectrum drenches has led to resistant worm populations, complicating control efforts. In parts of northern New South Wales and Queensland, for example, resistance is so severe that only narrow-spectrum treatments like closantel remain effective. This global issue highlights the need for integrated parasite management, combining chemical treatments with grazing management, biosecurity, and, where available, vaccination.

Barber's pole worm is a significant global challenge in sheep and goat farming, particularly in warm and wet climates where the parasite thrives. Effective management involves a combination of monitoring, strategic drenching, and preventative practices to reduce worm burdens and minimize losses. The emergence of drug resistance has complicated control efforts, making the development of new vaccines and integrated control strategies critical to safeguarding livestock health and productivity worldwide.

Itch Mite in Sheep

Itch mites (*Psorobia ovis*) are microscopic parasites that live on the skin surface of sheep, causing irritation, rubbing, and fleece damage in affected animals. Although infestations of itch mites are rare in many regions, they can still pose a challenge in managing sheep health. The impact of itch mite infestations varies, with many sheep showing no symptoms, but some exhibit noticeable discomfort and fleece damage. While the issue is not highly prevalent or economically significant in most regions, it is worth addressing due to its potential impact on animal welfare and fleece quality.

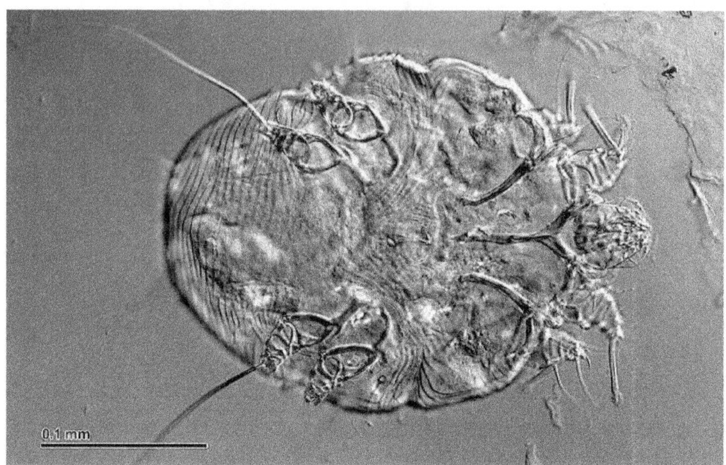

Figure 39: Itch mite; Magnification: 600x. Doc. RNDr. Josef Reischig, CSc., CC BY-SA 3.0, via Wikimedia Commons.

The itch mite is a small, barely visible parasite that resides on the skin of sheep. These mites do not burrow into the skin but cause irritation by their presence on the skin's surface, leading to scratching, rubbing, and chewing of the fleece. Female itch mites lay eggs on the sheep's skin, and after a period of five weeks, the eggs hatch, developing through several nymphal stages into adults. The adult mites, which are the only mobile stage, are responsible for spreading the infestation from one sheep to another, though the spread is slow.

The life cycle of *Psorobia ovis* typically increases during cooler months, with mite populations peaking in the winter and spring. Infestation is spread primarily through close contact, especially in yarded or shorn sheep, where transmission is more likely. However, it is rare for more than 5-10% of a flock to exhibit severe signs of infestation, and mites may persist in a flock for several years before being detected.

Sheep affected by itch mites may show signs of rubbing, scratching, and biting at their fleece, leading to tufted or ragged fleece appearance. These behaviours are primarily due to the irritation caused by the mites on the skin. Unlike lice infestations, where sheep often show more widespread itching, itch mites affect individual sheep more subtly, with very few sheep (typically less than 1%) displaying severe fleece damage.

Fine-wool breeds, such as Merinos, and sheep that are poorly nourished are more prone to infestation. The irritation caused by the mites may lead to skin flaking, visible as dry scurf or scaly patches on the skin. Severe infestations can reduce wool quality and yield, though the overall economic impact on the flock is typically minimal.

The financial implications of itch mite infestations in sheep are generally minor. Wool cuts may be slightly reduced in badly affected sheep, but this occurs in a small fraction of the flock. Infestations are primarily a concern for the appearance of the flock, especially for breeders managing stud animals. Sheep exhibiting visible signs of rubbing and damaged fleece may lead to a false suspicion of lice infestation, further complicating the situation. In general, the economic losses due to itch mites are not substantial compared to other parasitic diseases.

Diagnosing itch mite infestations is a challenge due to the microscopic size of the mites and the similarity of symptoms to other conditions, such as lice infestations. To confirm the presence of itch mites, skin scrapings must be examined under a microscope. However, mites may be difficult to detect even with multiple scrapings, and a negative result does not necessarily rule out their presence. Diagnoses are most reliable in spring when mite populations peak.

Figure 40: Photograph of infestation of sheep with parasitic mite. [[User:Alan R Walker (talk) 09:24, 28 January 2014 (UTC)]], CC BY-SA 3.0, via Wikimedia Commons.

Before investigating itch mites, other causes of rubbing and fleece damage should be ruled out, including lice, grass seeds, and fleece rot. These conditions can all lead to similar symptoms in sheep and must be differentiated from mite infestations through thorough examination and testing.

There is no known eradication method for itch mites, but certain treatments can significantly reduce their numbers and alleviate symptoms. Macrocyclic lactones (ML), such as ivermectin, abamectin, and moxidectin, are effective against itch mites. These treatments are commonly used for controlling internal parasites and ectoparasites in sheep, which may explain the low prevalence of itch mites in many regions.

It is unnecessary to treat an entire flock solely for itch mite unless the infestation is severe. In general, treatments are most effective when administered in spring when mite populations are highest. Infested sheep should be monitored regularly, and individual animals that continue to show signs of infestation may be culled to manage the condition.

Itch mites are found in sheep worldwide but are more prevalent in regions with certain environmental conditions, such as in parts of Australia and New Zealand. The use of modern parasite control treatments has significantly reduced the incidence of infestations in many areas. However, as with other parasitic conditions, farmers need to remain vigilant, particularly in managing flock health and nutrition, which can influence susceptibility to mites.

In summary, while itch mites are not a widespread or highly damaging parasite, they can cause discomfort and fleece damage in sheep. Managing these infestations involves maintaining good flock health, using effective parasite control treatments, and conducting regular monitoring to ensure that sheep are free from parasitic infections.

Johne's Disease (JD) in Cattle

Johne's disease (JD) is a chronic, incurable infectious disease that primarily affects ruminants, including cattle, sheep, goats, alpacas, and deer. It is caused by the bacterium *Mycobacterium avium* subspecies *paratuberculosis* (MAP), which leads to inflammation of the intestines, progressive weight loss, chronic diarrhea, and ultimately, death. Globally, JD is considered a significant animal health concern, particularly due to its potential economic impact on the livestock industry and its difficulty in diagnosis and management.

JD is caused by the bacterium *Mycobacterium paratuberculosis*, which thrives in the lymph nodes and intestines of infected animals. As it multiplies, the bacterium thickens the walls of the intestines, preventing proper absorption of nutrients and water. The long incubation period of JD, which can range from two to five years, means animals often harbor the disease without showing symptoms, making early detection challenging.

There are several strains of JD, including those specific to cattle (C-strain), sheep (S-strain), and bison. While each strain tends to infect the species it is named after, cross-species infection can occur, albeit rarely. For example, sheep strains of JD can infect cattle, and this has been observed in a limited number of cases.

JD spreads primarily through the ingestion of contaminated faeces, milk, or colostrum. The bacterium can survive in the environment, particularly in soil, for up to a year under cool, moist conditions. Calves are at the highest risk of infection, especially during their first 30 days of life, as they may ingest the bacteria through contaminated milk or by suckling udders covered in infected faeces. Cross-species transmission can also occur if cattle come into contact with infected sheep, goats, or deer. Infected animals can silently shed the bacteria in their faeces, contaminating the environment and posing a long-term risk to healthy livestock.

The symptoms of JD typically do not appear until cattle are at least three to four years old, though they are likely infected as calves. Once symptoms manifest, they are often triggered by stress, such as calving, poor nutrition, or a lack of feed. Common signs include:

- **Chronic diarrhea** that does not respond to treatment.

- **Weight loss** despite a normal or increased appetite and adequate feeding.

These visible signs indicate advanced stages of the disease, and infected animals may succumb within weeks or months of showing clinical symptoms. Even before these symptoms are evident, JD can silently reduce productivity by lowering milk yield, slowing weight gain, or decreasing fertility.

JD has a significant economic impact on the global cattle industry. Infected dairy herds experience reduced milk production, lower conception rates, and increased culling rates. Beef herds also suffer economic losses, though typically less severely than dairy herds. The prolonged incubation period of the disease means that the number of infected animals increases over time if the disease is not managed, progressively worsening its impact.

Additionally, JD poses a challenge for international trade. Some countries have stringent JD-related conditions for importing cattle or cattle products. The detection of JD in livestock on a property may also affect the farm's eligibility for certain export markets.

Managing JD on a global scale involves stringent biosecurity measures and herd management protocols. Since there is no cure for JD, prevention and early detection are critical to limiting its spread. Some key management strategies include:

- **Biosecurity:** Preventing the introduction of infected animals is the most effective way to control JD. This involves ensuring that new livestock come from herds with no history of JD, using national cattle health declarations, or ensuring a high Johne's Beef Assurance Score (J-BAS) or Dairy Assurance Score for cattle.

- **Vaccination:** In sheep, vaccination against JD can reduce the risk of transmission to other species such as cattle.

- **Environmental management:** Reducing livestock access to contaminated pastures, waterways, and drainage channels can help limit environmental exposure to the bacteria.

The diagnosis of JD is complex due to the prolonged incubation period and the limited effectiveness of some tests. The available diagnostic methods include:

- **Blood tests**: These detect the immune response to JD but are prone to false positives and are better suited for herd-level screening.

- **Faecal tests**: This method looks for MAP bacteria in faeces using culture or polymerase chain reaction (PCR). However, it may only detect infections in animals that are already shedding bacteria.

- **Post-mortem tests**: These involve examining the intestines and lymph nodes under a microscope for signs of infection. While this is more reliable, it is not always practical for live herd monitoring.

JD is a global issue, with different countries adopting various approaches to manage the disease. In Australia, for instance, JD is reportable and managed under a national framework. Livestock producers are encouraged to develop biosecurity plans, monitor herd health, and implement disease prevention strategies. In Europe, stringent testing and movement controls are enforced in countries where JD is prevalent. North America also focuses heavily on biosecurity, and the United States has implemented a voluntary control program for JD in cattle.

Managing Johne's Disease (JD) in a livestock production enterprise involves a combination of biosecurity measures, herd management practices, and, when available, disease surveillance and testing programs. The goal is to prevent the introduction of the disease, limit its spread, and minimize its economic impact. Here's a detailed outline of how JD can be effectively managed:

1. Biosecurity Measures

The most effective strategy for managing JD is preventing its introduction into the herd. This requires strong biosecurity protocols that focus on the careful management of incoming livestock, feed, and farm visitors.

- **Source New Animals from Low-Risk Herds**: Before introducing new animals, request detailed information on their health status, including a **Johne's Beef Assurance Score (J-BAS)** or Dairy Assurance Score. Ensure animals come from herds with no history of JD, particularly if importing from regions where the disease is prevalent.

- **Minimize Cross-Species Contact**: As JD can affect multiple ruminants (e.g., sheep, goats, alpacas, and deer), avoid co-grazing cattle with other species that may carry the disease. Grazing cattle on land previously used for sheep or other ruminants should also be avoided unless disease-free status is assured.

- **Use Quarantine**: All new stock should be quarantined for a minimum period, usually 30-60 days, before integrating them into the main herd. This allows time for disease symptoms to become visible if present and ensures they do not introduce JD to healthy stock.

- **Implement Hygiene and Disinfection Protocols**: Clean and disinfect equipment, particularly those shared between animals, such as feeders and waterers. Also, maintain strict visitor protocols to minimize the risk of JD introduction via contaminated boots, equipment, or clothing.

2. Herd Management Practices

Once JD is introduced, it is difficult to eradicate from a herd, so management practices should focus on reducing its impact and spread.

- **Calf Management**: Since calves are most susceptible to JD infection, it is critical to prevent contamination of their environment. Calves should be separated from the general herd early, avoiding suckling from infected cows. Use pasteurized milk or colostrum from JD-negative animals to reduce the risk of transmission.

- **Grazing Management**: JD bacteria can survive in the environment for up to a year, particularly in cool and moist conditions. Rotate grazing areas and avoid overstocking to prevent high levels of contamination in paddocks. Use "clean pastures" where the risk of JD contamination is low, particularly for young stock.

- **Monitor for Clinical Signs**: Regularly observe cattle for signs of JD, such as chronic diarrhea and weight loss despite normal appetite. Early identification of infected animals can prevent further spread and allow for targeted culling.

3. Testing and Surveillance

JD can be challenging to detect due to its long incubation period, but regular testing can help identify infected animals and prevent spread.

- **Routine Testing**: Blood, faecal, or post-mortem tests can be used to monitor JD in the herd. Blood tests are useful for screening, but faecal tests or tissue samples from post-mortem analysis are more definitive. Periodic testing helps detect animals that may be shedding the bacteria without showing symptoms.

- **Cull Infected Animals**: Once JD is confirmed, culling infected animals is the most effective way to reduce the risk of disease spread. Infected animals will continue to shed bacteria, contaminating the environment and posing a risk to other livestock.

- **Use of National Assurance Programs**: Programs like J-BAS (Johne's Beef Assurance Score) provide a framework for producers to manage the risk of JD. Participation in these schemes can help maintain a herd's health status, enhance market access, and demonstrate disease control efforts.

4. Vaccination (Where Applicable)

Vaccination is available for sheep in some regions to control JD, but a similar vaccine is not yet widely available for cattle. In countries where a JD vaccine is approved, it can be used to reduce disease spread, particularly in areas where JD is endemic.

- **Consider Vaccination for Sheep**: In enterprises that raise sheep alongside cattle, vaccination of sheep against JD can help reduce the risk of transmission between species.

5. Good Record-Keeping

Maintaining accurate records of animal health, testing results, and biosecurity measures is crucial for managing JD effectively.

- **Track Movements and Disease Status**: Record all animal movements on and off the farm, as well as testing and vaccination results. This will help trace potential sources of infection and manage the spread within the herd.

- **Maintain Biosecurity Logs**: Ensure that visitors and workers follow biosecurity protocols and record their entry onto the farm, particularly if they have been in contact with other livestock operations.

6. Economic and Trade Considerations

JD can have serious implications for trade and the economic viability of a livestock enterprise.

- **Maintain Export Market Eligibility**: Many countries have stringent requirements for JD status, especially in dairy and beef herds intended for export. Ensuring the herd remains free of JD helps maintain access to these markets and avoid restrictions.

- **Minimize Production Losses**: In dairy herds, JD can lead to reduced milk production and fertility. In beef herds, it can cause weight loss and lower conception rates. Preventative measures and early intervention help mitigate these economic losses.

Sheep-strain (S-strain) Johne's disease (JD) in cattle is a complex issue with significant implications for livestock management worldwide. JD is caused by the bacterium *Mycobacterium avium* subspecies *paratuberculosis* (MAP), which can infect multiple species, including cattle, sheep, goats, alpacas, and deer. There are three primary strains of JD: sheep/ovine (S-strain), cattle (C-strain), and bison strains. While the S-strain typically infects sheep, it can occasionally spread to other species, including cattle.

S-strain of JD is endemic in certain regions, such as Western Australia, where it affects sheep flocks. While primarily a sheep disease, it has been detected sporadically in cattle. In Australia, there have been a few confirmed cases of S-strain JD in cattle, with the first case in 2013. By 2019, there were three confirmed clinical cases in Western Australia. Although the transmission of S-strain from sheep to cattle is considered low-risk, it is still possible, particularly under certain conditions such as high stocking densities or close contact between infected sheep and susceptible cattle.

There is currently no evidence to suggest that the S-strain can be transmitted between cattle. The primary risk arises when cattle graze on pastures contaminated by sheep shedding the bacteria or come into indirect contact with infected environments, such as through water run-off.

In regions like Australia, all strains of JD, including the S-strain, are reportable diseases. However, if cattle are detected with S-strain JD, no regulatory actions are taken on the property. The detection is recorded against the property identification code (PIC), and this information is used when certifying property origin for JD-relevant markets. This detection, however, can affect a property's Johne's Beef Assurance Score (J-BAS). If a laboratory confirms S-strain JD on a property with a J-BAS 7 or 8 score, the property will be downgraded to J-BAS 6, provided other requirements are met. The property can retest in two years to regain a higher score after the removal of high-risk animals.

Several factors increase the likelihood of transmission of the S-strain from sheep to cattle:

- **Increased Stocking Densities**: High-density grazing environments raise the chance of transmission.

- **Co-grazing Practices**: When cattle, particularly calves under one year old, are grazed alongside sheep that are shedding the bacteria, the risk of infection increases.

- **Contaminated Pastures**: Cattle grazing on pastures previously contaminated by sheep with JD are at greater risk of infection.

- **Climatic Conditions**: Environments with higher rainfall allow the bacteria to survive longer in the soil, increasing the risk of infection.

- **Indirect Contact**: Cattle may also be exposed to S-strain through run-off from contaminated pastures, further heightening the risk.

- **Drought-feeding Practices**: Hand-feeding sheep and cattle together can lead to the ingestion of contaminated feed, especially when animals graze close to the ground where faecal contamination is more likely.

To reduce the risk of S-strain JD spreading from sheep to cattle, producers can take several precautions:

- **Avoid Co-Grazing**: It is advisable to avoid grazing cattle, especially calves under one year old, alongside sheep to minimize contact with infected faeces or contaminated environments.

- **Vaccination of Sheep**: Vaccinating sheep by 16 weeks of age is a proactive measure to minimize the risk of JD transmission. This can help reduce the number of infected sheep shedding the bacteria.

- **Testing and Monitoring**: On-farm testing and abattoir monitoring of sheep can help identify JD in the flock. If JD is detected, immediate steps like culling infected or high-risk animals should be taken.

- **Grazing Management**: Cattle, particularly young calves, should not be grazed on pastures previously grazed by infected sheep. Areas with high-risk environmental contamination, such as proximity to neighbours with JD-positive sheep, should also be avoided.

- **Separate Feeding**: Avoid hand-feeding sheep and cattle together to reduce the risk of cross-contamination through shared feed sources.

While the transmission of S-strain JD from sheep to cattle is relatively rare, it can occur under specific conditions. By implementing sound biosecurity practices, such as avoiding co-grazing, vaccinating sheep, testing and culling infected animals, and managing grazing areas, producers can significantly reduce the risk of JD transmission between species. Properly managing this risk is essential for maintaining cattle herd health and minimizing the economic impact of JD, particularly in regions where the disease is endemic in sheep flocks.

Anthrax

Anthrax is a highly infectious bacterial disease caused by *Bacillus anthracis*, affecting a wide range of animals, particularly livestock like cattle, sheep, and goats. The disease primarily spreads through spores, which can persist in soil for decades, sometimes up to

50 years or even 200 years in the bones of deceased animals. Outbreaks are often triggered by environmental changes, such as soil disturbance during construction, flooding, or drought, which brings dormant spores to the surface. This makes grazing animals particularly susceptible, as they ingest spores from contaminated soil, feed, or water.

Anthrax spores are the primary mode of transmission, with grazing animals contracting the disease from contaminated pastures. The bacteria form hardy spores under unfavourable conditions, and these spores are what animals ingest. While direct contact between animals or inhalation of spores is rare in the natural environment, it can happen under specific conditions. Cattle, sheep, goats, and horses are particularly vulnerable, though pigs, dogs, and even wildlife can be affected. Dogs can contract the disease by consuming infected carcasses, although such occurrences are rare.

The hallmark of anthrax in livestock is sudden death, often without prior visible illness. Postmortem, affected animals may display bloody discharges from natural orifices, such as the mouth, nose, and anus. Additional symptoms might include a sudden drop in milk production, red-stained milk or urine, weakness, staggering, and in some species like horses and pigs, colic or throat swelling. Anthrax is diagnosed through blood tests, which must be conducted with extreme caution to avoid exposure.

Although anthrax primarily affects animals, humans can become infected through contact with infected carcasses, animal products like wool, or by handling contaminated soil. Infection can occur through cuts or abrasions on the skin, inhalation of spores, or ingestion. The disease is serious and can be fatal in humans if left untreated, particularly when inhaled or ingested. Early symptoms in humans may resemble flu-like conditions, but the disease can rapidly worsen.

Anthrax outbreaks are rare but do occur. In Australia, outbreaks have been recorded mainly in Victoria and New South Wales, where vaccination of livestock is routine in anthrax-prone areas. Western Australia has only seen one recorded outbreak in 1994, which led to the death of 31 cattle and one horse [40]. Vaccination is highly effective in preventing anthrax, but treatment is usually not possible as affected animals die quickly [41] . In rare cases where the disease is caught early, high doses of penicillin administered by a veterinarian can be effective.

Anthrax is a bacterial disease caused by *Bacillus anthracis,* primarily affecting livestock such as cattle, sheep, and goats. It can also infect wildlife, as demonstrated by the recent outbreak in Wyoming, where dozens of cattle and a moose were found to have died from the disease. This outbreak is the first of its kind in the state in decades, highlighting the resilience and persistence of anthrax spores in the environment [42]. These spores can remain viable in the soil for decades, waiting to infect animals through contaminated feed or water, especially when environmental conditions, such as droughts or heavy rains, disturb the soil [42].

The disease's hallmark symptom in grazing animals is sudden death, often accompanied by bloody discharges from the mouth, nose, or anus. This rapid onset of symptoms means that treatment is rarely possible, as animals die before intervention is effective. When detected early, however, antibiotics like penicillin can sometimes save infected animals, and vaccinations are highly effective for preventing further cases [42].

Humans can contract anthrax through direct contact with infected animals, particularly when handling carcasses or contaminated animal products such as wool. There are four types of anthrax infections in humans—cutaneous, gastrointestinal, inhalation, and injection anthrax—each with varying degrees of severity. Cutaneous anthrax is the most common and least dangerous form, though it can still be fatal without proper treatment. Inhalation anthrax, which occurs when spores are breathed in, is much more deadly and is of particular concern for bioterrorism [42].

While anthrax outbreaks in livestock are not common in the U.S., such as the case in Wyoming, they pose significant risks both to public health and to the economy due to potential impacts on livestock industries. In cases like these, state and federal agencies swiftly respond to contain the outbreak, often advising strict biosecurity measures, proper disposal of carcasses, and precautions for hunters and those living near affected areas to avoid contact with infected animals or soil [42]. The Wyoming outbreak serves as a reminder of anthrax's continued threat, especially in regions where environmental conditions favour the persistence of spores.

Carcass disposal is crucial in preventing the spread of anthrax, as spores form only once the carcass is exposed to the environment. Therefore, infected animals should either be burned or buried intact, following strict guidelines to minimize contamination. Infected carcasses must not be opened, as this would release the spores into the environment, increasing the risk of further infection. Proper disposal methods are critical in controlling the spread of the disease, and advice from authorities, such as the Department of Primary Industries and Regional Development (DPIRD), should be followed.

Anthrax is a reportable disease in most countries, meaning any suspected cases must be reported to veterinary or biosecurity authorities immediately. Sudden death in livestock, particularly when accompanied by bloody discharges, should raise immediate concern.

Vaccination, proper biosecurity measures, and prompt reporting are essential in preventing the spread of this deadly disease, which poses both a risk to livestock health and a threat to public health through potential human infections.

Bovine Pestivirus or Bovine Viral Diarrhoea Virus (BVDV) and Mucosal Disease in Cattle

Bovine viral diarrhoea virus (BVDV), also known as bovine pestivirus, is a complex viral infection affecting cattle and other ruminants. It is a highly contagious disease, primarily impacting herd health and production through reproductive failures, immune suppression, and persistent infections. There are two strains of BVDV globally—BVDV Type 1 and BVDV Type 2—though only BVDV Type 1 is currently present in Australia, while BVDV Type 2 is more common in Europe and North America.

BVDV spreads primarily through persistently infected (PI) animals, which shed large amounts of the virus throughout their lives. These animals transmit the virus through bodily fluids such as saliva, urine, respiratory secretions, milk, and faeces. Calves are often infected in utero, leading to persistent infections that cause widespread virus dissemination within and across herds. Direct contact with PI animals, whether in a feedlot, pasture, or during transport, results in rapid viral spread. Naïve cattle, or those never exposed to the virus, are particularly susceptible.

The effect of BVDV varies depending on the timing of infection. Infected pregnant cows can suffer from early pregnancy losses, abortions, birth defects, or give birth to PI calves. These PI calves become lifelong carriers of the virus and may eventually develop mucosal disease, a fatal condition characterized by ulcers, lethargy, and persistent diarrhoea. Cows infected in the later stages of pregnancy often develop immunity, and their calves are born healthy and resistant to the virus. In feedlots, BVDV can trigger significant respiratory and gastrointestinal disease outbreaks due to the virus's immunosuppressive effects.

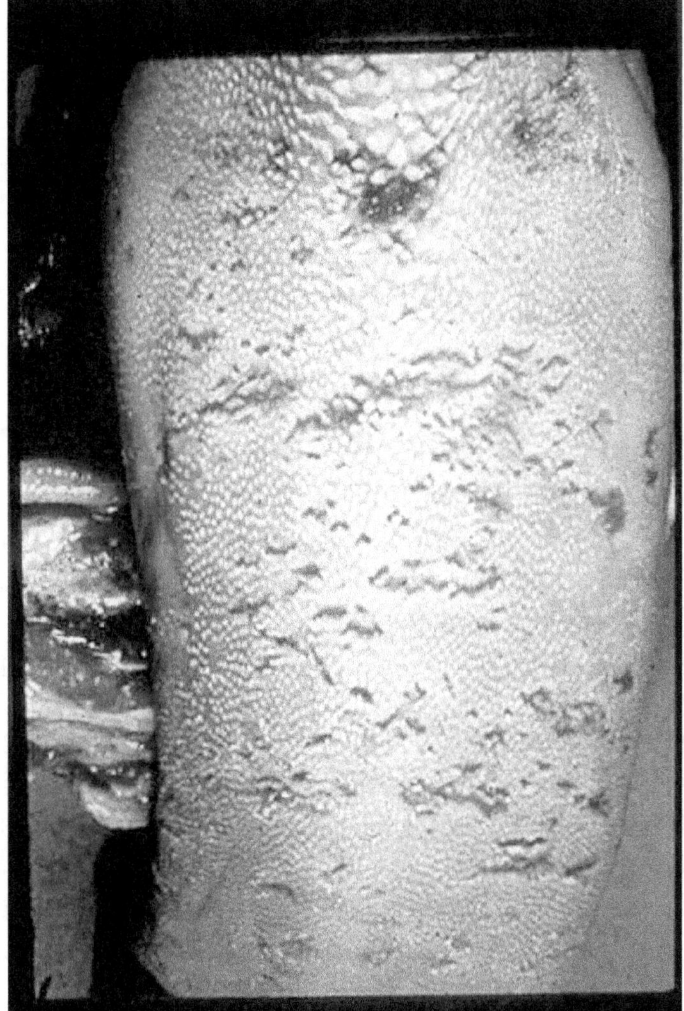

Figure 41: Tongue lesions on confirmed BVD/MD case (mucosal disease form). Lucien Mahin, CC BY-SA 3.0, via Wikimedia Commons.

Mucosal disease occurs only in animals persistently infected with BVDV when they acquire a more virulent strain of the virus. It is fatal, typically affecting cattle between six and 21 months of age. Symptoms include fever, lethargy, ulceration in the mouth and nose, and profuse diarrhoea. Once mucosal disease develops, the prognosis is poor, and affected animals must be euthanized.

BVDV has three key disease statuses that determine its progression and impact on cattle: *naïve, immune,* and *persistently infected (PI).* Each status corresponds to different stages of infection and immunity in cattle.

BVDV Disease Statuses:

1. **Naïve**: These cattle have never been exposed to the BVDV virus and thus have no immunity. Naïve cattle are highly vulnerable to infection if they come into contact with persistently infected animals or contaminated environments. When infected, naïve cattle often experience a mild illness but can suffer more severe consequences if they are pregnant, as BVDV can result in embryo death, abortion, or birth defects.

2. **Immune**: Immune cattle have either been exposed to the virus naturally or vaccinated against it. These cattle are not carriers and possess lifelong immunity to future BVDV infections. Natural exposure occurs when BVDV circulates through the herd, while vaccination helps create immunity in naive cattle. Immunity is crucial in reducing the spread of the virus, particularly among breeding cattle, as it prevents the development of persistently infected offspring.

3. **Persistently Infected (PI)**: PI animals are carriers of BVDV throughout their lifetime. They are the primary source of viral transmission within and between herds. These animals shed the virus in high quantities through bodily fluids such as saliva, urine, and faeces. Most PI cattle die within one to two years of birth, often due to the development of mucosal disease. PI animals that survive into adulthood pose a significant risk to the herd, particularly in pregnant cows, as they can transmit the virus to their offspring.

BVDV can lead to various reproductive and health issues in cattle, significantly impacting herd productivity. Some of the notable signs include:

- **Poor fertility**: Low calving rates, high rates of pregnancy testing indicating cows not in calf (PTNIC), and reproductive abnormalities such as stillbirths, abortions, and mummified foetuses.

- **Outbreaks of disease**: Infected cattle may show signs of fever, lethargy, and loss of appetite. The immune suppression caused by BVDV makes cattle more susceptible to secondary diseases such as pneumonia and scours, particularly in stressful environments like feedlots.

- **Mucosal disease**: This severe form of BVDV manifests in persistently infected animals and results in lethargy, ulcers, drooling, and profuse, persistent diarrhoea. Mucosal disease is fatal, and affected animals must be euthanized.

The timing of BVDV infection during pregnancy plays a critical role in determining the outcome for both the cow and the calf. Infection during different stages of gestation has varying consequences:

- **First trimester (1–4 months)**: Infection during this period often leads to embryo death, abortions, or the birth of persistently infected calves. These PI calves shed the virus throughout their lives and are a primary source of transmission within the herd.

- **Mid-pregnancy (4–5 months)**: Pregnant cows may abort or give birth to viable or non-viable calves, with some exhibiting congenital defects. Some calves may be born healthy with natural immunity to BVDV.

- **Late pregnancy (5–9 months)**: While abortion may still occur, calves born during this period are typically immune to BVDV.

BVDV is a complex and costly disease that affects the reproductive health, immunity, and productivity of cattle. Understanding the different disease statuses and stages of infection is essential for effective management and prevention. PI animals are the main drivers of disease transmission, and managing their presence in herds through biosecurity measures, vaccination, and testing is crucial to mitigating the impact of BVDV on cattle operations. Early detection and management are key to controlling the spread of BVDV, ensuring healthier herds, and preventing significant economic losses in the cattle industry worldwide.

BVDV is diagnosed through various laboratory tests, including blood samples, ear notches, or post-mortem tissues, to detect viral antibodies or persistent infections. Dairy herds can be tested using bulk milk samples. Early detection and diagnosis are crucial for implementing effective management strategies to prevent significant economic losses.

To manage the risk of BVDV in cattle, it is essential to establish a herd's immune status, control cattle movements, and identify PI animals. Vaccination is a key preventive measure, particularly for naïve animals entering the breeding herd or feedlot operations. Biosecurity measures such as proper fencing, separating naive animals from potential PI carriers, and using robust vaccination protocols can help control BVDV spread. In feedlots, managing cattle in groups with a common background and minimizing contact with unknown or unvaccinated cattle can help reduce the risk of outbreaks.

BVDV is a major concern for the cattle industry worldwide, causing significant economic losses due to reduced fertility, lower milk production, and increased susceptibility to other diseases. In Australia, BVDV is considered the second most costly disease after cattle tick infestation, with an estimated impact of $114 million annually [43]. Moreover, some international markets are sensitive to BVDV status, especially BVDV Type 2, and may require certification to ensure cattle are free from infection before export. Early detection of the disease is critical to maintain market access and ensure the health of livestock.

Furthermore, the introduction of BVDV into herds can lead to significant declines in milk production, as evidenced by studies showing that herds affected by BVDV experience lower yields compared to those implementing control measures [44, 45]. The economic implications of BVDV are not only limited to direct losses from reduced productivity but also include the costs associated with veterinary interventions and the management of secondary diseases that arise due to the immunosuppressive effects of BVDV [46]. Therefore, early detection and intervention are essential to minimize the economic impact of BVDV and to maintain the overall health and productivity of cattle herds globally.

BVDV poses a significant risk to cattle herds, impacting fertility, production, and overall herd health. Managing this disease requires comprehensive strategies, including vaccination, biosecurity, and monitoring for persistently infected animals. With careful herd management and proactive biosecurity measures, the economic impact of BVDV can be minimized, ensuring healthier livestock and protecting market access.

Pink Eye in Cattle

Pink eye, also known as infectious bovine keratoconjunctivitis (IBK), is a highly contagious eye disease affecting cattle worldwide. While it can affect cattle of any age, it is most common in calves and younger stock, leading to significant production losses and market implications. Globally, the disease presents challenges in terms of animal welfare and economic impact.

Pink eye is primarily caused by damage to the cornea (the outer layer of the eye), which may result from environmental factors such as dust, sunlight, or trauma. This damage allows bacteria from the genus Moraxella, most commonly Moraxella bovis, to infect the eye. The disease tends to flare up during the summer and autumn months when conditions are dry, and there are high populations of flies, which act as carriers of the bacteria, spreading the infection from animal to animal. The bacteria can persist in the herd because infected cattle may become carriers, ensuring that the disease remains a recurrent issue from year to year.

Pink eye can vary geographically due to environmental conditions and cattle management practices. In tropical and subtropical regions, where flies are prevalent, outbreaks tend to be more severe, while in colder climates, the disease might be less frequent.

The symptoms of pink eye are distinctive and include:

- Excessive blinking, watery eyes, and discomfort

- Ulcers on the surface of the cornea

- Cloudy or white spots on the eye due to the accumulation of pus and white blood cells

- Sensitivity to sunlight, leading to increased blinking or seeking shade

In severe outbreaks, up to 80% of a herd can be affected, making it a serious concern for cattle producers worldwide. The disease can significantly affect cattle welfare, as the discomfort and pain caused by pink eye can result in blindness or scarring in extreme cases.

Pink eye causes both economic losses and animal welfare concerns. Cattle affected by pink eye may have reduced market value due to scarring, blindness, or other eye issues. These animals may also be unsuitable for export or sale at saleyards, particularly if the disease is in an active stage. Internationally, some exotic diseases resemble pink eye, making accurate diagnosis crucial to avoid mistaken identity and prevent trade disruptions. Countries that prove freedom from such diseases can maintain better access to international markets.

Veterinary intervention is essential for diagnosing and managing pink eye, especially since some reportable diseases, such as malignant catarrhal fever (MCF) or infectious bovine rhinotracheitis (IBR), present with similar symptoms. Vets can perform early diagnosis through clinical signs and laboratory testing, helping rule out other serious or reportable diseases. Early intervention can reduce the severity of outbreaks and prevent widespread infection.

Treatment options for pink eye vary depending on the severity of the infection. Early stages may respond well to topical ointments applied directly to the affected eyes. More advanced cases might require antibiotic injections, sometimes combined with anti-inflammatory treatments to reduce pain and swelling. In extreme cases where the cornea is at risk of rupture, veterinary professionals may use eye patches or surgically close the eyelid to protect the eye and promote healing. It is important to treat both eyes, even if only one appears affected, as the likelihood of cross-infection is high.

Farmers and veterinarians should observe appropriate withholding periods (WHPs) and export slaughter intervals (ESIs) for any cattle treated with antibiotics or other medications to ensure compliance with safety standards.

Prevention is the most effective approach to managing pink eye. Key strategies include:

- Isolating affected cattle to prevent further spread within the herd

- Managing flies, using insecticides or encouraging dung beetles to reduce fly populations

- Reducing dust in cattle yards through environmental management, such as sprinklers in dusty areas

- Vaccination, where available, as a preventive measure before high-risk seasons like summer and autumn

Currently, in regions like Australia, only one vaccine is available to help prevent pink eye in cattle. It should be administered three to six weeks before the pink eye season begins, with annual booster shots recommended to maintain immunity.

Pink eye in cattle poses a global challenge, with significant economic and welfare implications for livestock producers. Proper diagnosis, treatment, and prevention are essential to minimize the impact of the disease. By implementing effective biosecurity measures, isolating infected animals, and controlling environmental factors like dust and flies, producers can reduce the incidence of pink eye. Vaccination, while helpful, should be used in combination with other management strategies for the best results.

Mastitis in Sheep

Mastitis, a bacterial infection of the udder, is a significant health concern for ewes, particularly those with high milk production or raising multiple lambs. It occurs globally, particularly in regions with large sheep populations, and is more common during the early stages of lactation or near weaning. The disease can lead to severe health issues in ewes and lambs, impacting the productivity of sheep farms worldwide.

Mastitis is an infection of the udder caused by bacteria, typically entering through injured teats. It results in inflammation, swelling, and, in severe cases, tissue necrosis. Ewes become more vulnerable to mastitis when their teats are damaged by mechanical injuries (such as cuts during shearing) or during lamb suckling, especially if the ewe is not producing sufficient milk. The infection can lead to significant economic losses, as it affects the health and productivity of the ewe and lambs, particularly in sheep industries in countries like Australia, New Zealand, and the UK.

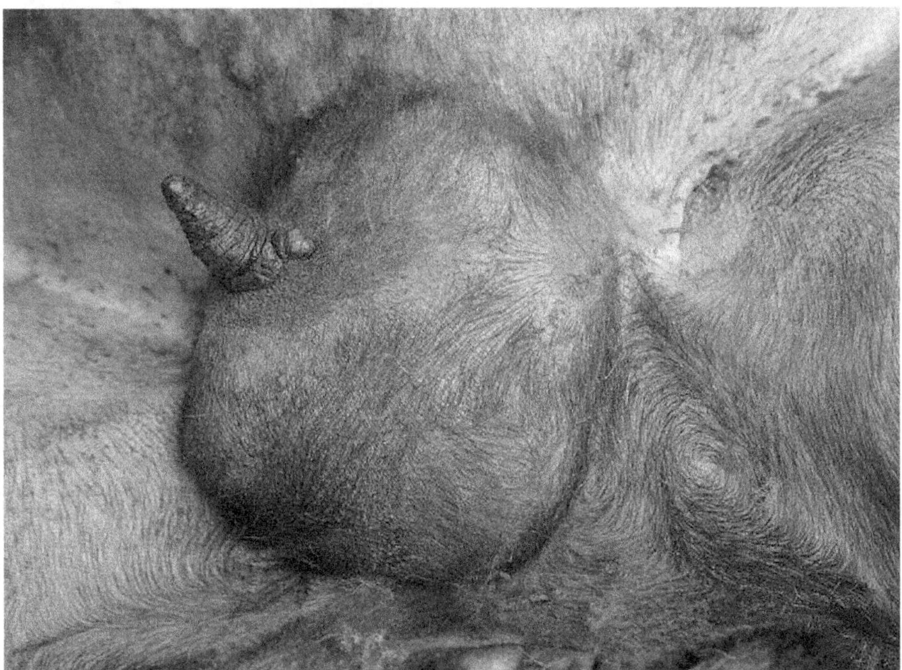

Figure 42: Udder of a of a Roux du Valais sheep (Walliser Landschaf) with only one teat after a healed mastitis. The second teat was lost due to the mastitis. Photographed by User:Bullenwächter, CC BY-SA 3.0, via Wikimedia Commons.

Certain breeds, like Poll Dorset and Suffolk, which are known for high milk production, are more susceptible to mastitis compared to other breeds like the Merino. Ewes nursing multiple lambs are also at a higher risk, especially in environments with poor hygiene, overstocking, or adverse weather conditions. Globally, environmental factors such as overcrowded lambing conditions or adverse weather events play a role in increasing mastitis cases, with wet, dirty conditions being particularly conducive to the disease's spread.

Mastitis is predominantly caused by two bacterial species: *Staphylococcus aureus* and *Mannheimia haemolytica*. These bacteria are often found in the skin and oral cavities of lambs and ewes. Poor hygiene, especially in crowded conditions or after adverse weather, can exacerbate the spread of these bacteria. The bacteria's resistance to several antibiotics complicates treatment, highlighting the importance of veterinary consultation for appropriate diagnosis and antibiotic selection. In many parts of the world, controlling mastitis requires stringent hygiene protocols, especially during lambing seasons.

Mastitis presents in two forms: clinical and subclinical. Subclinical mastitis is harder to detect, with no visible symptoms, but may be indicated by firm, hot udders and poor lamb growth. Clinical mastitis, on the other hand, shows more obvious signs, such as blackened, cold udders due to necrosis (often referred to as "black mastitis"). Lambs of infected ewes may suffer from poor growth or die from consuming infected milk. In severe outbreaks, the disease can cause death in ewes and lambs alike, making it a critical concern for sheep farmers globally.

Mastitis can be managed through appropriate antibiotic treatment, determined by laboratory testing. Veterinarians play a crucial role in diagnosing the disease and guiding treatment decisions. Humane euthanasia is often recommended for ewes with advanced cases of black mastitis that do not respond to treatment, and culling recovered ewes after weaning helps prevent future outbreaks in the flock. The global challenge of antibiotic-resistant bacteria has prompted veterinarians and farmers to be cautious about antibiotic use and to adopt more sustainable disease management practices.

Preventing mastitis involves a combination of good ewe nutrition, maintaining clean lambing paddocks, and minimizing physical injuries to the udder. Farmers worldwide are advised to ensure that sheep are not housed in unsanitary conditions or exposed to high stocking densities during critical periods like early lactation. Regular monitoring of ewes for signs of mastitis, coupled with proper hygiene practices, can significantly reduce disease incidence. In some regions, vaccination strategies have been developed to reduce the risk of mastitis, although these are not universally available.

Mastitis poses significant market challenges, particularly for sheep-exporting countries. In some cases, diseases that mimic mastitis symptoms, such as scabby mouth and contagious agalactia, can affect export markets. Ensuring that cases of mastitis are promptly reported and addressed helps maintain disease-free status for countries that rely on livestock exports, such as Australia and New Zealand. Early detection and management are crucial for minimizing the economic and social impacts of mastitis in these markets.

Mastitis in sheep is a widespread issue with serious health and economic consequences. Its management requires a combination of good hygiene, timely veterinary intervention, and, in some cases, culling of affected animals. With antibiotic resistance on the rise, sustainable farming practices and preventative measures, such as better nutrition and controlled environments, are becoming increasingly important in managing this disease on a global scale.

Managing mastitis in sheep within livestock production enterprises involves a multifaceted approach, focusing on prevention, early detection, treatment, and long-term management practices. These strategies aim to reduce the risk of infection, mitigate its impact on productivity, and promote animal welfare.

1. Prevention through Good Management Practices

Preventative measures are crucial for reducing the risk of mastitis in sheep. Livestock enterprises should focus on creating a clean and stress-free environment, particularly during key periods such as lambing and lactation.

- **Clean Lambing Environments:** Provide clean, dry, and sheltered lambing paddocks to minimize exposure to bacteria. Wet and dirty environments increase the likelihood of bacterial infections.

- **Good Nutrition:** Ensuring ewes have a balanced diet supports their overall health and immune function, making them less susceptible to infections. Proper nutrition is especially important for ewes nursing multiple lambs or those with high milk production.

- **Avoiding Overcrowding:** High stocking densities in paddocks or yards increase the risk of physical injury to the udders and the spread of bacteria. Managing stocking levels helps reduce teat damage and contamination.

2. Early Detection and Monitoring

Regular monitoring of ewes during the first few weeks after lambing and just before weaning is essential for detecting mastitis early.

- **Routine Udder Checks:** Regular inspection of the udder for signs of swelling, heat, or firmness helps identify cases of subclinical mastitis. Catching the disease early allows for quicker intervention, reducing the severity of the infection.

- **Monitoring Lamb Growth:** Lambs that are not growing as expected or seem weaker may indicate that their ewe has subclinical mastitis. Identifying these signs helps focus attention on potential udder infections.

3. Treatment Protocols

Treatment should be based on veterinary guidance, as bacteria causing mastitis can be resistant to certain antibiotics. Veterinary diagnostics, including bacterial cultures, can determine the most effective antibiotic treatment.

- **Antibiotic Treatment:** Antibiotics are often necessary to treat bacterial infections in the udder. However, because some bacterial strains are resistant to commonly used antibiotics, laboratory testing is essential for prescribing the correct treatment.

- **Supportive Care:** In severe cases of mastitis, particularly when the infection has led to tissue necrosis (black mastitis), supportive care like anti-inflammatories and fluids may be needed. In such cases, euthanasia may be considered if recovery is unlikely.

4. Long-term Management Strategies

Long-term strategies focus on preventing recurrence and maintaining the overall health of the flock.

- **Culling Affected Ewes:** Ewes that have suffered from mastitis may have permanent damage to their udders, making them prone to future infections. Culling these ewes after weaning helps prevent ongoing mastitis issues within the flock.

- **Biosecurity Measures:** Mastitis can spread within a flock through physical contact and contaminated environments. Isolating affected animals and maintaining biosecurity practices, such as regular disinfection of equipment and pens, can help limit the spread of infection.

- **Vaccination:** In areas where vaccines are available, vaccination can be a useful preventive tool to reduce the incidence of mastitis, especially in high-risk flocks.

5. Minimizing the Risk from Physical Injuries

Ewes with damaged teats from rough suckling or cuts during shearing are at a higher risk for mastitis. Livestock production enterprises can minimize this risk by:

- **Ensuring Proper Shearing Practices:** Avoid cutting or injuring the teats during shearing to prevent bacterial entry points.

- **Scabby Mouth Management:** Scabby mouth, a viral infection in sheep, can lead to sores around the mouth that increase the risk of mastitis when lambs suckle. Vaccination against scabby mouth can prevent its occurrence and thus reduce the associated mastitis risk.

6. Regular Veterinary Consultation

Veterinary input is vital for managing mastitis in a livestock production enterprise. Regular check-ups, laboratory tests for bacterial resistance, and tailored treatment plans help control the disease effectively.

Ovine Campylobacteriosis

Ovine campylobacteriosis, formerly known as ovine vibriosis, is a significant infectious disease in sheep, particularly affecting breeding ewes. It causes abortion in late pregnancy and is caused by the bacteria *Campylobacter fetus* subspecies *fetus*. The disease is a global concern in sheep-rearing regions but is less common in some areas, such as Western Australia (WA). Due to its potential impact on reproductive success in sheep, effective management and control measures are essential.

Ovine campylobacteriosis is not only a livestock issue but also poses risks to human health. The *Campylobacter fetus* bacteria can infect humans, particularly those with weakened immune systems. Infection in humans can lead to conditions like bacteraemia. Therefore, it is critical to use personal protective equipment such as gloves when handling aborted fetuses, placental materials, or contaminated areas to prevent zoonotic transmission. Proper sanitation, including washing hands and disinfecting equipment, is crucial after contact with infected materials.

In sheep, the disease manifests primarily in ewes during late pregnancy, leading to:

- **Vaginal discharge**: Initially, a brownish discharge may appear.

- **Sudden abortion**: Ewes typically experience abortion without much distress, but abortion rates can escalate quickly within a flock, sometimes affecting up to 70% of animals.

- **Post-abortion discharge**: Ewes may discharge infectious materials for up to two weeks after the abortion.

- **Foetal symptoms**: Lambs aborted due to *Campylobacter* infection often exhibit liver lesions, swelling, and fluid accumulation in body cavities. Recovered ewes tend to develop immunity, and repeat outbreaks in the same flock are uncommon in subsequent years.

Campylobacteriosis spreads through oral ingestion of bacteria from contaminated water or pasture. The bacteria typically enter the ewe's bloodstream and infect the placenta, leading to abortion. Contaminated environments, especially in cool, moist conditions, facilitate the survival and spread of the bacteria. Aborted foetuses and placental materials shed high numbers of bacteria, significantly contributing to the transmission within a flock. Wildlife and birds, particularly crows, are also known to spread the bacteria, especially when animals are hand-fed or concentrated in certain areas, such as around water points.

Ovine campylobacteriosis typically occurs in colder, wetter seasons when bacteria can thrive in the environment. However, in arid regions, outbreaks have been linked to sheep gathering around water and feed points, particularly during handfeeding in dry seasons.

Once *Campylobacter* bacteria enter a flock, the source is often difficult to trace. The bacteria can persist in the intestines of livestock like sheep and cattle or be spread by wildlife, including foxes and birds. Outbreaks can occur when infected animals introduce the bacteria into a previously healthy flock.

Diagnosis of ovine campylobacteriosis is typically made through laboratory testing of aborted foetal tissues, particularly liver samples, as well as placental materials. Post-mortem examination of foetuses may reveal liver lesions, which are a hallmark of the disease. Blood tests can also detect antibodies in ewes, though single tests may not be definitive in proving infection.

There is no specific treatment for preventing abortion once infection has occurred, as the damage to the placenta is irreversible. However, antibiotics can help reduce losses from secondary uterine infections (metritis). Infected ewes should be isolated from the flock to prevent further transmission.

Control measures for ovine campylobacteriosis focus on limiting the spread of infection and preventing contamination of feed and water sources:

- **Isolation**: Infected ewes should be immediately isolated to prevent transmission to other animals.

- **Sanitation**: Aborted foetuses and placental materials should be collected and burned to avoid further environmental contamination.

- **Biosecurity**: Flocks that have experienced an outbreak should not be used for breeding, as carrier animals may continue to spread the bacteria.

- **Vaccination**: A vaccine is available and can provide immunity to susceptible flocks, especially before periods of high risk. Vaccination is an important preventive tool in flocks with a history of the disease.

Ovine campylobacteriosis is a widespread disease in sheep-producing regions around the world, with varying prevalence depending on local conditions. In parts of the United States, New Zealand, and Europe, the disease is a major concern for sheep farmers. In these areas, vaccination and strict biosecurity protocols are critical to managing outbreaks. In contrast, regions like Western Australia report fewer cases, though occasional outbreaks still occur, particularly under favourable conditions for bacterial survival and spread.

Ovine campylobacteriosis presents a significant risk to sheep reproduction worldwide. Managing the disease requires a combination of good farm management practices, early detection, and the use of vaccination to protect vulnerable flocks. The zoonotic potential of the bacteria further emphasizes the need for careful handling and sanitation to protect both animals and humans.

Grass Tetany in Cattle

Grass tetany, also known as hypomagnesemia, is a fatal metabolic disorder affecting beef cattle, primarily caused by low levels of magnesium in the bloodstream. This disease is most prevalent during winter and spring and poses a global challenge to livestock production enterprises, especially in regions where cattle graze on rapidly growing pastures with low magnesium content.

The main cause of grass tetany is the low absorption of magnesium from the animal's diet. While cattle store magnesium in their bones and muscles, they cannot easily access these reserves during periods of high demand, particularly in lactating cows. Peak lactation, which occurs six to eight weeks after calving, significantly increases magnesium loss through milk. If dietary magnesium intake is insufficient, blood levels can drop dangerously low.

Several factors exacerbate the risk of grass tetany:

- **Pasture Composition**: Cool-season grasses and cereals, common in winter and spring, have lower magnesium levels than legumes. Pastures grown on leached, acidic soils also provide insufficient magnesium.

- **Fertilizer Use**: The application of nitrogen and potassium fertilizers can reduce magnesium availability in pastures, compounding the problem.

- **Environmental Stress**: Bad weather, such as winter storms, stress from transport or yarding, and high moisture content in grass, which speeds up digestion, further reduce magnesium absorption.

Cattle affected by grass tetany are often found dead, with signs of distress like leg paddling (indicative of convulsions before death). Early symptoms include muscle twitching, stiffness, and excitability. Affected animals may become aggressive and show an exaggerated response to external stimuli, which can progress to galloping, bellowing, and staggering. In severe cases, animals may collapse and die rapidly if untreated.

Immediate treatment is critical to save cattle showing signs of grass tetany. The best treatment involves intravenous administration of calcium and magnesium solutions, which quickly restore blood magnesium levels. In emergencies, cattle producers can inject a calcium-magnesium solution subcutaneously if veterinary assistance is not immediately available. Following treatment, providing oral magnesium supplements to the herd is essential to prevent relapses. Options include magnesium oxide powder, lick blocks, slow-release capsules, and magnesium-enriched feeds.

Prevention of grass tetany focuses on managing magnesium intake and reducing risk factors. Key strategies include:

- **Magnesium Supplementation**: Providing magnesium-enriched feeds, licks, or hay treated with magnesium sulphate can prevent deficiencies. Spraying pastures with magnesium-containing solutions is another effective measure, particularly during high-risk periods.

- **Grazing Management**: Moving lactating cows to pastures with higher legume content or dry matter can reduce the risk of magnesium deficiency. Reducing stocking density and minimizing environmental stress, such as sheltering cattle from bad weather, also help.

- **Soil and Pasture Management**: Long-term preventive measures include correcting soil acidity with lime or dolomite, limiting nitrogen and potassium fertilizer use, and establishing clover in pastures to improve magnesium availability.

Grass tetany shares clinical signs with several other diseases, making accurate diagnosis essential. Conditions such as nitrate poisoning, lead toxicity, and various types of grass staggers (caused by phalaris, ryegrass, or paspalum) can mimic grass tetany. In rare cases, exotic diseases like bovine spongiform encephalopathy (BSE) or Aujeszky's disease may present similar symptoms, underscoring the need for vigilance and prompt veterinary diagnosis.

Grass tetany is a worldwide issue affecting cattle in regions with magnesium-deficient soils and cool-season pastures, such as the United States, Europe, and Australia. The disease has significant economic implications, causing losses through cattle deaths, reduced milk production, and treatment costs. In Australia alone, grass tetany is responsible for substantial financial losses to the beef industry during peak risk seasons. Managing magnesium deficiencies in cattle is a global challenge, requiring a combination of nutrition management, pasture maintenance, and timely intervention to safeguard herd health.

Eperythrozoonosis (E. ovis) in Sheep

Eperythrozoonosis, now known as *Mycoplasma ovis* infection, is a bacterial disease affecting sheep and goats, leading to ill-thrift, anemia, and in severe cases, death. This disease, caused by the bacterium *Mycoplasma ovis* (formerly *Eperythrozoon ovis*), poses a significant risk to livestock, especially in regions where environmental conditions or farm practices facilitate its spread.

Eperythrozoonosis is observed worldwide, with varying prevalence depending on geographical location, climate, and farm management practices. It is notably more common in regions where sheep are raised in large numbers, such as Australia, the United States, Europe, and South America. The disease can lead to significant economic losses due to reduced productivity, ill health, and, in some cases, carcass condemnation in abattoirs due to jaundice. In Western Australia, the disease is most commonly diagnosed during spring in areas such as the northern agricultural region surrounding Geraldton and the great southern agricultural regions.

The primary signs of eperythrozoonosis in sheep include:

- **Ill-thrift and anemia**: Pale gums and overall lethargy in affected animals.

- **Jaundice**: Yellowing of the gums and mucous membranes due to the breakdown of red blood cells.

- **Dark red urine**: Indicating hemolysis, a hallmark of the disease.

- **Death**: Often following a stressful event such as yarding or marking.

The bacteria *Mycoplasma ovis* infects the red blood cells of the animals, prompting the spleen to destroy the infected cells, leading to anemia, jaundice, and eventual death. Disease outbreaks may last for up to four weeks, and in some cases, animals that recover remain carriers of the bacterium, serving as a reservoir for future infections.

Eperythrozoonosis tends to affect:

- **Young sheep**: Particularly those with compromised immune systems.

- **Sheep with concurrent infections or poor nutrition**: Malnourished sheep or those suffering from parasitic infestations like worms are more vulnerable.

- **Pregnant ewes**: They are particularly susceptible due to the stress of pregnancy and their reduced immunity.

The disease is typically spread through the transfer of infected red blood cells between animals. This can happen during common farming practices such as shearing, marking, or mulesing, especially when hygiene protocols are not strictly followed. Blood-sucking insects like mosquitoes and flies can also act as vectors, transmitting the bacterium from infected to healthy animals.

Eperythrozoonosis is diagnosed through laboratory testing, often by examining blood samples for the presence of *Mycoplasma ovis* bacteria. However, by the time clinical signs appear, many infected red blood cells have already been cleared, making diagnosis more challenging. To confirm the disease, it is essential to collect blood samples from both symptomatic and asymptomatic animals early in the course of the infection.

Treatment options are limited. While tetracyclines have been used in the past to control outbreaks, they are not recommended as a long-term solution. The best management strategy is supportive care, ensuring that affected animals receive adequate nutrition and are not stressed further. Over time, infected animals can recover naturally and develop immunity to the disease.

Preventing eperythrozoonosis in a flock requires strict hygiene measures and good farm management practices, such as:

- **Hygiene during marking, mulesing, and shearing**: Disinfecting instruments and avoiding the reuse of needles between animals.

- **Controlling biting insects**: Reducing the presence of flies and mosquitoes, which can act as vectors.

- **Management of stress**: Avoiding yarding or transporting sheep within six weeks of stressful events like marking or mulesing.

- **Nutritional support**: Ensuring that sheep have a balanced diet with sufficient trace elements and are free from parasitic infestations to minimize the risk of severe disease.

The prevalence of eperythrozoonosis varies across regions. Surveys in Western Australia suggest that up to 5% of weaner sheep may carry the infection, while in Victoria, 90% of farms tested positive for *Mycoplasma ovis*. In other countries such as the United States and parts of Europe, the disease also poses a significant concern, particularly in densely stocked or poorly managed herds. In some areas, infection rates in flocks can vary from as low as 2% to as high as 80%.

Eperythrozoonosis is a globally prevalent disease that requires careful management to prevent economic losses and animal suffering. Although there is no vaccine or highly effective treatment, good farm management practices, such as maintaining hygiene, reducing stress, and controlling biting insects, can significantly reduce the risk of outbreaks. Regular veterinary oversight and early intervention are critical in managing the disease and maintaining flock health.

Salmonellosis of Sheep

Salmonellosis is a serious infectious bacterial disease that affects sheep, causing illness and potentially death. It is caused by the proliferation of *Salmonella* bacteria, most notably *Salmonella typhimurium*, which has been identified as the primary serotype causing the disease in sheep in various parts of the world, including Western Australia. This disease can result in symptoms like profuse diarrhea, dehydration, fever, and in pregnant ewes, it often leads to abortion. The condition poses significant risks not only to sheep but also to humans, as *Salmonella* is zoonotic, meaning it can spread from animals to humans.

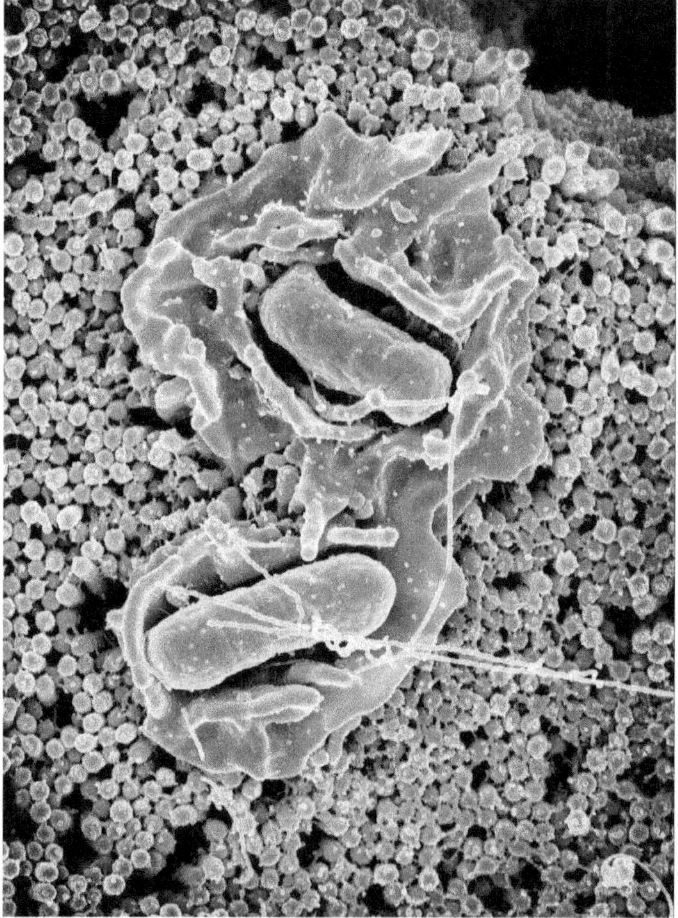

Figure 43: Salmonella Typhimurium. NIAID, CC BY 2.0, via Flickr.

Salmonellosis is found in sheep across the globe, although the prevalence and serotypes can vary by region. While *Salmonella* has over 2,000 serotypes globally, *S. typhimurium* is particularly problematic in sheep herds. In many countries, including Australia and the UK, this serotype is responsible for outbreaks in livestock.

The disease tends to be more common during winter and spring, particularly in areas with poor pasture quality, high sheep densities, or adverse weather conditions. High stocking densities—often seen in feedlots, during supplementary feeding, or when animals are crowded due to limited feed or water—significantly increase the likelihood of an outbreak. Other risk factors include stress caused by transportation, shearing, and poor weather conditions, along with heavy worm burdens, which make sheep more vulnerable to infection.

Sheep can become carriers of the bacteria and shed it in their faeces, contaminating their environment and thus infecting other sheep. The disease often spreads when stressed animals shed large quantities of bacteria into water or feed sources, infecting the rest of the flock. Common symptoms in sheep include:

- Sudden death in previously healthy animals

- Fever and reluctance to move

- Profuse, foul-smelling diarrhea

- Dehydration and abortion in pregnant ewes

- Jaundice and dark red urine in severe cases

Diagnosis of salmonellosis is confirmed through laboratory analysis of samples from deceased animals. Veterinarians typically conduct post-mortems to find evidence of infection, such as inflamed intestines, enlarged lymph nodes, or a characteristic khaki-coloured diarrhea. A rapid diagnosis is essential for limiting the spread of the disease and preventing further losses.

Treatment focuses on minimizing further stress, improving hygiene, and isolating infected animals. Veterinarians may recommend the use of fluids and antibiotics to treat infected animals, although in severe cases, humane euthanasia may be necessary.

There is no vaccine available for salmonellosis in sheep, making prevention a critical component of managing the disease. Preventative strategies include maintaining good nutrition and parasite control, avoiding high stocking densities, and minimizing stressors like prolonged periods without feed or water. Cleaning and disinfecting the environment, particularly water troughs and feed sources, can help reduce the risk of bacterial spread.

In terms of zoonotic risk, humans can contract salmonellosis through contact with infected animals, carcasses, or contaminated materials. Therefore, strict hygiene protocols are recommended, including wearing gloves when handling animals, washing hands frequently, and disinfecting equipment and clothing.

Salmonellosis is recognized as a significant health issue in livestock worldwide, and its zoonotic potential makes it a public health concern as well. Outbreaks can have devastating economic impacts due to livestock losses, decreased productivity, and restrictions on trade. Managing salmonellosis in sheep is therefore important not only for animal welfare but also for maintaining global food safety standards and ensuring the viability of agricultural industries across regions.

Lupinosis in Sheep

Lupinosis is a severe liver disease primarily affecting sheep that consume lupin stubble colonized by the fungus *Diaporthe toxica* (formerly *Phomopsis leptostromiformis*). This disease can manifest in two forms: an acute condition leading to rapid death or a chronic liver dysfunction syndrome. It is particularly prevalent during the summer and autumn months in regions where lupins are grown, especially in areas with summer rains.

Lupinosis occurs when sheep ingest lupin plants contaminated by *Diaporthe toxica*. Lupin stubbles, often left after harvesting, become the primary source of this toxin. While newer varieties of narrow-leafed lupins bred for resistance to this fungal infection have reduced the incidence of lupinosis, the disease remains a threat, especially under certain environmental conditions. It is most commonly associated with blue lupins, particularly following rains in summer.

The liver damage caused by lupinosis impairs the body's ability to metabolize food and toxins effectively, leading to a range of symptoms that may appear rapidly in acute cases or develop slowly over time in chronic cases.

Lupinosis can cause both acute and chronic conditions:

1. Acute Lupinosis:

- **Marked appetite reduction**: Affected sheep lose interest in food, often before other signs become evident.

- **Severe depression and lethargy**: Animals may isolate themselves from the flock, and in severe cases, jaundice (yellowing of the eyes and mucous membranes) becomes apparent.

- **Rapid mortality**: Some animals may die suddenly, with signs of liver swelling and jaundice seen in post-mortem examinations.

2. Chronic Lupinosis:
- **Loss of condition**: Affected sheep experience significant weight loss and weakness.

- **Stiff gait and disorientation**: Sheep may wander, appear disoriented, or get caught in fences.

- **Abortions and reduced lambing percentages**: A reduction in fertility and lamb survival rates is common in flocks affected by chronic lupinosis.

In both acute and chronic cases, secondary symptoms such as photosensitization, reduced wool production, and overall poor animal health can occur.

Veterinarians can diagnose lupinosis through a combination of clinical signs, post-mortem findings, and laboratory tests. Jaundice and liver damage are key indicators during physical exams. Blood samples and tissue tests from the liver and brain can confirm the presence of the fungal toxin. Early identification is essential to prevent further losses within a flock.

There is no specific cure for lupinosis, but prompt action can save affected animals and reduce losses:
- **Immediate removal from lupin stubble**: Once symptoms are observed, all sheep should be moved to a safe paddock with shade and water.

- **Nutritional support**: Providing high-quality hay or oats can help restore energy levels. Lupin seeds and high-protein feeds should be avoided due to the liver's compromised ability to process protein.

- **Avoid stress**: Stress exacerbates the disease, so minimizing handling, yarding, or any further stressors is critical during the recovery phase.

- **Veterinary intervention**: Severely affected animals that do not recover their appetite may need to be euthanized.

Sheep that recover from lupinosis often take several months to fully regain their health, and long-term liver damage may predispose them to other health issues such as pregnancy toxaemia.

Preventing lupinosis is mainly focused on good grazing management and close observation of sheep during risk periods:

- **Graze lupin stubbles early**: Grazing lupin stubble before the toxicity increases after summer rains can reduce the likelihood of exposure.

- **Monitor stock**: Regular checks for lethargy, appetite reduction, and signs of illness are crucial.

- **Manage lupin seed feeding**: Only feed lupin seed that is free from fungal contamination, especially avoiding discoloured or mouldy seeds.

- **Limit flock size**: Keeping weaner sheep in smaller groups can help manage grazing behaviour and reduce the risk of overconsumption of lupin stems.

- **Water management**: Providing multiple watering points encourages even grazing across paddocks, reducing the concentration of sheep in areas with potentially toxic lupin stubble.

Lupinosis is a concern in sheep-producing regions worldwide, particularly in areas where lupins are used as a fodder crop. The disease has historically been a major cause of sheep deaths in countries like Australia, particularly in Western Australia, where lupin farming is common. However, the development of resistant lupin varieties and improved grazing management strategies have significantly reduced its prevalence. Nevertheless, outbreaks can still occur in regions with susceptible lupin varieties or where environmental conditions favour fungal colonization.

The international livestock industry must remain vigilant about lupinosis, especially as climate change may alter rainfall patterns and create conditions conducive to fungal growth. Maintaining resistant lupin varieties, educating farmers on prevention, and promoting research into better management strategies remain key to controlling this disease globally.

Brucellosis in Animals

Brucellosis in Animals: A Global Overview

Brucellosis is a significant bacterial disease affecting livestock and causing economic losses, particularly in cattle, sheep, goats, pigs, and dogs. It is caused by various species of the *Brucella* bacterium and is recognized for its potential to infect humans, making it a zoonotic disease. Human infection occurs through contact with infected animals or consumption of contaminated animal products. Due to its economic impact and zoonotic nature, many countries have strict control measures to prevent and eradicate brucellosis from livestock populations.

Brucellosis is found worldwide, with different *Brucella* species affecting different animal hosts:

1. **Brucella abortus**: Mainly affects cattle, causing late-term abortions, infertility, and undulant fever in humans. Australia and many other countries have eradicated *B. abortus* through rigorous control and surveillance programs. However, it is still prevalent in regions of Africa, Asia, and Latin America.

2. **Brucella melitensis**: Primarily affects goats and sheep, causing abortions, stillbirths, and weak offspring. It is the most virulent *Brucella* species for humans, causing severe illness through contact with infected animals or consumption of unpasteurized dairy products. This strain is widespread in the Middle East, Africa, southern Europe, and parts of Latin America.

3. **Brucella suis**: Affects pigs and is a known cause of reproductive failure, including abortions and stillborn piglets. Infected pigs may also display lameness due to swollen joints. *B. suis* is present in parts of Australia, particularly in feral pig populations in Queensland and New South Wales, and it is found in other regions such as North and South America.

4. **Brucella canis**: Known to infect dogs, *B. canis* causes reproductive issues like abortions and inflammation of the epididymis. While zoonotic, human infections are rare. This strain has been reported in many countries, including the Americas and parts of Asia, but it has not been found in Australia.

5. **Brucella ovis**: This strain mainly affects sheep, causing reproductive issues in rams and early embryonic death in ewes. Unlike other *Brucella* species, *B. ovis* is not zoonotic, meaning it does not infect humans. The disease is found in most sheep-producing regions worldwide, including Australia.

                    RICHARD SKIBA

Brucellosis is transmitted between animals and to humans through direct contact with infected tissues (such as the placenta or foetus), ingestion of contaminated food (especially unpasteurized dairy products), or inhalation of aerosols. Animals become highly infectious after giving birth or aborting, which is when the bacteria are most likely to spread.

The disease spreads more easily in conditions where animals are densely stocked or have inadequate biosecurity measures. Poor farm management, lack of vaccination, and inadequate animal health monitoring can exacerbate outbreaks. In humans, brucellosis is often seen in farmers, veterinarians, and slaughterhouse workers due to their close contact with animals.

Brucellosis leads to significant economic losses in affected livestock industries due to reduced fertility, abortion storms, reduced milk production, and the need for culling infected animals. Countries free from brucellosis must maintain strict surveillance and control measures to ensure disease-free status, which is crucial for international trade. Livestock products from brucellosis-free regions are in high demand, as importing countries often require proof of freedom from the disease.

For example, Australia, which is free from most strains of brucellosis, conducts regular surveillance to demonstrate disease-free status. This is vital for protecting its export markets for livestock and livestock products. However, *Brucella suis* remains a concern in feral pig populations, posing a risk to pig producers and hunters.

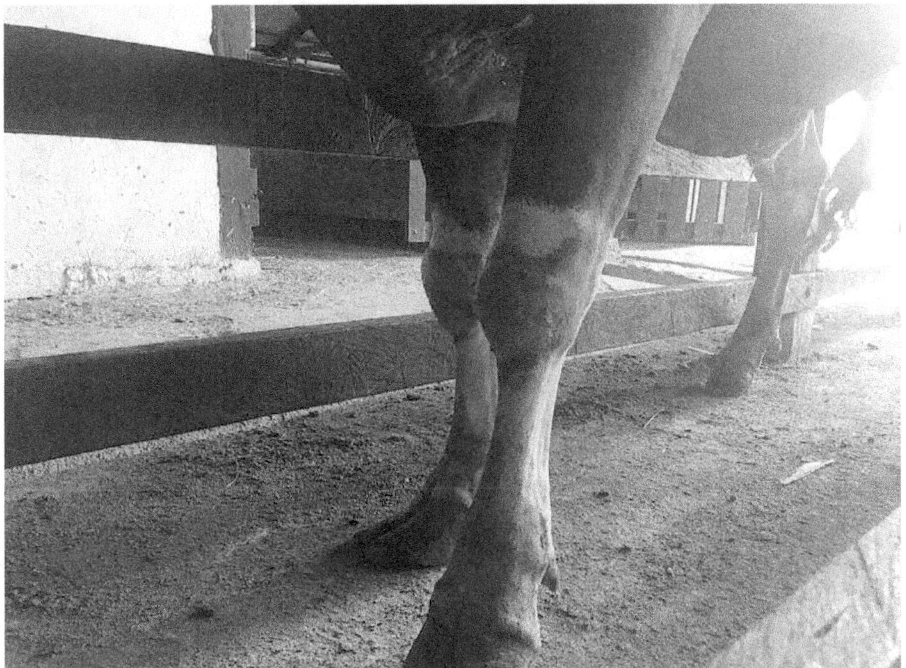

Figure 44: Swelling of joints due to hygroma which is a clinical sign of cattle brucellosis. RianHS, CC BY-SA 4.0, via Wikimedia Commons.

Identifying brucellosis in animals requires a combination of clinical observation, laboratory testing, and an understanding of the disease's epidemiology. Brucellosis can affect various livestock species such as cattle, sheep, goats, pigs, and dogs, and it is caused by different strains of the *Brucella* bacterium, including *B. abortus*, *B. melitensis*, *B. suis*, and *B. canis*. Accurate identification is essential due to the economic impact of the disease and its zoonotic potential.

1. Clinical Signs in Animals

The clinical signs of brucellosis can vary depending on the species of animal, but some general symptoms across species include:

- **Reproductive Issues**: The most prominent clinical feature of brucellosis is reproductive failure, such as late-term abortions, stillbirths, and weak offspring. Abortion storms in livestock herds are a strong indicator.

- **Epididymitis in Males**: Infected males, particularly rams and bulls, may develop epididymitis, characterized by swelling and inflammation of the epididymis, leading to infertility.

- **Reduced Fertility**: Animals that have been infected may exhibit a reduction in fertility, leading to lower calving, lambing, or farrowing rates.

- **Other Symptoms**: Some animals may show signs of arthritis, joint inflammation, and lameness, especially in pigs affected by *B. suis*.

In rams infected with *Brucella ovis*, the disease is often subclinical (with few visible symptoms), but infertility is a common outcome. In dogs, infections with *B. canis* can lead to reproductive disorders like abortion and epididymitis, as well as systemic issues like discospondylitis (inflammation of the spine).

2. Laboratory Diagnosis

Laboratory tests are crucial for the definitive diagnosis of brucellosis, as clinical signs alone can be nonspecific and similar to other diseases. Diagnostic tests include:

- **Serological Tests**: Blood tests are commonly used to detect antibodies against *Brucella* species. These include the Rose Bengal Test, complement fixation tests, and ELISA (enzyme-linked immunosorbent assay), which are effective in detecting an immune response.

- **Bacterial Isolation**: Culturing the bacteria from samples (such as fetal membranes, placental tissues, or aborted fetuses) can provide a conclusive diagnosis. However, culturing *Brucella* is labour-intensive and requires biosafety precautions because of the zoonotic risk.

- **Polymerase Chain Reaction (PCR)**: PCR tests can detect the DNA of *Brucella* species and are highly sensitive. This test is particularly useful when other diagnostic methods fail or are inconclusive.

- **Milk Testing**: In dairy cattle, bulk milk testing can be conducted to screen for brucellosis in a herd, especially for *Brucella abortus*. This is important for countries that are brucellosis-free and conduct routine surveillance.

3. Post-mortem Examination

In the event of animal death, post-mortem examinations can help confirm brucellosis:

- **Lesions**: Affected animals may have lesions on reproductive organs such as the placenta, liver, and spleen. The placenta may appear inflamed with thickened and necrotic cotyledons (fetal membranes).

- **Enlarged Lymph Nodes**: Affected animals often have enlarged lymph nodes, particularly in the abdominal and pelvic regions.

- **Abscesses**: Some species, like pigs infected with *B. suis*, may develop abscesses in various organs including joints, reproductive organs, and the liver.

4. Transmission History and Risk Factors

The identification process should also involve tracing the potential transmission pathways. Brucellosis is primarily transmitted via direct contact with infected animals or their fluids, particularly during or after parturition (birth). In regions where brucellosis is common, veterinarians will look at the following factors to identify the disease:

- **Abortions in the herd**: If multiple animals are experiencing late-term abortions, brucellosis may be a suspected cause.

- **Introduction of New Animals**: Brucellosis often spreads when infected but asymptomatic animals are introduced into a previously brucellosis-free herd.

- **Environmental Conditions**: Poor biosecurity measures, high stocking densities, and lack of vaccination increase the risk of brucellosis spreading within a herd.

5. Surveillance and Reporting

In many countries, brucellosis is a reportable disease due to its economic and zoonotic importance. National programs monitor herds and flocks for the presence of *Brucella* species to maintain the country's disease-free status. Regular testing of animals, particularly in high-risk populations, is vital to identify and contain outbreaks.

6. Differential Diagnosis

Brucellosis shares clinical signs with other diseases, so it is crucial to rule out similar conditions. These include:

- **Leptospirosis**: Another bacterial infection that can cause abortions in livestock.

- **Q Fever**: Caused by *Coxiella burnetii*, it can result in reproductive failure in animals and is also a zoonotic disease.

- **Listeriosis**: Can cause late-term abortions and neurological symptoms in affected livestock.

- **Campylobacteriosis**: This bacterial infection can cause abortions in sheep and goats and should be differentiated from brucellosis.

Brucellosis is a zoonotic disease, meaning it can spread from animals to humans. In humans, the disease typically causes flu-like symptoms, including fever, weakness, and joint pain. The more severe forms, such as those caused by *B. melitensis* and *B. abortus*, can lead to chronic conditions and severe complications if left untreated. Human infections occur through contact with infected animals or their secretions, ingestion of unpasteurized dairy products, or inhalation of aerosols in high-risk environments like abattoirs.

Preventing human infection relies heavily on good hygiene practices, including wearing protective gear when handling animals or animal products, pasteurizing milk, and proper disposal of infected carcasses and birthing materials.

Control measures for brucellosis vary depending on the region and the specific *Brucella* species involved. Strategies include:

- **Vaccination**: Vaccines are available for certain strains, such as *B. abortus*, which are crucial in preventing the spread of the disease in livestock.

- **Testing and Culling**: Regular testing of livestock herds and culling of infected animals are essential for disease control, particularly in eradication programs.

- **Biosecurity**: Implementing strict biosecurity measures on farms, such as isolation of new animals, proper sanitation, and controlling wildlife or feral animal populations that may carry the disease.

- **Surveillance**: Continuous surveillance is vital for countries that have eradicated the disease to ensure it does not re-enter.

Brucellosis remains a global challenge due to its impact on both animal and human health. The disease can cause severe economic losses, particularly in livestock industries, and poses a risk to human health in regions where brucellosis is still prevalent. The most effective means of controlling the disease are through vaccination, strict biosecurity, regular surveillance, and proper management of infected animals. Worldwide efforts continue to minimize the impact of this disease on both agriculture and public health.

Thiamine Deficiency Induced Polioencephomalacia (PEM) of Sheep and Cattle

Thiamine deficiency-induced polioencephalomalacia (PEM) in sheep and cattle is a neurological disease caused by a lack of thiamine (vitamin B1), which plays a vital role in energy metabolism, especially in the brain. In ruminants, thiamine is normally produced by bacteria in the rumen when animals are on a well-balanced diet rich in roughage. However, under certain conditions, the rumen environment can shift, leading to thiamine deficiency, which results in a serious neurological disorder called polioencephalomalacia.

The root cause of PEM lies in the disruption of thiamine production or its absorption. This can happen for various reasons:

- **Proliferation of Thiaminase-Producing Bacteria**: Under certain conditions, like diets high in carbohydrates or sudden dietary changes, certain rumen bacteria can produce enzymes called thiaminases. These enzymes break down thiamine before it can be absorbed by the animal, leading to a deficiency.

- **Sulphur Toxicity**: In some cases, PEM can be induced by sulphur poisoning, especially in feedlots where high levels of sulphur may come from feed supplements or water sources. Excess sulphur disrupts normal thiamine metabolism and can mimic the symptoms of thiamine deficiency.

- **Thiaminase-Containing Plants**: Certain plants, such as bracken fern and horsetail, contain thiaminases, which can prevent the absorption of thiamine in ruminants, although this is a rare cause of PEM in many regions.

PEM typically affects only a small number of animals in a herd or flock, but in some cases, the mortality rate can reach as high as 10%. In Western Australia and globally, the disease is more prevalent during periods of rapid dietary change, such as during the spring or autumn when animals may be moved to different pastures or feeding regimens. Feedlots are particularly vulnerable to outbreaks due to high carbohydrate, low-fibre diets that encourage the growth of thiaminase-producing bacteria.

The disease can affect animals of any age, though younger, rapidly growing animals and those under stress are more susceptible. Animals in good body condition are not immune and can develop the disease suddenly, often within a few days of changes to their diet or environment.

The clinical signs of PEM are predominantly neurological, given that thiamine deficiency severely affects brain function. Common symptoms include:

- **Initial Signs**: Agitation, anxiety, and muscle twitching. Animals may hold their heads in abnormal positions, such as with the head tilted upwards, and walk with a stiff, high-stepping gait.

- **Neurological Deterioration**: As the disease progresses, affected animals become blind, may press their heads against objects, and develop severe muscle spasms and seizures.

- **Terminal Stages**: Animals often lie down, with convulsions (paddling of the legs), and die within 24 to 48 hours if untreated.

Diagnosing PEM involves both clinical observation and laboratory tests. Vets may collect blood samples to measure thiamine levels and submit brain tissue samples for histopathology to detect the characteristic lesions of PEM. Post-mortem examination can reveal softening of the brain tissue, a hallmark of the disease.

Early intervention is critical for treating PEM. If diagnosed early, thiamine can be administered either orally or via injection. A single oral drench of thiamine may correct the deficiency and restore balance in the rumen. For severe cases, repeated injections of thiamine (up to three times a day) may be required over several days. Animals that are treated in the early stages of the disease typically recover fully within 48 hours, though some may have lasting neurological deficits such as blindness.

In feedlots or during outbreaks where multiple animals are affected, prevention strategies include:

- **Dietary Adjustments**: Increasing the fibre content in the diet to at least 50% to reduce the proliferation of thiaminase-producing bacteria.

- **Thiamine Supplementation**: Adding thiamine to the feed for two to three weeks helps stabilize thiamine levels while the animals' rumen environment adjusts.

- **Environmental Management**: Reducing stress factors such as overcrowding, sudden transport, or dietary changes also helps minimize the risk of PEM.

Prevention of PEM revolves around good dietary and environmental management. Ensuring that animals have consistent access to roughage, avoiding rapid changes in feed composition, and monitoring animals closely during periods of high risk (e.g., during spring/autumn transitions or in feedlots) are crucial. When animals are exposed to diets low in thiamine or high in thiaminases, adding thiamine to their diet can help prevent the onset of the disease.

PEM is a global concern, particularly in regions where cattle and sheep are fed on high-carbohydrate diets or where there is limited access to roughage. While the disease is relatively well-understood in countries like Australia, similar cases have been reported worldwide, particularly in intensive farming systems where sudden dietary changes are common. Countries with large livestock industries, especially in North and South America, Europe, and parts of Asia, have reported PEM outbreaks, especially in feedlot cattle.

Additionally, as thiamine deficiency may mimic other serious diseases such as rabies, listeriosis, and lead poisoning, early diagnosis and treatment are crucial, not only to protect livestock health but also to maintain market access and ensure food safety in global trade.

Botulism in Cattle

Botulism is a highly fatal disease in cattle caused by the ingestion of the botulinum toxin, produced by the bacterium *Clostridium botulinum*. This bacterium thrives in low-oxygen environments like rotting animal carcasses, decaying plant material, or poorly prepared silage, where it produces toxins that lead to muscle paralysis in affected animals. Once cattle ingest the toxin, the disease progresses rapidly, often leading to death within days.

Clostridium botulinum spores are widely present in soil and water worldwide. Under favourable conditions—such as in decomposing carcasses or improperly preserved silage—these spores can germinate and release toxins. The toxin interferes with the neuromuscular junction, preventing the release of acetylcholine, which is essential for muscle contraction. As a result, affected animals experience progressive paralysis, beginning with the hind limbs and eventually leading to respiratory failure.

Different types of botulinum toxins (labelled A through G) exist, with types C and D being most common in cattle in regions like Australia and other parts of the Southern Hemisphere. In contrast, type B toxin is more frequently observed in Europe and North America.

The primary sources of botulism outbreaks in cattle include:

- **Bone chewing**: In regions where protein and phosphorus are scarce, especially during the dry season, cattle may chew on animal bones or carcasses, inadvertently ingesting the botulinum toxin.

- **Contaminated feed**: Carrion-contaminated feed rolls or silage can lead to widespread botulism outbreaks in feedlots, dairy farms, or intensive beef farms.

- **Spoiled silage**: Silage that has not been properly fermented (with a pH above 4.5) can also harbor the bacteria, increasing the risk of botulism in cattle consuming it.

- **Toxicoinfectious botulism**: Though rare, this form occurs when live *C. botulinum* bacteria proliferate in an infected wound or in the intestines and release the toxin into the bloodstream.

Botulism typically presents as an outbreak, with multiple cattle exhibiting signs of weakness and paralysis. Early signs include:

- Hindlimb weakness and stumbling

- Flaccid paralysis, starting with the hindquarters and progressing to the forequarters and head

- Difficulty swallowing, drooling, and inability to retract the tongue

- Drooping eyelids and shallow breathing

- Complete paralysis, leading to death within 24 to 48 hours if not treated

Diagnosis of botulism is based on clinical signs, as laboratory detection of the toxin in blood or tissues can be challenging. Veterinarians may take samples of suspected feed or carcass material to test for botulinum toxin. Other diseases with similar signs—such as milk fever, ketosis, spinal cord injuries, or tick paralysis—must be ruled out.

Once botulism sets in, treatment options are limited, and most cattle die from respiratory failure. Mildly affected animals may recover with intensive supportive care, including fluids and nutritional support. Vaccination is the most effective method for preventing botulism, especially in high-risk environments such as feedlots, dairies, and pastoral operations.

Both short-acting and long-acting vaccines are available, providing protection against the C and D types of botulinum toxin. Long-acting vaccines can offer protection for up to three years, making them ideal for extensive cattle operations. Ensuring proper silage preparation, removing carcasses, and limiting cattle access to decaying organic material are also key preventive measures.

Botulism is a global concern in cattle-rearing regions, particularly where environmental and feeding practices increase the risk of exposure. In countries like Australia, botulism is a significant issue in pastoral regions where cattle chew bones due to nutrient deficiencies. In contrast, in parts of Europe and North America, botulism tends to be linked more to silage contamination and feedlot outbreaks.

In developing regions, botulism remains a public health risk due to the potential contamination of food sources and inadequate animal health infrastructure. Vaccination programs, improved animal husbandry practices, and monitoring of feed quality are essential strategies to mitigate the risk of botulism worldwide.

Beef Measles

Beef Measles (Cysticercus bovis) is a parasitic infection in cattle caused by the larval stage of the human tapeworm *Taenia saginata*. This condition is of significant concern in both veterinary and human medicine due to its public health implications and economic impact on the beef industry worldwide.

Beef measles refers to the presence of small, fluid-filled cysts, typically the size of a pea, that are found in the muscles of cattle. These cysts contain the immature form of *Taenia saginata* and are primarily located in the jaw, heart, diaphragm, and tongue muscles. Although less frequently, they can also be found in other muscles of the cattle.

Humans become infected with *Taenia saginata* by consuming raw or undercooked beef that contains these viable cysts. Once ingested, the cyst releases the immature tapeworm, which attaches to the human's intestinal wall and grows into an adult tapeworm within 2-3 months. This adult tapeworm can reach a length of 4 to 10 meters and may live in the human intestine for many years, often without causing noticeable symptoms. Infected individuals unknowingly pass tapeworm segments in their feces, which contain thousands of eggs, perpetuating the cycle when the feces contaminate cattle grazing areas.

Cattle become infected when they graze on pasture contaminated with human feces containing *Taenia saginata* eggs. Once inside the cattle's digestive system, the eggs release larvae that burrow through the intestinal wall, enter the bloodstream, and lodge in the muscles where they develop into cysts. These cysts, commonly referred to as "beef measles," are not harmful to cattle and typically die within 4 to 6 months, leaving behind calcified lumps. However, they become problematic when the infected beef is consumed by humans.

Beef measles are a global issue, but certain regions with inadequate sanitation practices and areas where human feces contaminate cattle grazing land are at higher risk. In developed countries, infections are more likely in regions where cattle are exposed to human fecal contamination through septic tank drainage, improper waste disposal near campsites, or through agricultural practices involving untreated sewage.

Routine inspections in abattoirs help detect beef measles. Carcasses showing evidence of cysts undergo strict processing protocols, including freezing at temperatures below -12°C for at least 10 to 20 days to kill any viable cysts. Heavily infected carcasses may be condemned. The economic consequences of beef measles can be severe, as infected meat may be downgraded or destroyed, leading to significant financial losses for cattle producers.

Effective control of beef measles focuses on both reducing infection in cattle and preventing human infection. Measures include:

1. **Sanitation Practices**: Preventing cattle from grazing on pastures contaminated with human waste is critical. This involves proper sewage management, particularly around areas where humans frequent, such as campsites or rural areas without proper waste disposal systems.

2. **Meat Inspection**: Beef consumed by humans should undergo proper meat inspection at abattoirs to ensure it is free from cysts. Home-killed beef, if not properly inspected, may pose a significant risk to human health.

3. **Proper Cooking**: To eliminate the risk of human infection, beef should be cooked thoroughly, ensuring the meat reaches a uniform temperature of at least 56°C for a minimum of 5 minutes.

Beef measles represents a zoonotic disease with significant public health and economic concerns. It underscores the importance of proper hygiene, waste management, and meat inspection to break the life cycle of *Taenia saginata* and protect both cattle and humans from infection. Ongoing vigilance and education about safe meat-handling practices, combined with regular abattoir inspections, remain crucial in managing and reducing the prevalence of this disease worldwide.

Lumpy Jaw

Lumpy Jaw is a chronic bacterial infection primarily affecting cattle, caused by the bacterium *Actinomyces bovis*. It is characterized by swelling and distortion of the jaw, a condition that is typically associated with bone infections. While cattle are the most commonly affected species, pigs and horses can also develop lumpy jaw. In a global agricultural context, this disease represents a significant health concern for cattle farmers, as it can lead to economic losses due to the deterioration in animal health and productivity.

Lumpy jaw is caused by *Actinomyces bovis*, a bacterium that normally inhabits the mouths of healthy cattle without causing disease. The infection often begins when the lining of the mouth is punctured by sharp objects like sticks, rough forage, or abrasive plants. This allows the bacteria to invade deeper tissues, where they infect the jawbone, causing the bone to swell and deteriorate. The infection can also result from tooth eruption in young cattle, which creates an entry point for the bacteria. In some rare cases, the infection can spread to the digestive tract, but the mouth and jaw are the primary areas affected.

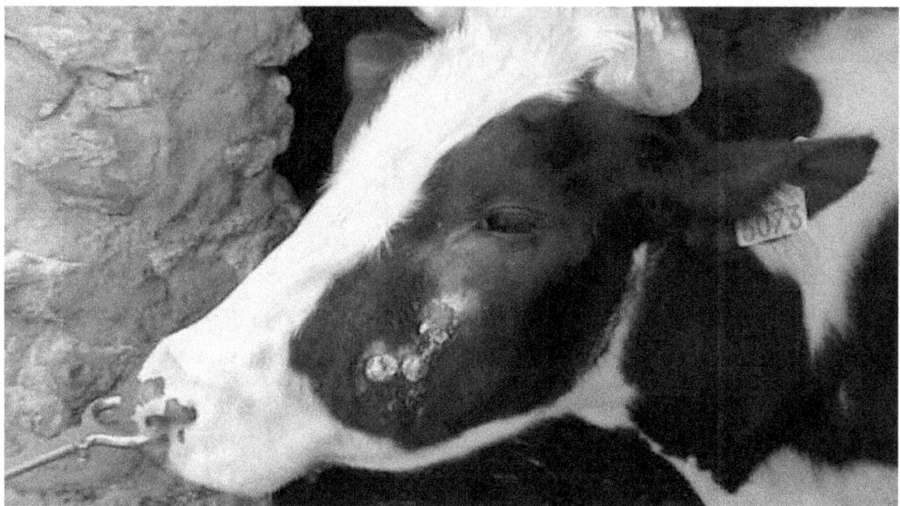

Figure 45: Bovine actinomycosis, 3-years-old bull, 2-month evolution. L. Mahin, CC BY-SA 3.0, via Wikimedia Commons.

Lumpy jaw develops gradually over weeks or months, so the initial signs may be subtle and often go unnoticed. The first symptom is usually a hard, immovable swelling on the jawbone, which can become progressively larger. As the disease progresses, the swelling may produce a discharge of sticky, honey-like material containing small granules. Because the infection causes pain, the affected cattle may have difficulty eating, leading to weight loss and gradual wasting. In advanced cases, the condition may also cause "bottle jaw," a swelling of the soft tissue under the jaw.

Cattle of all ages can be affected, but younger animals are more prone to the disease due to their developing teeth, which can provide an entry point for bacteria. While the disease is not directly fatal, most untreated cattle eventually succumb to malnutrition and secondary infections.

Several other conditions may be mistaken for lumpy jaw due to similar signs of swelling or jaw deformities:

- **Grass Seed Abscesses**: Caused by grass seeds, these abscesses appear as lumps around the face but are usually located in soft tissue rather than the bone. They can be cured with proper treatment.

- **Woody Tongue**: Caused by a bacterium closely related to *Actinomyces bovis*, this condition affects only the tongue, causing it to become hard and swollen. It is generally easier to treat than lumpy jaw.

- **Bottle Jaw**: This condition is a soft, fluid-filled swelling under the jaw, often associated with parasitic infections like liver fluke or conditions like Johne's disease.

- **Bone Tumours and Foreign Bodies**: Rare bone tumours and trapped foreign objects can also cause jaw swelling and should be considered during diagnosis.

Treatment of lumpy jaw is most effective when initiated in the early stages of the disease. The primary approach is the administration of iodide solutions, either orally or intravenously, to reduce bone swelling and inflammation. However, because infections in bone are challenging to treat, success rates are generally low, especially if the condition has advanced. Long-term treatment may be necessary, and in some cases, euthanasia is recommended for animals that do not respond to therapy or are suffering from severe malnutrition.

Preventing lumpy jaw involves minimizing the risk of mouth injuries that could allow bacterial entry. This includes providing cattle with high-quality forage and avoiding sharp or abrasive plants. Regular dental care for younger animals, particularly during the teething stage, can also reduce the risk of infection. In areas where the disease is common, early detection and prompt veterinary intervention can help limit the spread and severity of the condition.

Lumpy jaw is a worldwide issue that affects cattle farmers across different regions. While the disease is more commonly observed in cattle, pigs and horses can also be affected. In areas with poor-quality forage or inadequate veterinary care, the disease can be more prevalent, leading to significant economic losses due to the culling of affected animals and reduced productivity. It is essential for cattle producers worldwide to maintain good oral hygiene for livestock, manage pastures to minimize injury risk, and seek veterinary care at the first sign of infection to prevent the disease from advancing.

CHAPTER EIGHT

Prevention and Treatment Strategies to Resolve Health Issues

To effectively manage and prevent common diseases and disorders in cattle, it is crucial to understand the potential health risks that can affect your beef enterprise. Implementing appropriate management practices and corrective treatments requires a thorough understanding of local disease risks, herd conditions, and veterinary advice.

Identifying Disease Risks in Beef Cattle

In managing cattle, producers must assess various factors that contribute to disease outbreaks. Some of the common causes of diseases in cattle include [47]:

- **Infectious agents** such as bacteria, viruses, or fungi.

- **Parasite infestations**, both internal (like gastrointestinal worms) and external.

- **Nutritional deficiencies or imbalances**, which can lead to metabolic disorders such as grass tetany, milk fever, or ketosis.

- **Reproductive diseases** such as vibriosis, trichomoniasis, and leptospirosis, which affect cattle fertility and pregnancy rates.

By evaluating local grazing conditions, husbandry practices, cattle age groups, and disease status of introduced animals, producers can assess the risk of these diseases. Knowing the common cattle diseases in a specific locality enables farmers to be proactive in disease prevention.

Developing a Disease Management Plan

Consulting with local veterinarians, neighbouring farms, and agricultural departments is essential in developing a comprehensive disease management plan. A proactive approach to disease management often proves more cost-effective than reactive treatment, especially when it comes to vaccination programs that can prevent high-risk diseases. Once a disease risk is identified, producers should either [47]:

1. Take immediate action to implement preventive measures, such as vaccinations and biosecurity protocols, or

2. Monitor herd health and respond promptly when disease symptoms appear.

For example, the control of internal parasites in cattle often involves an integrated approach that includes strategic drenching at key points in the production cycle. This practice is particularly important in young cattle, which are most susceptible to parasitic infections. Monitoring through faecal worm egg counts (WECs) and using effective drenches is vital to ensure optimal herd health and prevent production losses.

Managing Specific Health Issues

Certain diseases require specific treatments and management strategies. For instance [47]:

- **Calf scours** (neonatal diarrhea) needs early intervention with supportive fluids and, if necessary, antibiotics.

- **Grass tetany** and **milk fever** require careful nutritional management to ensure adequate magnesium and calcium levels in the diet.

- **Pinkeye** requires both preventative measures (such as fly control and maintaining clean, dust-free environments) and treatment with antibiotics when symptoms appear.

Example: Integrated Approach for Internal Parasites

Managing internal parasites requires a combination of strategic treatments and environmental controls. Young stock, especially cattle up to 12 months of age, are most at risk for internal parasites. Best practices for managing internal parasites include [47]:

- **Strategic drenching** at weaning and during peak worm seasons (such as winter).

- **Monitoring pasture contamination** to prevent worm pickup.

- **Grazing management** to ensure young cattle graze on low-risk pastures (e.g., pastures previously grazed by sheep or mature cattle).

- **Nutrition management** to maintain good body condition and minimize the impact of parasitic infections.

It is also essential to be vigilant about drench resistance by regularly testing the efficacy of drenches through faecal worm egg count reduction trials.

Drench resistance is a growing concern in livestock management, particularly in cattle, sheep, and other ruminant operations. Drench resistance refers to the reduced effectiveness of anthelmintic treatments (drenches) in controlling parasitic worms due to the development of resistant worm populations. Over time, repeated use of the same drench can lead to selective pressure, where susceptible worms are killed off and resistant ones survive, reproduce, and proliferate.

Testing the efficacy of drenches is essential to ensure that treatments remain effective in managing worm burdens. Regular testing helps detect early signs of resistance and allows farmers to adjust their parasite management strategies before resistance becomes widespread. One of the most effective ways to test for drench resistance is through faecal worm egg count reduction trials (FECRT).

FECRT is a method used to measure the effectiveness of a drench by comparing the number of worm eggs in the faeces before and after treatment. The process involves:

1. **Initial Faecal Sampling:** A sample of faeces is collected from a group of animals before drenching. This sample is analysed to determine the number of worm eggs present (measured as eggs per gram of faeces).

2. **Administration of Drench:** The animals are then treated with a specific anthelmintic or drench.

3. **Post-treatment Sampling:** After 10 to 14 days, another faecal sample is col-

lected and analysed to determine the new worm egg count.

4. **Calculating Reduction:** The difference between the pre- and post-treatment egg counts is calculated to assess the efficacy of the drench. A reduction of 95% or more is considered effective, whereas anything below this indicates a possible resistance issue.

FECRT provides a clear indication of whether a drench is still effective in reducing worm populations. A low reduction rate suggests that the worm population has developed resistance to the treatment. The results of these trials enable producers to:

- **Identify resistant worm species**: Some worms may be resistant to certain classes of drenches, requiring a change in treatment strategy.

- **Evaluate the efficacy of different anthelmintics**: FECRT allows for comparisons between different products or drug classes to determine the most effective treatment for the current worm burden.

- **Optimize parasite control programs**: By understanding the level of resistance, producers can rotate drenches or implement integrated pest management (IPM) practices, such as combining grazing management with drenching.

Failure to regularly test for drench resistance can lead to significant production losses due to poor parasite control. Worm burdens can cause reduced weight gain, lower milk production, and even death in severe infestations. Over-reliance on ineffective drenches not only worsens animal health but also accelerates the spread of resistance across the farm.

To reduce the risk of drench resistance, it is essential to use effective drenches and rotate between different classes of anthelmintics. Other strategies include:

- **Strategic Drenching**: Only drench animals at critical times, such as before high-risk periods, rather than routine or calendar-based drenching.

- **Targeted Selective Treatment (TST)**: Treat only animals showing signs of a worm burden, leaving healthier animals untreated to maintain genetic diversity among worms.

- **Grazing Management**: Reducing worm contamination on pastures by rotating animals between high- and low-risk pastures can minimize the need for drenches.

By regularly conducting FECRTs and integrating other management practices, producers can slow the development of drench resistance, protect animal health, and ensure the continued efficacy of anthelmintic treatments in their operations.

Vaccinations For Beef Cattle

Vaccinations are a critical part of maintaining the health and productivity of beef cattle herds by preventing diseases that can cause severe economic losses. Implementing an effective vaccination program is essential to safeguard against a range of bacterial, viral, and parasitic diseases. When developing a vaccination strategy, it's important to work with a veterinarian to ensure it is tailored to the specific health risks prevalent in your locality and production system [48].

Importance of Vaccinations in Disease Prevention

Vaccinations work by stimulating the animal's immune system to recognize and combat pathogens before they can cause illness. In cattle, various vaccines are available to prevent both bacterial diseases, such as those caused by clostridial bacteria (e.g., blackleg, tetanus), and viral diseases (e.g., bovine viral diarrhoea virus, pestivirus). Combining vaccination with good management practices, such as proper nutrition, biosecurity, and parasite control, can maximize the effectiveness of disease control programs [48].

Key Vaccinations for Beef Cattle

5-in-1 and 7-in-1 Vaccines:

- **5-in-1** protects against five clostridial diseases: pulpy kidney (enterotoxaemia), black disease, tetanus, blackleg, and malignant oedema.

- **7-in-1** extends this protection to include Leptospira hardjo and Leptospira pomona, which are significant causes of reproductive losses and infections in humans.

Clostridial Diseases (Blackleg, Tetanus, Pulpy Kidney):

- Clostridial bacteria are commonly found in soil and can cause sudden death in cattle if ingested. Vaccination is critical in areas with a high risk of these diseases. Calves should receive two doses of the 5-in-1 vaccine, 4-6 weeks apart, followed by annual boosters.

Leptospirosis:

- Leptospirosis can cause reproductive problems such as abortions and stillbirths in cattle. A two-dose vaccination schedule is recommended for maiden heifers before mating, with an annual booster during mid-pregnancy. Bulls should also be vaccinated, especially in areas where leptospirosis is prevalent.

Three Day Sickness (Bovine Ephemeral Fever):

- Bovine ephemeral fever, or three-day sickness, is a viral disease that causes fever, lameness, and reduced productivity. Vaccination is recommended for valuable animals, such as bulls and sale cattle, but not the entire herd due to the cost. Vaccination is administered in two doses, with a booster as needed.

Botulism:

- Caused by toxins from Clostridium botulinum, botulism can be fatal if cattle ingest contaminated feed or animal carcasses. Vaccination is particularly important in northern Australia where phosphorus deficiency increases the risk of bone chewing and exposure to botulism. Long-acting vaccines provide up to three years of protection, but annual vaccination is often recommended in high-risk areas.

Tick Fever:

- Tick fever, transmitted by ticks, causes significant losses in cattle herds by leading to high fever, anaemia, and death. Vaccination is crucial for all cattle, particularly young stock and those introduced from tick-free areas. A single dose of the tick fever vaccine provides protection.

Vibriosis:

- Vibriosis is a common cause of infertility in cattle, especially affecting maiden heifers. Bulls in all breeding herds should be vaccinated, and maiden heifers should receive a single dose one month before mating to prevent the spread of the disease.

Pestivirus (Bovine Viral Diarrhoea Virus - BVDV):

- Pestivirus can cause reproductive losses and immunosuppression in cattle. Vaccination is recommended for all breeding cattle and bulls, especially if pestivirus is endemic in the herd. Testing for antibodies in older cows can help determine whether vaccination is necessary.

Considerations for Vaccination Programs

- **Timing and Age Groups**: It is important to vaccinate cattle at key points in their life cycle, such as branding, weaning, or before mating, to ensure optimal protection.

- **Consultation with Veterinarians**: Misdiagnosis and inappropriate vaccination strategies can lead to economic losses. Consulting with veterinarians can ensure accurate disease identification and effective prevention.

- **Following Manufacturer Instructions**: Always adhere to the vaccine manufacturer's instructions regarding dosage, administration, and storage to maintain vaccine efficacy.

A well-structured vaccination program, tailored to the specific health risks in your region, is a critical component of disease management in beef cattle production. By vaccinating against key diseases such as clostridial infections, leptospirosis, botulism, and tick fever, cattle producers can prevent costly outbreaks and improve herd health and productivity. Regular consultation with veterinarians and keeping abreast of the latest vaccination guidelines is essential for effective disease prevention.

Parasite Immunity to Different Classes of Chemicals

The emergence of chemical resistance, especially in livestock parasites, presents a growing concern for the long-term sustainability of animal health management. Chemical resistance refers to the ability of parasites, bacteria, or other organisms to survive treatments that previously would have controlled or killed them. This problem often arises from over-reliance on chemicals such as drenches, pour-ons, and antibiotics, without addressing other factors that contribute to parasite infestations and diseases [49].

Parasites develop resistance over time when chemical treatments are overused or misapplied. For instance, parasites exposed to frequent low-dose treatments or incomplete treatments can adapt, rendering those chemicals ineffective. In particular, drench resistance—where internal parasites in sheep, goats, and cattle become resistant to anthelmintics—is widespread, especially in southern Australia. Over time, the same chemical formulations used repeatedly will no longer be effective, resulting in the need for alternative control strategies.

Producers are encouraged to adopt an integrated parasite management (IPM) approach. This approach recognizes that parasites are not the only cause of livestock health problems; animal susceptibility, environmental conditions, and farm management practices also play significant roles. Successful parasite control programs focus on improving animal resilience through good nutrition, reducing parasite exposure via grazing management, and strategically using chemicals when needed.

Specific strategies to minimize chemical resistance include [49]:

- **Quarantining and treating introduced livestock** to prevent the introduction of resistant parasites.

- **Using combination drenches** from different chemical classes to avoid over-reliance on a single treatment.

- **Rotating chemical treatments** based on resistance patterns identified through regular **drench resistance tests** every 2-3 years.

- **Monitoring parasite loads** through tools like faecal worm egg counts (WECs), which can help determine the necessity and timing of treatments.

- **Enhancing grazing management** to reduce exposure of young livestock to contaminated pastures, thereby minimizing the risk of heavy parasite infestations.

Antimicrobial resistance (AMR) poses a serious threat to both human and animal health, as bacteria, viruses, and parasites evolve to withstand previously effective treatments. The misuse and overuse of antibiotics—in both human medicine and livestock production—are major contributors to AMR. In livestock, AMR can lead to the spread of resistant organisms through meat and other animal products, and through direct contact with animals [49].

Australia, with its extensive livestock systems, low use of antimicrobials, and strict biosecurity measures, has maintained relatively low levels of AMR in food-producing animals compared to other regions globally. However, vigilance is needed to prevent resistance from increasing. Livestock producers must use antibiotics judiciously, following all veterinary and label instructions precisely and avoiding unregistered chemicals unless prescribed by a veterinarian.

The presence of chemical residues in livestock products such as meat, milk, and wool can jeopardize access to domestic and international markets. Stringent market demands ensure that Australian red meat products are free from unacceptable levels of chemical residues. To meet these requirements, producers must follow strict guidelines on the use of veterinary chemicals, including adhering to withholding periods (WHPs), export slaughter intervals (ESIs), and other safety measures.

The key to combating chemical resistance and preventing residue contamination lies in adopting integrated disease management practices. By combining strategic use of chemicals with improved management, nutrition, and biosecurity practices, producers can protect their livestock and maintain market access. Additionally, efforts to curb antimicrobial resistance will benefit both animal welfare and public health, ensuring that antibiotics remain effective for future generations.

In the United States, managing Antimicrobial Resistance (AMR) in livestock production is a priority, with multiple strategies and programs in place to address the issue. The U.S. Department of Agriculture (USDA), alongside other federal agencies like the Centers for Disease Control and Prevention (CDC), plays a central role in overseeing these efforts.

Key management strategies include:

1. **Veterinary Oversight**: U.S. livestock producers are required to work closely with veterinarians to ensure antibiotics are used appropriately. Veterinarians prescribe antibiotics only when necessary and provide guidance on how to administer them safely and effectively. This includes determining the correct dosage, route of administration, and treatment duration to prevent misuse, which could contribute to AMR [50, 51]

2. **Judicious Use of Antibiotics**: One of the core strategies in AMR management is ensuring that antibiotics are used exactly as prescribed, avoiding overuse or misuse. Producers are encouraged to follow strict biosecurity measures, improve animal welfare (such as ensuring proper nutrition and reducing overcrowding), and use vaccines to prevent disease, thereby reducing the need for antibiotics [50].

3. **Integrated Disease Management**: In addition to antibiotic use, integrated

management approaches focus on improving general livestock health through better nutrition, pasture management, and parasite control. This reduces the need for antibiotics by minimizing disease outbreaks [50].

4. **Surveillance and Monitoring**: USDA programs like the National Antimicrobial Resistance Monitoring System (NARMS) and other initiatives focus on tracking antimicrobial use and resistance patterns in livestock. These efforts help identify emerging threats and guide future regulations [51].

5. **Research and Education**: Agencies like the USDA Agricultural Research Service (ARS) conduct research to better understand the drivers of AMR in livestock systems. This research informs educational programs aimed at producers, helping them adopt sustainable practices that curb the spread of resistant organisms [51].

Drench resistance in livestock, particularly in sheep, arises when internal parasites (worms) develop an inherited tolerance to commonly used deworming chemicals, making these treatments less effective over time. This resistance develops gradually as a result of genetic mutations within the worm population. When animals are treated with a drench, most worms are killed, but those with resistance genes survive and reproduce. Over time, as these resistant worms proliferate and reinfect other animals, their offspring make up a larger proportion of the worm population, eventually leading to a population that is largely resistant to the particular drench [52].

Drench resistance often occurs due to repeated and incorrect usage of the same class of dewormers over time. Over-reliance on chemical drenches, incorrect dosing, and frequent treatments contribute to the rapid development of resistance. Parasites with genetic mutations that help them survive exposure to a particular chemical can reproduce, creating a resistant population.

Management Strategies to Slow Drench Resistance [52]:

1. **Rotating Drench Classes**: Switching between different classes of drenches helps reduce the pressure on worms to develop resistance to any single class. Using drenches with different active ingredients ensures that no single group of parasites builds a full resistance.

2. **Targeted Selective Treatment (TST)**: TST involves selectively drenching only animals that show signs of heavy worm infestation. By treating only the

most affected animals, a portion of the worm population remains susceptible to treatment, slowing the development of resistance.

3. **Quarantine Drenching**: New animals introduced to a farm should be quarantined and treated with a combination of drenches before they join the main flock. This helps prevent the introduction of resistant parasites from other farms.

4. **Refugia**: This method involves leaving a portion of the worm population untreated, allowing susceptible worms to survive and reproduce. Refugia helps maintain a population of non-resistant worms, diluting the resistant gene pool in future generations.

5. **Pasture Management**: Strategic grazing practices, such as rotational grazing, reduce the buildup of worm larvae on pasture, thus lowering the overall worm burden in livestock. By reducing the number of larvae ingested, the need for frequent drenching is minimized.

6. **Monitoring Worm Burdens**: Regular monitoring of worm egg counts in livestock faeces helps identify when drenching is necessary, avoiding unnecessary treatments. Monitoring also allows for the early detection of resistance in worm populations, enabling more informed decisions on drench use.

Drench resistance is a significant challenge in livestock farming, particularly for sheep. To slow the development of resistance, farmers need to adopt an integrated parasite management strategy that includes a combination of chemical and non-chemical control measures. This helps preserve the effectiveness of available drenches, ensuring that they remain a useful tool in managing internal parasites.

Sheep Worm Monitoring

To avoid drench resistance and ensure effective parasite management in sheep, farmers must adopt a strategic and proactive approach. Drench resistance develops when internal parasites, such as worms, evolve to survive common deworming treatments. To manage and prevent this, an integrated plan that combines strategic chemical use with other non-chemical methods is critical.

Key Strategies for Reducing Drench Resistance:

1. **Use Effective Drenches**: It's essential to only use drenches that have proven effective on your property. Farmers should conduct regular drench resistance tests (ideally every 2-3 years) to determine the efficacy of the products they use. Worm egg count reduction tests (FECRT) can identify whether a drench is reducing the worm burden by at least 98%, which is the preferred level of effectiveness.

2. **Combination Drenches**: The use of combination drenches—products that contain two or more active ingredients targeting different types of parasites—reduces the risk of resistance. By combining drench groups, farmers can ensure that parasites are attacked from multiple angles, slowing the development of resistance. This can be achieved by using multi-active products or administering two different products sequentially.

3. **Short-Acting Treatments**: Short-acting drenches should be the default choice in most situations, as they allow for the parasite population to be exposed to the treatment for a limited time, reducing the selection pressure for resistance. Persistent products, which remain in the animal's system for extended periods, should be reserved for high-risk periods or specific purposes, such as during seasonal peaks in worm activity.

4. **Quarantine Drenching**: Any sheep introduced to the farm should be given a quarantine drench containing at least four unrelated active ingredients to prevent introducing resistant parasites. This should be followed by keeping the sheep in a yard or a small paddock for 24–48 hours to allow any resistant worms to pass out in the faeces. Afterward, they should be moved to a "wormy" paddock where they can interact with existing worm populations, promoting refugia.

5. **Refugia**: Refugia is the practice of allowing a proportion of the worm population to remain untreated, ensuring that susceptible worms survive and dilute the resistant genes in the overall population. This can be achieved by leaving some animals in a treated mob undrenched or by relying on worms in the environment (such as pasture) that haven't been exposed to treatment. This method is crucial

for slowing the spread of resistance.

Monitoring and Managing Parasites:

1. **Worm Egg Counts (FEC)**: Regular faecal egg counts are essential to assess the worm burden in the flock. This should be done at key points in the production cycle, including:

   ○ 4–6 weeks after weaning drench.

   ○ Prior to summer drenches in November or December.

   ○ Pre-lambing, and during times of suspected parasitic challenge. These tests not only help decide when to drench but also ensure that the drench used is effective.

2. **Strategic Drenching**: Drenching should be reserved for specific times, such as when the worm burden is high (evidenced by FEC results), or when environmental factors make sheep more vulnerable to parasite infections. Avoiding unnecessary drenching reduces the selection pressure for resistant worms.

3. **Pasture Management**: Proper grazing management can reduce worm exposure. Young, susceptible sheep should be grazed on lower-risk pastures, such as those grazed previously by mature sheep or other species. This lowers the chances of the animals encountering a heavy worm burden.

Identifying Animals Affected by Infection or Parasites or Requiring Treatment under the Animal Health Plan

Parasites are a major concern for livestock production worldwide, as they can significantly impact animal health, welfare, and productivity. Livestock such as cattle, sheep, and goats are affected by both internal parasites (e.g., worms, flukes, protozoa) and external parasites (e.g., flies, ticks, lice). These infestations can lead to economic losses and a range of health issues for animals. In regions such as Australia, parasitic diseases have been identified as some of the most costly diseases affecting farm productivity. For example, a study by Meat & Livestock Australia (MLA) identified that parasitic diseases like worms, flystrike, and lice in sheep, along with cattle ticks and buffalo flies, are among the top causes of financial loss in the livestock sector [53].

Impact on Animal Health and Welfare

Parasites can have severe effects on the health and welfare of livestock [53]:

- **Blood Loss**: Some parasites, like certain types of worms and ticks, cause blood loss, leading to anaemia and, in severe cases, death.

- **Diarrhea**: Gastrointestinal parasites often cause diarrhea, which can lead to dehydration and death if untreated.

- **Debilitated Animals**: Infestations can reduce an animal's appetite, weakening them and making them more vulnerable to other diseases.

- **Disease Vectors**: Some parasites, such as ticks and flies, act as vectors, transmitting diseases from one animal to another.

- **Irritation and Stress**: External parasites, especially flies, cause irritation, leading to reduced grazing and stress, which further impacts productivity.

Impact on Livestock Productivity

The negative effects of parasitic infestations include [53]:

- **Reduced Growth Rates**: Infested animals often experience reduced weight gain due to diminished appetite and energy.

- **Lower Reproductive Rates**: Health-compromised animals are less likely to conceive or carry pregnancies to term.

- **Carcass and Product Condemnation**: Severe infestations can result in carcasses or parts being condemned during slaughter, reducing the financial return for producers.

- **Milk and Wool Production**: Parasites can lead to a reduction in milk yield and a decline in fleece quality, including reduced fibre diameter and staple strength, directly affecting the wool industry.

Parasites of Livestock and their Global Significance

Several parasites impact livestock globally:

- **Gastrointestinal Worms**: Affect cattle, sheep, and goats, causing loss of condition and digestive issues.

- **Coccidiosis**: A protozoal parasite prevalent in young animals, often leading to bloody diarrhea and anaemia.

- **Ticks**: These external parasites not only cause blood loss but are also vectors for diseases like babesiosis and anaplasmosis in cattle.

- **Flies**: Flies such as the buffalo fly in Australia or the screwworm fly (found in other regions) cause irritation and can lead to wounds or flystrike, which require prompt treatment.

- **Liver Fluke**: A global parasite, especially in regions with wet conditions, impacting the liver and reducing the animal's productivity.

Management Strategies

Parasite management in livestock includes three essential components: **identification**, **control**, and **resistance management**. A robust strategy must focus on:

1. **Identification**: Regular monitoring, including faecal egg counts and physical inspections for external parasites, is critical to early detection.

2. **Control**: Various chemical and non-chemical methods are used. However, the overuse of chemicals can lead to **chemical resistance** in parasites, making them harder to control.

3. **Integrated Parasite Management**: Producers are encouraged to combine chemical treatments with pasture and animal management strategies. This might involve rotating grazing pastures, strategic drenching (deworming), and ensuring good nutrition to reduce the animals' susceptibility to parasitic infestations.

Chemical Resistance and Residues

The overuse or misuse of chemicals to treat parasites can lead to drench resistance (in the case of worms) and the development of antimicrobial resistance in some cases. In the long term, this can reduce the effectiveness of treatments, requiring stronger and more costly interventions. Monitoring for resistance and rotating treatments is crucial to slow the onset of resistance. Additionally, ensuring that chemical residues from treatments are within acceptable levels is vital for maintaining market access, as consumers and regulators demand that meat and dairy products are free from harmful residues.

Special Considerations: Floods and Parasitic Surges

Flood events, which are becoming more frequent in many parts of the world due to climate change, can lead to a surge in parasitic populations. Standing water provides breeding grounds for biting insects like mosquitoes and ticks, leading to higher infestation rates. During such times, producers must be particularly vigilant and employ timely intervention strategies to prevent outbreaks that can severely impact livestock productivity.

Flies and Their Impact on Livestock

Flies are a significant issue in livestock management globally, causing discomfort to animals, reducing productivity, and sometimes spreading diseases. In Australia, the flies of most concern to livestock producers are buffalo flies, nuisance flies in feedlots, and sheep blowflies. Each of these flies requires specific control measures to mitigate their impact on livestock health and welfare.

Figure 46: Flies Perched on Cows Faces. Maurice Engelen, Public Domain, via Pexels.

Buffalo Flies

Buffalo flies (*Haematobia irritans exigua*) are primarily a problem in northern Australia, where the hot and humid climate creates ideal breeding conditions. These flies are a serious pest for cattle, as they feed on blood, causing skin irritation and stress. Affected cattle will often rub against fences or trees to relieve the irritation, which disrupts grazing and leads to damaged hides. Some cattle, particularly dark-coated cattle, bulls, and older animals, are more prone to attracting buffalo flies. Cattle that are allergic to the bites may also develop skin ulcers.

Effective control of buffalo flies relies on an **integrated approach**:

- **Assessing the Need for Treatment**: Not all cattle need to be treated for buffalo flies, so it's important to monitor fly populations before administering treatments.

- **Culling Allergic Animals**: Sensitive or allergic cattle may require culling to reduce overall susceptibility in the herd.

- **Buffalo Fly Traps**: These can be used to physically capture and reduce the fly population.

- **Dung Beetles**: By breaking down cattle dung, dung beetles help reduce fly breeding grounds.

- **Chemical Control**: Ear tags, sprays, pour-ons, back rubbers, and dust bags are commonly used, but should be rotated to prevent resistance.

Nuisance Flies in Feedlots

In feedlots, flies are a persistent issue due to the large concentrations of animals and manure, which provide ideal breeding conditions. The four most common types of nuisance flies in Australian feedlots include the house fly (*Musca domestica*), bush fly (*Musca vetustissima*), stable fly (*Stomoxys calcitrans*), and various blowflies (*Chrysomyia spp.*, *Calliphora spp.*, and *Lucilia spp.*). These flies can cause annoyance, stress, and, in some cases, transmit diseases among livestock.

Managing nuisance flies in feedlots involves several strategies:

- **Reduce Breeding Sites**: Ensuring proper manure management and drainage can reduce the areas where flies lay eggs.

- **Selective Use of Insecticides**: Rather than blanket spraying, insecticides should be used selectively to prevent resistance and reduce chemical usage.

- **Feedlot Design**: Feedlots should be designed to minimize fly breeding, with good drainage and manure management systems in place.

- **Biological Control Agents**: Enhancing populations of natural predators, such as parasitic wasps, can help keep fly populations under control.

- **Systematic Monitoring**: Regular monitoring of fly populations allows producers to take timely action before infestations become severe.

Sheep Blowflies

Blowflies, particularly the Australian sheep blowfly (*Lucilia cuprina*), are a major cause of flystrike, a condition in which flies lay eggs in the wool or skin of sheep, leading to maggot infestations. This is particularly problematic in warm, moist conditions, where flies thrive. Flystrike not only causes pain and suffering in sheep but can also lead to significant production losses due to decreased wool quality and weight loss.

The control and prevention of flystrike involve a combination of management techniques:

- **Genetic Selection**: Breeding sheep with plainer breeches (fewer folds and wrinkles in the skin) reduces the risk of flystrike.

- **Chemical Control**: Strategic use of insecticides and dips during high-risk periods can protect sheep from blowflies.

- **Shearing and Crutching**: Regular shearing and crutching (removing wool around the breech) can help reduce the likelihood of flystrike by keeping susceptible areas clean and dry.

- **Monitoring Fly Activity**: Understanding the lifecycle of blowflies and monitoring weather conditions can help predict high-risk periods for flystrike and enable timely interventions.

While these examples focus on Australia, the management of flies in livestock is a global challenge. In tropical and subtropical regions, flies can spread vector-borne diseases such as trypanosomiasis (transmitted by tsetse flies in Africa) and leishmaniasis (spread by sandflies). In Europe and North America, stable flies and house flies are more common, causing irritation and transmitting diseases like pinkeye. Integrated pest management, combining biological control, chemical treatments, and environmental management, is key to reducing the impact of flies on livestock worldwide.

Gastrointestinal Worms in Livestock

Gastrointestinal worms are a significant health challenge for cattle, sheep, and goats in Australia and around the world. The presence of these parasites can result in substantial economic losses due to reduced growth rates, poor animal condition, and even death in severe cases. The types of gastrointestinal worms and their impact can vary between regions, influenced by climate, livestock management practices, and the immune status of the animals.

Several environmental and management conditions increase the likelihood of gastrointestinal worm infections. High rainfall regions (more than 500–600 mm annually) and areas with irrigation are particularly prone to infestations due to the moist conditions that favour worm larval survival on pastures. High stocking rates or set-stock grazing situations also lead to increased pasture contamination, as does the grazing of short pastures where livestock are more likely to ingest worm larvae. Young animals, especially those recently weaned, are at higher risk due to their developing immune systems. Additionally, bulls, rising 2-3-year-old cows, and ewes or does around lambing or kidding are more susceptible to worms due to lower immunity, often exacerbated by stress.

In Australia, Barber's pole worm (*Haemonchus contortus*) is common in hot, humid regions, particularly in the summer rainfall areas, while black scour worm (*Trichostrongylus spp.*) and brown stomach worm (*Ostertagia spp.*) are more problematic in cooler, winter rainfall regions. Identifying which worms are common in a region is crucial for targeted worm control strategies.

Gastrointestinal worm infections can often be identified through a combination of signs and diagnostic tools. Poor performance on pasture, despite adequate nutrition, may indicate a parasitic burden. Common clinical signs include scouring, weight loss, pale gums, and bottle jaw (fluid swelling under the jaw). In severe cases, worms may also be visible in the faeces of infected animals. Worm egg counts from faecal samples (faecal egg count, or FEC) and larval cultures are common diagnostic tools used to confirm the presence of worms. For certain worms like type 2 ostertagiasis in cattle, a blood test measuring serum pepsinogen levels can help detect infections.

The best approach to controlling gastrointestinal worms is through an integrated parasite management strategy. This includes measures beyond drenching (deworming), as over-reliance on chemical treatments can lead to the development of drench resistance. Preventing worm infections requires:

- **Understanding regional worm risks**: Knowing which types of worms are prevalent on a property and the seasons in which they pose the highest threat.

- **Monitoring**: Regularly testing livestock for worm burdens, particularly during high-risk seasons or in vulnerable groups such as young or lactating animals.

- **Quarantine drenching**: Any new animals introduced to the property should be treated with a quarantine drench to prevent the introduction of new or resistant worm strains.

- **Improved nutrition**: Adequate nutrition strengthens the immune response, helping animals combat worm infections.

- **Grazing management**: Young animals should be placed on pastures with lower worm contamination levels, often achieved by rotational grazing or grazing mature animals on worm-contaminated pastures first to reduce larvae levels.

- **Strategic drenching**: While drenching remains a key tool, it should be used judiciously. Knowing the drench resistance status of worms through regular resistance testing ensures that only effective drenches are used. Using combination drenches (containing more than one active ingredient) can also slow the development of resistance.

Drench resistance is a growing concern globally, especially in regions with intensive livestock production. Resistance occurs when worms survive a previously effective treatment, and their resistant offspring become the dominant population over time. To combat this, producers should avoid unnecessary drenching and rotate drench groups to prevent resistance buildup. Conducting regular resistance tests (every 2-3 years) helps determine which drenches are still effective on a property.

Using refugia—the practice of leaving a portion of the worm population untreated—helps slow the development of resistance. By maintaining a population of susceptible worms in the environment, refugia reduces the selective pressure for resistance genes to dominate. Additionally, improving grazing management, selecting animals for worm resistance, and focusing on better nutrition can all contribute to more sustainable parasite control.

Managing gastrointestinal worms in livestock requires a comprehensive approach that integrates chemical treatments with better management practices, regular monitoring, and improved animal nutrition. By taking a proactive and informed approach to worm control, livestock producers can minimize the impact of parasitic infections on animal health and productivity, while also mitigating the risks of drench resistance. Global efforts to manage parasitic infections are critical to ensuring the sustainability of livestock production, protecting both animal welfare and farm profitability.

Lice Infestations

Lice are a common external parasite that affects various livestock species, including cattle, sheep, and goats. These pests can cause irritation, reduced productivity, and welfare concerns in affected animals. Two main types of lice infest livestock: biting lice and sucking lice.

Biting Lice

Biting lice primarily feed on skin debris, including dead skin cells, scurf, and bacteria found on the animal's skin surface. They are commonly found in livestock with longer coats during the colder months, as the thicker hair provides an ideal environment for lice proliferation.

Sucking Lice

Sucking lice, in contrast, feed by penetrating the animal's skin with their mouthparts and drawing blood from capillaries or the exuded serum. This type of lice can cause more severe health problems due to blood loss and irritation, potentially leading to anaemia in heavily infested animals.

Host Specificity

Lice are generally **host-specific**, meaning the species of lice found on cattle are different from those found on sheep or goats. For instance, sheep are mainly infested with **biting lice**, whereas cattle and goats can harbor both **chewing and sucking lice**. This specificity emphasizes the importance of tailored lice management strategies for each livestock species.

Conditions Favouring Infestations

Lice infestations are more likely to occur under the following conditions:

- **Winter to Early Spring**: The colder months often see the most severe infestations, as lice thrive in the dense coats animals grow during this time.

- **Poor Condition of Animals**: Animals in suboptimal health or malnourished are more prone to lice infestations, likely due to their weakened immune responses.

- **Long Hair or Wool**: Animals with longer hair or fleece, especially during the colder months, provide a more suitable environment for lice to multiply.

Lice infestations are often first suspected when animals begin exhibiting signs of discomfort, such as rubbing, scratching, or biting at themselves. This behaviour can lead to secondary issues like skin infections, scaly skin, raw patches, hair loss, or fleece derangement in sheep. Lice can be detected by parting the hair or fleece, where they can be seen moving away from the light.

In sheep, a lice detection test can be performed by testing wool grease in shearing cutters to confirm the presence of lice. Early detection is key in preventing the spread of lice within a herd or flock.

A robust management plan can prevent lice infestations and their negative impacts. Key strategies include:

- **Source Lice-Free Animals**: Purchase animals only from lice-free herds or flocks and thoroughly inspect them before introduction. Request an **animal health statement** from the vendor to ensure transparency about the disease status of the livestock.

- **Secure Fencing**: Maintain secure boundary fences to prevent the introduction of lice from stray or neighbouring animals.

- **Post-Shearing Treatment**: Treat sheep after shearing to eliminate lice before they can spread to other animals. It's important to only introduce lice-free animals into existing flocks.

- **Shearing Equipment Hygiene**: Shearers should sterilize their equipment, including microwaving moccasins or other footwear, to prevent lice from being transmitted between flocks.

- **Personal Hygiene**: After handling infested animals, handlers should change clothing before interacting with uninfested animals.

- **Animal Condition**: Ensure that cattle, sheep, and goats are maintained in good condition during challenging seasons, as animals in poor health are more vulnerable to lice infestations.

- **Exclusion from Contaminated Facilities**: If animals are treated for lice with a non-residual chemical, consider a 2-3 week exclusion period from contaminated facilities to prevent re-infestation.

Liver Fluke in Livestock

Liver fluke (Fasciola hepatica) is a parasitic flatworm that infects cattle, sheep, goats, and several other animals, including humans. It can significantly impact livestock productivity and the economy due to health issues and the condemnation of livers in slaughterhouses. Understanding its lifecycle, risk factors, and prevention strategies is crucial for managing its effects on farm operations.

The liver fluke has two primary life phases:

1. **On pasture and in snails**: The fluke eggs are released from infected animals via faeces and hatch in water. They develop into larvae (miracidia), which infect specific freshwater snails (e.g., *Lymnea tomentosa*). Inside the snail, the larvae multiply and are released as free-swimming cercariae.

2. **In animals**: The cercariae encyst on vegetation as metacercariae and are ingested by livestock while grazing. Once inside the host, the immature flukes migrate to the liver, where they mature into adults and begin to damage the liver tissue.

Liver fluke infections can reduce animal productivity, causing weight loss, poor growth rates, and reduced milk production. Infected livers are often condemned at slaughter, leading to significant economic losses for the meat industry. Severe infestations can also result in anaemia, scouring, and bottle jaw (swelling beneath the jaw due to fluid accumulation). Chronic infections, especially in cattle, may lead to long-term ill-thrift.

Liver fluke thrives in specific environmental conditions:

- **Habitat suitable for snails**: Liver fluke requires snails for part of its lifecycle. Thus, areas with **marshy, swampy grounds, watercourses, springs, or irrigated pastures** are high-risk areas for infections.

- **Regions of high prevalence**: In Australia, liver fluke is most prevalent in **south-eastern areas**, including the tablelands, coastal regions, and irrigation zones of New South Wales and Victoria.

- **Properties with a history of liver fluke**: Farms that have dealt with liver fluke in the past are at higher risk of recurrence, especially if appropriate preventive measures aren't implemented.

Identifying and Diagnosing Liver Fluke

Liver fluke infections can manifest as either **acute or chronic disease**:

- **Acute disease**: Most common in sheep, acute liver fluke infection typically occurs from **late summer to autumn**. It is characterized by severe liver damage due to the migration of immature flukes through the liver.

- **Chronic disease**: This form of the disease is more common in cattle, and while it can occur year-round, it is most prevalent from **autumn to spring**. Chronic liver fluke infection results from the accumulation of adult flukes in the bile ducts.

Clinical signs of liver fluke include:

- **Bottle jaw** (oedema under the jaw).

- **Pale gums and membranes around the eyes**, indicative of anaemia.

- **Scouring (diarrhea)**.

- **Weight loss and general ill-thrift**, leading to decreased production and poor body condition.

To confirm the presence of liver fluke, diagnostics may involve:

- **Faecal egg counts** to detect fluke eggs in animal waste.

- **Blood tests** to measure specific liver enzymes indicative of fluke damage.

- **Liver condemnation reports** from abattoirs, which detail the extent of liver damage in slaughtered animals.

Strategies to Prevent Liver Fluke

Prevention and control of liver fluke are essential for maintaining livestock health and minimizing economic losses. Key strategies include:

1. **Quarantine and Drenching**: Always quarantine and treat animals coming from fluke-endemic areas. This helps prevent the introduction of liver fluke onto your property. Drenching with effective flukicides is essential to clear liver fluke from the animals before they are integrated into your herd or flock.

2. **Monitor Liver Fluke Status**: Regularly monitor the liver fluke status of your livestock through faecal egg counts, blood tests, and abattoir reports. This can help detect fluke infections early and prevent widespread issues.

3. **Avoid 'Flukey' Areas**: Where possible, fence off marshy or swampy areas, watercourses, or irrigation zones to limit the exposure of livestock to liver fluke habitats.

4. **Use Strategic Drenching**: **Fluke drenches** are critical in managing liver fluke populations within livestock. Depending on the season and the risk level, strategic drenching can help reduce the fluke burden and improve animal health.

5. **Animal Health Statements**: Always request an animal health statement when purchasing new stock. This helps ensure transparency regarding the disease status of the animals, reducing the risk of introducing fluke-infected animals into your herd or flock.

Ticks

Ticks are a significant concern for livestock producers around the world due to their ability to transmit diseases, cause economic losses, and affect the health and welfare of animals. In Australia, the cattle tick (*Rhipicephalus microplus*) is the most economically important tick species, particularly in the northern and coastal regions. The presence of ticks can lead to direct production losses and incur high costs for tick control measures.

The cattle tick primarily affects cattle but can also infest sheep, goats, horses, and other livestock. European cattle breeds, in particular, are highly susceptible. Cattle ticks are notorious for spreading diseases like tick fever (also known as "red water"), which is caused by blood parasites like *Anaplasma marginale*, *Babesia bovis*, and *Babesia bigemina*. Tick fever can lead to anemia, loss of condition, and even death, especially in susceptible animals.

A study by Meat and Livestock Australia (MLA) estimates that the annual cost of ticks to the Australian cattle industry is about $146 million. This includes both the production losses from affected cattle and the costs of tick control measures. Additionally, there are costs associated with maintaining the "tick line", a regulatory boundary that controls the spread of ticks between New South Wales and Queensland as well as within Queensland. This "tick line" involves inspection points and management efforts to minimize infestations in regions where ticks are not endemic.

Cattle tick infestations are most common in the tropical and subtropical regions of northern Australia, including Queensland, the Northern Territory, and parts of Western Australia. Ticks thrive in warm, humid conditions, with infestations peaking in late spring and summer. While cattle ticks primarily occur within these endemic regions, sporadic infestations can appear in non-endemic areas, prompting stringent control measures when detected.

Producers can identify cattle tick infestations through the following clinical signs:

- **Engorged female ticks** (pea- to blueberry-sized) are visible on the neck, brisket, flanks, and hind legs of affected animals.

- **Licking and rubbing** at tick bites, causing discomfort known as tick worry.

- **Tick sores and ulceration** at bite sites.

- **Anaemia**, evident through pale gums and membranes around the eyes.

- **Lack of energy, weight loss, and death** in severe cases of infestation or secondary tick fever infections.

Effective control of cattle ticks involves an integrated approach that incorporates resistant cattle breeds, chemical treatments, vaccinations, and management of pastures. Here are some key strategies:

- **Use of Resistant Cattle Breeds**: Certain cattle breeds exhibit greater resistance to ticks. In Australia, Brahman and Brahman-cross cattle are known to be more resistant to ticks compared to European breeds like Herefords and Angus.

- **Chemical Treatments**:

  - Acaricides (tick-killing chemicals) are commonly used to control cattle tick populations. These chemicals can be applied as sprays, pour-ons, backline treatments, or in cattle dips.

  - Producers must carefully follow label instructions to prevent chemical resistance from developing in tick populations. Rotating chemical groups is also recommended to minimize resistance.

- **Vaccination**: There is a tick fever vaccine available, which can protect cattle against the diseases spread by ticks, especially in areas where tick fever is common. This vaccine helps prevent the development of tick fever but does not directly kill the ticks themselves.

- **Pasture Management**: Rotational grazing and the use of "tick-safe" pastures

(pastures with lower tick populations) can help reduce tick infestations. Ticks tend to survive in moist, humid environments, so pastures that are dry or well-drained can lower the risk of infestation.

Ticks, especially cattle ticks, represent a significant challenge for livestock producers in Australia and other parts of the world. Their ability to transmit diseases like tick fever and cause direct harm to animals makes them a priority for control efforts. Through a combination of resistant cattle breeds, chemical control, vaccination, and pasture management, producers can minimize the negative impacts of ticks on livestock health and productivity. Maintaining awareness of regional tick infestations and following best practices for tick control are essential for managing this costly pest effectively.

Trichomoniasis in Cattle

Trichomoniasis is a contagious venereal disease that significantly impacts reproductive efficiency in cattle herds. It is caused by the protozoal parasite Tritrichomonas foetus, which primarily affects the reproductive tract of cows and heifers, leading to early embryonic losses, vaginal discharges, and abortions. This disease is especially problematic in natural breeding herds and is more prevalent in regions where controlled breeding practices are not rigorously applied.

The risk of trichomoniasis increases when there is movement of infected bulls or cows between herds. This movement can occur during bull sales, herd dispersals, or when cattle are co-grazed with neighbouring herds. Bulls are the main vectors of this disease, as they can carry and spread the infection without showing any clinical signs. Older bulls, in particular, are more likely to harbor the parasite in their preputial folds and pass it to cows during mating. Once a cow is infected, the protozoan colonizes her reproductive tract, leading to embryonic death or early-term abortion, often unnoticed by the producer.

Diagnosis of trichomoniasis can be challenging because the disease's primary clinical signs—early-term abortion and low calving rates—are not specific and can be confused with other reproductive issues. However, the following clinical signs may lead producers to suspect trichomoniasis:

- **Early embryonic losses or abortions**, typically within the first 60 to 90 days of gestation.

- **Vaginal discharge** in infected cows.

- **Extended calving intervals** or a significant reduction in calving rates, as cows may take longer to conceive after initial pregnancy losses.

Definitive diagnosis requires veterinary intervention. Diagnostic procedures include the collection of samples from the reproductive tract of bulls and cows, such as preputial fluid from bulls or vaginal mucus from cows, followed by laboratory culture or PCR testing to detect the presence of *Tritrichomonas foetus*. Testing bulls is particularly crucial, as they are often asymptomatic carriers of the disease.

Preventing trichomoniasis relies on a combination of biosecurity measures, controlled mating practices, and herd management. Key strategies include:

1. **Culling Infected Bulls**: Bulls that test positive for *Tritrichomonas foetus* should be culled from the herd. Older bulls, especially those used for natural breeding, are at higher risk of harbouring the parasite and transmitting it to cows during mating. It is advisable to replace these bulls with younger ones, as younger bulls are less likely to carry the disease.

2. **Reducing the Age of Bulls and Improving Bull Control**: Implementing a system where younger bulls are used in the breeding herd can help reduce the incidence of trichomoniasis. This is because older bulls are more likely to carry the protozoan in their reproductive organs.

3. **Controlled Mating**: Restricting mating to a shorter breeding season (about 3 to 6 months) helps manage reproductive health more effectively. By limiting the breeding period, producers can monitor pregnancies closely and identify cows that fail to conceive or abort early. Controlled mating also allows for **sexual rest** in cows, facilitating better disease control and recovery.

4. **Segregating Heifers and Mating with Young Bulls**: Keeping **heifers separated** from the main herd and mating them with **young, disease-free bulls** can reduce the risk of spreading the infection. Heifers are especially vulnerable to reproductive diseases like trichomoniasis, so it's essential to manage their breeding carefully to minimize exposure to the parasite.

5. **Biosecurity and Testing**: Preventing the introduction of infected animals into a herd is critical. New bulls or cows entering the herd should be quarantined and tested for trichomoniasis before being allowed to breed. Herds should also maintain proper fencing to avoid accidental contact with neighbouring herds, which could be a source of infection.

Trichomoniasis remains a significant reproductive disease in cattle, leading to economic losses through reduced calving rates, abortion, and extended breeding seasons. Prevention and control of the disease rely heavily on proactive herd management practices such as regular testing, the use of younger bulls, culling of infected animals, and strict biosecurity measures. In regions where the disease is prevalent, these measures are crucial to maintaining reproductive efficiency and minimizing the impact on cattle production.

Diagnosing Disease and Deficiencies

Diseases and parasites can severely impact livestock production, sometimes leading to death if not properly managed. Livestock owners need to be vigilant, not only about common diseases in their region but also about exotic diseases that may pose future risks. Monitoring the health of livestock by recognizing **key symptoms**, understanding normal cattle functions, and taking prompt action when issues arise is crucial for maintaining animal health and productivity.

The ability to diagnose diseases in livestock requires professional expertise because many diseases exhibit similar symptoms. **Veterinary intervention** is essential for accurate diagnosis and treatment. The table provided earlier outlines key symptoms and potential causes, ranging from sudden death (which can be caused by **clostridial diseases**, botulism, or poisoning) to reproductive problems (such as those caused by **leptospirosis**, **trichomoniasis**, or **pestivirus**). Other symptoms like **diarrhea**, **fever**, and **scours** can also be attributed to specific parasites, deficiencies, or diseases.

One of the most critical factors in managing livestock health is **adequate nutrition**. Malnourished animals are more susceptible to diseases and parasites due to weakened immune systems. **Inadequate nutrition** could be caused by:

- Insufficient feed

- A lack of specific nutrients

- Nutritional imbalances

Providing animals with balanced diets can prevent many health issues and improve their capacity to resist infections and parasites.

External parasites such as **ticks, buffalo flies**, and **lice** are easy to detect by the experienced farmer. These pests cause irritation, reduce productivity, and in severe cases, lead to skin damage or secondary infections. Internal parasites like worms, however, require **faecal testing** to identify the types of worms and the level of infestation. **Gastrointestinal worms**, in particular, are a significant cause of reduced growth rates, poor condition, and scouring, especially in younger livestock or animals grazing in high-risk environments.

A successful disease prevention program includes:

- **Regular monitoring of livestock health**, particularly through **faecal testing** to detect internal parasites.

- **Strategic drenching** to control worm burdens, while also taking care to avoid the development of **drench resistance**.

- Vaccination programs to protect against diseases like **blackleg, tetanus, pestivirus**, and **leptospirosis**.

- Improved **biosecurity practices** such as quarantine and screening of new stock before introduction to the herd.

The following provides a summary that focuses on diagnosing various cattle diseases, highlighting the key diagnostic approaches for each in a global context:

Black Disease (Clostridial Disease)

Diagnosis: Black disease is linked to liver fluke infestations and is typically identified after sudden death in affected cattle. It can be diagnosed through post-mortem examinations, which often reveal severe liver damage associated with *Clostridium novyi* infection. In regions where liver fluke is prevalent, such as areas with marshy pastures, veterinarians look for liver damage during necropsies to confirm black disease. Blood and tissue samples may also be analysed for the presence of clostridial bacteria.

Bovine Johne's Disease (BJD)

Diagnosis: Diagnosing Johne's disease is challenging, especially in the early stages, as it may remain asymptomatic for years. Diagnosis relies on a combination of blood tests, faecal cultures, and PCR (Polymerase Chain Reaction) tests to detect *Mycobacterium avium subsp. paratuberculosis*. Infected animals often show chronic weight loss and diarrhea. In countries where BJD is prevalent, surveillance programs using antibody testing and herd-level faecal examinations help identify infected herds.

Bloat

Diagnosis: Bloat is generally diagnosed based on clinical signs such as distension of the left upper flank, discomfort, and restlessness in affected cattle. In more severe cases, cattle may be found dead with evidence of frothy bloat in the rumen. Diagnosis is typically clinical, based on history and symptoms observed in animals grazing on high-risk, legume-dominant pastures. Globally, veterinarians and farmers monitor grazing behaviour and dietary intake to prevent the disease, using pasture management strategies to control the risk.

Grass Tetany (Hypomagnesemia)

Diagnosis: Diagnosing grass tetany involves recognizing signs of magnesium deficiency, such as nervousness, muscle tremors, and eventual collapse. Blood tests measuring magnesium levels are used to confirm the condition. Globally, veterinarians diagnose grass tetany through serum magnesium concentrations and clinical signs in cattle grazing on pastures with low magnesium and calcium. For herds grazing lush, rapidly growing pasture in autumn and winter, magnesium supplementation is often implemented as a preventive measure.

Gastrointestinal Parasites

Diagnosis: Diagnosing gastrointestinal worm infestations involves faecal egg counts (FECs) to identify worm burdens in cattle, sheep, and goats. This method helps detect worm eggs and larvae, such as *Ostertagia*, *Cooperia*, and *Haemonchus*, which are commonly found in livestock worldwide. Veterinary professionals use larval differentiation tests and FEC reduction trials to assess the effectiveness of deworming treatments. In high rainfall areas where parasitic infestations are common, FECs are regularly used to monitor worm populations and guide treatment strategies.

Trichomoniasis (Reproductive Disease)

Diagnosis: Trichomoniasis is diagnosed through laboratory testing of preputial scrapings from bulls or vaginal discharges from cows. These samples are examined for the presence of *Tritrichomonas foetus* parasites. In regions where this venereal disease is a concern, diagnostic screening of breeding bulls is a routine practice to prevent its spread. Producers globally work closely with veterinarians to conduct testing, especially in herds with unexplained reproductive failures, low calving rates, or frequent abortions.

Ketosis

Diagnosis: Ketosis is typically diagnosed through blood tests measuring ketone levels or by testing milk or urine for ketone bodies. Cows with ketosis often exhibit weight loss, lethargy, and a sweet-smelling breath due to the presence of acetone. Veterinarians worldwide diagnose the condition by testing for elevated levels of ketones, particularly in late-pregnancy cows. Early detection is crucial to prevent metabolic collapse, and nutritional management practices are commonly implemented in high-risk herds.

Pinkeye (Infectious Bovine Keratoconjunctivitis, IBK)

Diagnosis: Pinkeye is diagnosed based on the clinical signs of eye irritation, such as excessive tearing, squinting, and corneal ulceration. Bacterial cultures or PCR tests may be performed to confirm the presence of *Moraxella bovis*, the most common cause of pinkeye. In environments where dust and flies are common, veterinarians diagnose pinkeye by examining the eye for characteristic lesions and inflammation, particularly in young cattle kept in close quarters.

Leptospirosis

Diagnosis: Leptospirosis is diagnosed through blood tests or PCR assays that detect antibodies to *Leptospira* species. In regions where leptospirosis is endemic, veterinarians monitor for reproductive issues such as abortions, stillbirths, and weak calves. Diagnostic blood tests and urinalysis are crucial for confirming the disease and preventing its spread to humans. Worldwide, leptospirosis is managed through vaccination and proper sanitation practices to reduce environmental contamination from infected animals.

Selenium Deficiency

Diagnosis: Diagnosing selenium deficiency involves blood tests measuring selenium levels or the activity of the enzyme glutathione peroxidase. Selenium-deficient cattle often exhibit symptoms like poor growth, muscle weakness, and reproductive issues. In regions where selenium levels are low in the soil, veterinarians recommend testing young, growing cattle to ensure adequate selenium levels, especially in areas with a history of deficiency.

Pulpy Kidney (Clostridial Disease)

Diagnosis: Pulpy kidney is diagnosed based on clinical signs of sudden death in young cattle, particularly those on rich diets. Post-mortem examination reveals a soft, pulpy kidney, which is characteristic of this disease. Vaccination is the most effective preventive measure, and veterinarians worldwide rely on vaccination schedules to protect cattle from this and other clostridial diseases.

Each disease discussed requires specific diagnostic methods, often involving clinical signs, laboratory tests, and post-mortem examinations. Globally, veterinarians and livestock producers utilize these diagnostic approaches to manage herd health, prevent economic losses, and ensure food safety.

Severity of Infection

Determining the severity of infection in livestock involves employing approved diagnostic methods and often requires expert veterinary advice. Livestock producers worldwide should prioritize accurate diagnosis to manage diseases effectively and avoid further spread or economic losses.

Veterinary advice can be provided remotely via phone, fax, or email, and it is often critical when formal inspections are completed or when an animal becomes injured after a veterinarian has departed. In cases of severe infection or injury, veterinarians can assess whether euthanasia or transportation is the best option. Accurate records of veterinary consultations, including the information shared and advice given, are crucial for both the veterinarian and the person in charge of the animals.

To manage sick or injured animals, it's essential to house them in a calm, group environment to reduce stress, as livestock separated from their herd can experience distress. Livestock handlers should observe and sample animals effectively to monitor their health. In cases where the disease is suspected, specific testing, such as faecal tests for gastrointestinal worms or blood tests for mineral deficiencies, should be conducted under veterinary guidance.

Common Livestock Health Issues:

1. **Pinkeye**: Handle carefully to prevent eye damage, and treat with antibiotics like procaine penicillin or oxytetracycline. Severe cases might require additional procedures, such as suturing eyelids to prevent corneal rupture.

2. **Lameness**: For conditions like footrot or solar abscesses, parenteral antibiotics and early aggressive treatment can often resolve the issue. Proper hoof care and providing soft, dry ground for affected animals aid recovery.

3. **Nervous Diseases**: Conditions such as transit tetany or polioencephalomalacia require prompt veterinary intervention with treatments like calcium borogluconate injections to manage electrolyte imbalances.

4. **Respiratory Diseases**: Heat stress and pneumonia can be managed with hydration, antibiotics, and anti-inflammatory drugs. Ensuring animals are housed in well-ventilated, shaded areas helps prevent respiratory distress.

5. **Skin and Abscesses**: Conditions such as ringworm and abscesses require local treatment. Abscesses may need surgical intervention for drainage, while ringworm can be managed with topical antifungal treatments. Timely intervention

is key to preventing infection from spreading or worsening.

6. **Lumpy Jaw and Wooden Tongue**: These bacterial infections, common in cattle, can be difficult to treat but may respond to long-term antibiotic treatments if caught early. For severe or chronic cases, euthanasia or slaughter may be necessary to avoid further suffering.

Determining the Type and Severity of Infestation, where applicable, through Faecal Egg Counts or Other Tests

To determine the type and severity of internal parasitic infestations in livestock, the primary diagnostic tool used worldwide is the faecal egg count (FEC), also known as the worm egg count (WEC). This method involves analysing the faeces of infected animals to assess the worm burden by counting the eggs shed by parasitic worms. It helps identify the type and extent of the infestation and guides appropriate treatment.

Types of Internal Parasites:

Internal parasites can be categorized into several major groups, each with unique characteristics and diagnostic needs:

1. **Nematodes (Roundworms)**: Common across various animal species, nematode infections are often diagnosed through FEC. In livestock like sheep and cattle, species such as the barber's pole worm, black scour worm, and brown stomach worm are prevalent. In poultry, large roundworms such as *Ascaridia galli* are commonly found, and faecal testing is an effective way to monitor their presence.

2. **Trematodes (Flukes)**: Liver fluke (*Fasciola hepatica*) is the most significant fluke in livestock, especially in areas with wet conditions conducive to the fluke's snail intermediate host. Faecal sedimentation rather than flotation is required for diagnosis, as fluke eggs are dense and don't float in typical solutions used for other parasites.

3. **Cestodes (Tapeworms)**: Tapeworm infestations in ruminants, such as *Moniezia* in sheep, can be detected via faecal egg counts. Although they are less commonly pathogenic, knowing their presence helps manage overall livestock health.

4. **Protozoans**: *Coccidia* are common protozoan parasites affecting the intestines

of livestock, and faecal samples are essential for identifying species-specific in-festations. They are particularly a concern in young animals.

Diagnostic Methods:

FEC is the most widely used method to determine parasitic load. It involves mixing a sample of faeces with a saturated salt solution, which causes the lighter parasite eggs to float. These eggs are then counted under a microscope. By calculating the eggs per gram (epg) of faeces, veterinarians and livestock producers can estimate the worm burden.

- **Direct Life Cycles**: Many nematodes have a direct life cycle, where the parasite's eggs hatch in the environment, and larvae are ingested by the host. These cycles are straightforward and can be easily monitored through FEC.

- **Indirect Life Cycles**: Some parasites, like tapeworms, have an intermediate host in their life cycle, such as a beetle or flea, complicating diagnosis and manage-ment.

Bulk vs. Individual Testing:

- **Individual Testing**: Faecal samples are collected from individual animals to determine their specific parasitic load. This is especially important in cases where certain animals need targeted treatment or when assessing the breeding value (ASBV) for worm resistance.

- **Bulk Testing**: For larger herds, bulk faecal testing is common. Samples from 20–40 animals are combined and tested to give an average worm burden for the group, which is a cost-effective approach when dealing with large numbers of livestock.

Worldwide Relevance:

Globally, FEC is crucial in managing parasitic infestations in various climates and agricultural settings. In high-rainfall regions, for instance, gastrointestinal worms can proliferate, making regular FECs essential for maintaining herd health. In dry areas, parasites like liver flukes may become more prevalent when animals graze near water sources.

Safety and Quality Control:

Due to the infectious nature of faecal samples, strict hygiene protocols must be fol-lowed. Regular submission of samples to certified laboratories ensures the accuracy of

DIY faecal egg counts. This enables livestock managers worldwide to effectively control parasitic infections, reducing the risk of severe infestations and improving overall productivity.

Sheep Worms - Faecal Worm Egg Counts

Sheep worms and the associated worm control strategies, particularly using faecal worm egg counts (WEC), are critical components of sustainable livestock management globally. Worm egg counts help in determining the worm burden in sheep and other livestock by counting the number of worm eggs in a faecal sample, expressed as eggs per gram (epg). This method plays a vital role in understanding and mitigating the negative impacts of parasitic infestations.

Worm control in sheep is essential to prevent diseases caused by gastrointestinal parasites and to ensure efficient production. By performing WECs, livestock managers can determine whether the worm burden is high enough to require treatment with drenches or other anti-parasitic measures. WECs offer several benefits:

1. **Drenching Decisions**: WECs help producers decide when to administer drenches, preventing unnecessary treatments and reducing drench resistance. This is particularly important for regions where the overuse of drenches has led to the development of resistant parasite populations.

2. **Monitoring Drench Effectiveness**: By performing WECs before and after drenching (ideally 10-14 days later), producers can evaluate how effective a particular drench has been. If the egg count remains high after treatment, it indicates possible resistance, signalling a need for alternative strategies.

3. **Pasture Management**: WECs conducted in the autumn help assess pasture contamination and the risk of worm outbreaks later in the year, especially during high-risk seasons such as winter and spring. Treating animals based on WEC results prevents the build-up of worm larvae in pastures, minimizing the likelihood of significant infestations in grazing animals.

When performing a WEC, accurate sampling is key to obtaining reliable results. Typically, 20 faecal samples are recommended from each mob to provide a representative worm egg count. In cases of drench resistance testing, 10 samples per group are sufficient.

Sample Collection: Samples must be fresh and can be collected directly from the animal's rectum or from freshly deposited faeces in the paddock. It's crucial to avoid contamination, and samples should be stored in cool conditions to prevent eggs from hatching before testing. Maintaining proper handling protocols ensures the validity of the WEC results.

Drench resistance is a growing concern worldwide, particularly in areas where frequent use of chemical drenches has led to resistant parasite populations. WECs are fundamental to identifying the effectiveness of drenches. Through resistance testing, sheep are divided into groups and treated with different drenches. Post-treatment WECs help measure the percentage reduction in worm egg counts, indicating the level of resistance to each product.

For instance, a significant reduction (greater than 95%) indicates effective treatment, whereas lower reductions signal potential resistance. This testing allows farmers to adjust their worm control strategies, ensuring they use the most effective treatments for their specific situation.

In regions where worm infestations are a major challenge, genetic selection for worm resistance has gained attention. Ram breeders use WEC data to select rams with low worm egg counts, contributing to a more resistant flock over time. This approach reduces the reliance on chemical treatments and enhances the long-term sustainability of worm control.

Worm control and the use of WECs are not unique to any one country. In countries with intensive sheep farming, such as Australia, New Zealand, and the United Kingdom, WECs are a routine part of parasite management. Each region faces different challenges based on climate, pasture conditions, and parasite species, making WECs adaptable and essential tools worldwide.

Determining the Need for Treatment and the Type and Scope of Treatment

Determining the need for treatment and identifying the type and scope of treatment for cattle diseases requires careful assessment based on disease risk factors, herd health history, and veterinary advice. Vaccines and treatments for cattle diseases should always be administered following manufacturer specifications or professional veterinary guidance to ensure efficacy and safety.

Vaccination strategies for cattle should account for local disease prevalence, animal age, herd management practices, and environmental conditions. Disease prevention strategies involve planning and timing vaccinations to protect cattle during high-risk periods or before specific events such as calving, breeding, or seasonal changes. In addition, proper storage, handling, and administration of vaccines are critical to maintaining their potency and ensuring effective immunization.

Common Cattle Disease Vaccines and Treatment Strategies

Clostridial diseases, such as tetanus, black leg, and pulpy kidney, are caused by toxins produced by Clostridium species. Vaccination is a primary preventive measure, with vaccines like Ultravac 5 in 1™ and Websters 5 in 1™ being commonly used. Vaccination strategies recommend administering vaccines to cows 2-6 weeks before calving to ensure that calves receive maternal antibodies [54, 55]. Calves born to unvaccinated cows should be vaccinated early, with a booster given 4-8 weeks later. For adult cattle, an annual booster is advised, particularly before high-risk periods such as drought or grain feeding [56]. The importance of these vaccines is underscored by their role in preventing sudden and severe diseases that can lead to rapid mortality in cattle, especially in endemic areas [57].

Clostridial Diseases (e.g., Tetanus, Black Leg, Pulpy Kidney)

- **Vaccines**: Common vaccines include Ultravac 5 in 1™ and Websters 5 in 1™.

- **Strategy**:

  - Vaccinate cows 2-6 weeks before calving to protect young calves through maternal antibodies.

  - Calves born to unvaccinated cows should be vaccinated early with a booster 4-8 weeks later.

  - For adult stock, an annual booster is recommended, especially before high-risk periods like drought or grain feeding.

  - Clostridial vaccines are essential for preventing sudden, severe diseases that can lead to rapid death in cattle, especially in areas where these diseases are common.

Botulism, caused by the neurotoxins of Clostridium botulinum, poses a significant risk to cattle, particularly in regions where cattle may consume contaminated feed or carcasses [58]. Vaccines such as Ultravac Botulinum Vaccine™ and Singvac 3 Year™ are available, with the latter requiring only one booster every three years, thus providing extended protection [59]. For high-risk properties, an initial vaccination followed by a booster in 4-6 weeks and then annual boosters is recommended [60]. The economic impact of botulism outbreaks can be severe, necessitating proactive vaccination strategies to mitigate risks (Soares et al., 2018). Furthermore, the efficacy of recombinant vaccines against botulism has been demonstrated, showing promise in inducing protective immunity [61, 62].

Botulism

- **Vaccines**: Options include Ultravac Botulinum Vaccine™ and Singvac 3 Year™.

- **Strategy**:

  - High-risk properties with a history of botulism deaths should vaccinate cattle, with an initial vaccination followed by a booster in 4-6 weeks and then annually.

  - Singvac 3 Year™ requires only one booster every three years, offering longer protection.

  - Botulism vaccination is especially relevant in regions where phosphorus deficiency in soils contributes to cattle ingesting carcasses or decayed matter, leading to the disease.

Vibriosis, primarily affecting bulls and causing reproductive issues in females, can be managed through vaccination with Vibriovax™. The vaccination strategy involves administering two initial doses to bulls four weeks apart, followed by annual boosters [56]. For females, vaccination is only recommended if a vibriosis infection is confirmed by a veterinarian, as the disease can lead to infertility and extended calving intervals [56]. This targeted approach is essential for effective breeding management.

Vibriosis

- **Vaccines**: Vibriovax™.

- **Strategy**:

  - Vaccination focuses primarily on bulls, with two initial doses given 4 weeks apart, followed by annual boosters.

  - For females, vaccination is only recommended if vibriosis infection is confirmed, as diagnosed by a veterinarian.

  - This disease can cause infertility, early embryonic death, and long calving intervals, making vaccination crucial for breeding management.

Leptospirosis is a zoonotic disease that can lead to significant health issues in both cattle and humans. Vaccines such as Cattlevax LC 7 in 1™, Leptovax™, and Ultravac 7 in 1™ are utilized to prevent this disease. The vaccination strategy includes administering a priming dose to calves at four weeks, followed by a booster after 4-6 weeks, with annual boosters recommended before calving or during wet conditions that favour disease spread [56]. Vaccination not only protects cattle but also reduces the risk of transmission to farm workers [56].

Leptospirosis

- **Vaccines**: Cattlevax LC 7 in 1™, Leptovax™, Ultravac 7 in 1™.

- **Strategy**:

  - Calves should receive a priming dose at 4 weeks old, followed by a booster after 4-6 weeks.

  - Annual boosters are given before the calving season or during wet conditions, which favor the spread of leptospirosis.

  - The disease can lead to abortions, reduced milk yield, and serious health issues in humans (zoonotic disease), so vaccination helps protect both cattle and farm workers.

Bovine Pestivirus (BVDV) can cause significant losses in cattle herds. The vaccine Pestigard™ is recommended, with two doses administered 4-6 weeks apart, followed by annual boosters [56]. Vaccination is critical before the breeding season to protect the fetus from infection. Additionally, identifying and removing persistently infected animals from the herd is vital for controlling the spread of BVDV [56].

Mucosal Disease (Bovine Pestivirus/BVDV)

- **Vaccine**: Pestigard™.

- **Strategy**:

  - Two doses, 4-6 weeks apart, followed by annual boosters. Vaccination is vital before the breeding season for foetal protection.

  - Identifying and removing persistently infected animals from the herd helps control the spread of BVDV.

Pinkeye, often exacerbated by environmental factors, can be managed through vaccination with Coopers Piliguard™. It is recommended to vaccinate at-risk calves about four weeks before conditions that may promote pinkeye outbreaks, such as hot and dusty environments [56]. Alongside vaccination, environmental management, including reducing fly populations and dust, is essential for preventing disease outbreaks [56].

Pinkeye

- **Vaccine**: Coopers Piliguard™.

- **Strategy**:

    ○ Vaccinate at-risk calves about four weeks before conditions that may promote pinkeye (e.g., hot, dusty environments).

    ○ Managing the environment (reducing flies and dust) and quarantining infected mobs helps prevent disease outbreaks.

Vaccination schedules should be followed rigorously to ensure that cattle are protected against diseases at the right time. Vaccines generally take several weeks to confer full protection, so they should be administered well before high-risk periods, such as calving or weaning. Additionally, new animals should be integrated into the herd only after a thorough health check and appropriate vaccination.

It is essential to follow veterinary advice and the manufacturer's specifications regarding vaccine dosage, storage, and handling. This ensures not only the effectiveness of the vaccine but also the safety of the cattle and handlers. Veterinarians also play a crucial role in diagnosing diseases and recommending the right treatment protocols, especially in regions with high disease burdens or where resistance to treatments may be a concern.

The appropriate vaccination and treatment strategy depends on the disease risks specific to the region, herd health history, and ongoing management practices. Consulting with a veterinarian and adhering to vaccine protocols ensures the optimal health and productivity of cattle worldwide.

CHAPTER NINE

Treating Livestock

Zoonotic diseases are infectious diseases that animals can transmit to humans, posing a significant global public health challenge. Leptospirosis and Q-fever are two notable examples that can spread through contact with infected animals or their by-products. Zoonoses can be caused by bacteria, viruses, parasites, or fungi, and they affect a wide range of domestic, wild, and farm animals. The transmission routes for these diseases include direct contact with animals, contaminated environments, and ingestion of contaminated food or water [63].

Parasite control in animals often involves using chemicals that can be hazardous to humans. Exposure to these chemicals can occur through skin absorption, inhalation of vapours, or accidental ingestion. The improper handling of these chemicals, such as storing them in unlabelled containers or failing to follow safety guidelines, can increase the risk of exposure. Workers who handle veterinary products like antibiotics also face risks, such as needle-stick injuries or chemical splashes. It's essential to follow manufacturer specifications, maintain safety data sheets (SDS), and ensure that competent staff handle these substances to reduce health risks.

Certain animals are more likely to shed harmful microorganisms, thus increasing the risk of zoonotic transmission. These include pregnant animals, newborn animals (such as calves and chicks), stressed or sick animals, reptiles, and amphibians. For instance, the bacteria *Leptospira*, which causes leptospirosis, can be found in the urine of infected animals, including rodents, pigs, and cattle. The bacterium *Coxiella burnetii*, which causes Q-fever, can be spread via aerosols during animal birthing, making direct exposure to birthing products or contaminated dust hazardous [63].

Certain populations are more susceptible to zoonotic infections or may experience more severe symptoms. These groups include [63]:

- **Young children**, whose immune systems are still developing, making them vulnerable to infections from animal contact.

- **Pregnant women**, who may be more susceptible to zoonotic diseases like toxoplasmosis, which can cause birth defects or miscarriage.

- **Immunocompromised individuals**, such as those with HIV/AIDS or undergoing chemotherapy, are at a heightened risk due to their weakened immune systems.

- **Older adults**, who, despite often having immunity to certain diseases, may face more severe consequences from zoonotic infections due to the natural decline in immune function with age.

To prevent zoonotic disease transmission, people at higher risk, such as young children and pregnant women, should avoid direct contact with high-risk animals and contaminated environments. Supervising children around animals and enforcing hand hygiene after animal contact are essential preventive measures. Pregnant women should avoid handling cat litter due to the risk of toxoplasmosis and limit contact with rodents, which can carry the LCMV virus.

Handwashing is the most critical practice to prevent zoonotic disease transmission, especially in environments where people come into contact with animals. Washing hands thoroughly with soap and water after touching animals, animal enclosures, or contaminated surfaces is vital. Facilities providing animal contact opportunities should ensure the availability of proper handwashing stations equipped with running water, soap, and disposable paper towels. Signage promoting hand hygiene should be placed prominently, especially around animal enclosures.

Zoonotic diseases can spread through several routes [63]:

- **Ingestion**: Contaminated food, water, or animal products, such as unpasteurized milk, can spread diseases like salmonella.

- **Inhalation**: Diseases like Q-fever can be spread through aerosols from animal birthing products or contaminated dust in animal enclosures.

- **Skin contact**: Infections can enter the body through bites, scratches, or contact with contaminated animal excreta or surfaces.

- **Urine**: Leptospirosis is an example of a disease that can spread through contact with animal urine.

Operators of facilities that offer animal contact should take precautions to minimize the risk of zoonotic transmission. Healthy animals should be used for public contact, and areas should be regularly cleaned and disinfected. Operators should communicate zoonotic risks to visitors and provide adequate supervision, especially for high-risk groups like children and pregnant women. Waste from animals, including birth products, should be safely removed from public areas to prevent contamination.

Zoonotic diseases represent a significant risk in many areas of human-animal interaction, but proper hygiene practices, safe chemical handling, and awareness of high-risk groups can reduce these risks effectively.

Use of Personal Protective Equipment

Zoonotic diseases, which can be transmitted from animals to humans, pose a significant occupational health risk, particularly for people working closely with animals such as farmers, veterinarians, abattoir workers, and animal handlers. Zoonoses are caused by a range of pathogens including bacteria, viruses, fungi, parasites, and protozoa, and their transmission can occur through direct or indirect contact with animals, contaminated surfaces, or through inhalation and ingestion of harmful agents.

To mitigate the risks of zoonotic diseases, proper use of Personal Protective Equipment (PPE) is essential. The selection of appropriate PPE is influenced by the type of zoonotic agent, the nature of the work, and the level of exposure risk. Basic PPE includes gloves, safety eyewear, protective clothing, and respiratory masks when exposure to aerosols or dust is expected. For example, disposable gloves and respiratory masks should be worn when dealing with animals showing signs of infection, assisting in animal birthing, or cleaning animal environments.

Key Zoonotic Diseases and Precautions

1. **Bacterial Infections**: Diseases like *Leptospirosis, Brucellosis,* and *Anthrax* can be contracted through direct contact with infected animals, contaminated environments, or inhaling contaminated aerosols. Workers dealing with cattle, pigs, and other livestock are particularly vulnerable. Using gloves, aprons, and respiratory masks can significantly reduce the risk of exposure. For instance, dairy farmers should be vigilant in preventing contact with urine aerosols that may carry *Leptospira* bacteria.

2. **Viruses**: *Q-Fever*, caused by *Coxiella burnetii*, can be contracted through inhalation of airborne particles from infected animals, such as sheep, goats, and cattle. Vaccination is available and recommended for people in high-risk occupations. PPE, such as face masks and protective eyewear, should be worn when working in dusty or birthing environments where aerosolized particles are likely to contain pathogens.

3. **Parasites and Protozoa**: Diseases like *Cryptosporidiosis* and *Giardiasis* can be transmitted through contact with infected animal faeces or contaminated water. Protective gloves are vital for handling animals, particularly in environments where faecal contamination is possible. Practicing good hygiene, such as handwashing after handling animals or cleaning enclosures, is essential to prevent these infections.

4. **Fungal Infections**: Fungi such as those causing *Ringworm* can be transmitted through contact with infected animals or contaminated surfaces. PPE like gloves and coveralls are necessary to prevent skin contact with infectious agents. Regular cleaning and disinfection of workspaces, animal bedding, and tools can help minimize transmission.

Certain populations, including young children, pregnant women, the elderly, and immunocompromised individuals, are more susceptible to zoonotic diseases and may suffer more severe symptoms. For instance, pregnant women should avoid contact with cat litter to prevent exposure to *Toxoplasma gondii*, a parasite that can cause serious harm to the foetus. Immunocompromised workers should avoid handling animals with known infections and must strictly adhere to PPE protocols to minimize their risk.

In addition to using PPE, several preventive measures should be implemented in workplaces to minimize zoonotic disease transmission:

- **Regular cleaning and disinfection**: Workspaces and equipment should be regularly cleaned with appropriate disinfectants, especially areas where animals are housed or handled.

- **Vaccination programs**: Vaccinating livestock and workers (for diseases like Q-fever) is a critical preventive measure in high-risk environments.

- **Proper hand hygiene**: Workers should wash hands thoroughly with soap and water after handling animals or cleaning enclosures, and hand sanitizers should be available when handwashing is not feasible.

By using the appropriate PPE and following safety guidelines, the risk of zoonotic disease transmission in occupational settings can be significantly reduced, ensuring both worker safety and public health.

Handling and Restraining Animals Without Causing Harm or Injury to Animal or Handler

Handling livestock on farms presents significant risks, both in terms of physical injury and potential disease transmission. Animals like cattle, pigs, horses, and sheep can be unpredictable and may act aggressively, particularly during mating seasons. Farmers and farm workers must exercise caution, plan tasks ahead of time, and maintain safe practices to minimize risks. For example, working too closely with large animals or attempting to lift or move them improperly can result in severe physical injury, including strains, sprains, and fractures.

Before engaging in any task involving livestock, it is crucial to assess factors such as the breed, temperament, gender mix, and size of the animals. Both male and female animals can exhibit heightened aggression during mating seasons, increasing the risk of injury. Workers should be well-trained in animal handling techniques and familiar with the temperament of the specific animals they are managing. Ensuring that yards, fences, and enclosures are properly maintained is also critical for minimizing injury risk. Well-designed facilities reduce the chances of being trapped or injured by animals.

Farm workers must receive appropriate training to handle animals safely and effectively. Training should include understanding animal behaviour, especially herd dynamics, and how stress impacts both animal welfare and product quality. For example, cattle, sheep, and goats are herd animals that tend to follow each other; separating them from the herd can cause stress and unpredictable reactions. Low-stress handling techniques not only improve worker safety but also have productivity benefits. By minimizing stress, farmers can enhance livestock health, boost production, and improve meat quality.

Wearing suitable protective clothing—such as boots, gloves, and pants—plays a crucial role in preventing injury when handling animals. Animal-handling aids, including cradles, crushes, and head rails, should be used to restrain animals during essential procedures. Using these tools minimizes direct physical contact with animals and reduces the likelihood of being injured.

Farm animals can transmit zoonotic diseases, such as leptospirosis and Q fever, through contact with blood, saliva, urine, faeces, and other bodily fluids. To prevent the spread of diseases between animals and humans, it is important to vaccinate livestock, maintain good personal hygiene, and treat signs of illness promptly. Additionally, covering cuts and open wounds, washing hands after contact with animals, and practicing good sanitation are critical measures to reduce the risk of disease transmission.

Proper handling techniques vary depending on the species. For instance, cattle handling requires care when approaching, as cows may charge to protect their calves. Bulls should be separated during mating seasons to avoid aggression. Similarly, when handling pigs, boars should be kept apart, and lifting pigs should be avoided unless necessary. Sheep and rams can be unpredictable, so using sheepdogs for control and avoiding the isolation of individual animals are important safety strategies.

Administering Treatment

In livestock production environments around the world, the administration of vaccines is typically conducted by a variety of professionals and trained individuals, depending on local regulations, industry practices, and the type of livestock. Here are some of the key groups that can administer vaccines:

1. Veterinarians

- **Role**: Veterinarians are the most qualified professionals to administer vaccines to animals. They are responsible for diagnosing diseases, determining the appropriate vaccination protocols, and ensuring that vaccines are administered properly. In many countries, veterinarians may be legally required to administer specific vaccines, especially for diseases that pose public health risks or are regulated by government authorities.

- **Global Practice**: In countries like the United States, Australia, and across Europe, veterinarians are often required to administer vaccines for certain regulated diseases, such as foot-and-mouth disease or rabies.

2. Licensed Veterinary Technicians/Assistants

- **Role**: Veterinary technicians or assistants work under the supervision of veterinarians and are often tasked with administering vaccines. Their role is especially prominent in larger livestock operations where routine vaccination programs are necessary.

- **Global Practice**: In regions such as North America and Europe, veterinary technicians can administer vaccines after being properly trained, especially in larger-scale operations where veterinarians may not always be present.

3. Farmers and Livestock Producers

- **Role**: In many parts of the world, farmers and livestock producers who have received appropriate training can administer vaccines to their own animals. This is common in large-scale farming operations where routine vaccinations are part of herd management. These individuals are typically responsible for basic vaccinations that prevent common diseases, but they may still rely on veterinarians for guidance on complex issues.

- **Global Practice**: In countries with large livestock industries such as Australia, New Zealand, Brazil, and the U.S., farmers often vaccinate animals themselves. They are generally trained through agricultural extension services, government programs, or private sector initiatives to safely administer vaccines.

4. Paraprofessionals and Animal Health Workers

- **Role**: In developing countries or rural areas, paraprofessionals (such as community animal health workers) play an essential role in vaccinating livestock, especially in areas where veterinarians are not easily accessible. These workers often receive specific training through government or NGO programs.

- **Global Practice**: In countries across Africa and parts of Asia, paraprofessionals are commonly used to manage livestock vaccination programs due to the scarcity of veterinarians in rural areas. These paraprofessionals are often involved in national disease control campaigns, such as those for foot-and-mouth disease or brucellosis.

5. Government or Regulatory Officials

- **Role**: In some cases, government-employed veterinarians or officials are tasked with administering vaccines during large-scale national or regional disease control programs. This is common for the control of contagious diseases with public health implications, where strict oversight is necessary.

- **Global Practice**: Countries facing outbreaks of highly contagious livestock diseases, such as in Europe and Asia during foot-and-mouth outbreaks, may deploy government veterinarians to oversee or directly administer vaccines to contain the spread of the disease.

Legal and Regulatory Context

- **In the U.S.**: The U.S. Department of Agriculture (USDA) has strict guidelines about who can administer vaccines, especially for diseases of national importance. In most cases, vaccines for common livestock diseases can be administered by farmers, but in regulated cases, veterinarians are required.

- **In Australia**: The Australian Pesticides and Veterinary Medicines Authority (APVMA) provides guidelines on vaccine use, and veterinarians typically oversee vaccinations, though farmers often administer routine vaccines themselves.

- **In Europe**: The European Union has strict rules under its animal health laws, and while farmers may administer certain vaccines, veterinarians are often required for regulated diseases.

United States: In the United States, the U.S. Department of Agriculture (USDA) enforces strict guidelines regarding who can administer vaccines to livestock, especially for diseases of national importance. These regulations are designed to protect both animal and public health, ensuring that vaccinations are administered properly and by qualified individuals.

For many common livestock diseases, such as those caused by clostridial bacteria (e.g., blackleg or tetanus), livestock owners, including farmers and ranchers, are often allowed to administer vaccines themselves. These diseases do not pose significant public health risks, and vaccines are available over the counter. Farmers typically receive training through agricultural extension programs, veterinarians, or industry organizations on how to safely handle and administer these vaccines. Administering these vaccines regularly is crucial for maintaining herd health and productivity, and is a routine part of animal husbandry.

For diseases that are of national or public health importance, such as rabies, tuberculosis, and brucellosis, the USDA mandates that only licensed veterinarians administer vaccines. This is because these diseases can spread rapidly, potentially crossing species barriers and affecting humans, or causing widespread economic losses due to livestock deaths or trade restrictions. For example, rabies vaccination is closely regulated, and in most states, only veterinarians can legally administer the rabies vaccine to livestock.

Similarly, the control of diseases like brucellosis, which can be transmitted to humans (a zoonotic disease), is governed by state and federal regulations under the USDA's Animal and Plant Health Inspection Service (APHIS). APHIS oversees disease eradication and control programs, and vaccines for these diseases must be administered under the supervision of a veterinarian to ensure proper handling and recording.

In the event of an outbreak of a highly contagious disease, such as foot-and-mouth disease or avian influenza, the USDA may deploy special emergency response teams. In these cases, the administration of vaccines becomes part of a coordinated federal and state effort to control the spread of the disease. In such situations, farmers may not be authorized to vaccinate their animals; instead, government veterinarians or authorized personnel may be dispatched to administer vaccines, and there may be quarantine or other movement restrictions in place.

The USDA also mandates that vaccines must be stored and handled according to manufacturer specifications to ensure their effectiveness. Improper storage, such as allowing vaccines to be exposed to extreme temperatures, can render them ineffective. Both veterinarians and farmers are responsible for maintaining proper storage conditions, but for regulated vaccines, veterinarians are typically required to ensure compliance with storage and handling guidelines.

In addition to federal regulations, each state has its own guidelines regarding livestock vaccination. Some states may have more stringent rules than others, requiring veterinarian involvement in a broader range of vaccinations or specific documentation for vaccine administration. These regulations can vary based on the prevalence of certain diseases in a given region and the local agricultural economy.

These state-level rules are influenced by factors such as local disease prevalence, the type of livestock raised, and agricultural economics. Below are examples of states with differing regulations:

1. California

California has stringent vaccination regulations due to its large agricultural economy and diverse livestock population. For diseases such as brucellosis, only licensed veterinarians are authorized to administer vaccines, particularly for cattle. The state mandates extensive record-keeping and reporting for certain zoonotic diseases, such as rabies and tuberculosis, which are considered public health threats. In certain counties, rabies vaccination for livestock, especially horses and cows, is mandated due to the proximity to urban areas.

2. Texas

Texas, known for its large cattle industry, has specific requirements for diseases like anthrax, which is endemic in certain regions of the state. The Texas Animal Health Commission (TAHC) oversees vaccination programs, and veterinarians are required to supervise and report vaccinations for diseases such as brucellosis. For common vaccines such as those for clostridial diseases, farmers are allowed to administer them, but for regulated diseases, veterinarians must ensure compliance.

3. Florida

Florida's regulations focus on preventing the spread of diseases like equine encephalitis, rabies, and bovine tuberculosis due to its subtropical climate and the high risk of disease transmission in livestock and horses. The Florida Department of Agriculture mandates that veterinarians administer vaccines for certain diseases, and in some cases, proof of vaccination must be provided for animal sales or movement across state lines. The state also requires vaccinations for cattle against diseases such as brucellosis and tuberculosis in certain counties.

4. New York

New York, with its large dairy and livestock industry, has strict vaccination guidelines for diseases like rabies, which is prevalent in wildlife that may come into contact with livestock. In certain high-risk areas, rabies vaccination is required for all livestock that interact with the public. New York State Department of Agriculture and Markets oversees these programs, requiring that vaccines for regulated diseases, such as rabies and bovine viral diarrhea (BVD), be administered by a licensed veterinarian.

5. Nebraska

Nebraska, a state with a significant cattle industry, allows farmers to administer common vaccines but imposes stricter regulations on diseases that can impact interstate trade, such as brucellosis. The Nebraska Department of Agriculture mandates that certain vaccines, especially for diseases that could affect the local or national economy, be administered under the supervision of veterinarians. The state also enforces stricter regulations for vaccines during disease outbreaks, such as avian influenza in poultry.

6. Minnesota

Minnesota, with its large swine and poultry industries, has specific guidelines that require veterinarian involvement for diseases that could have a significant impact on public health, such as avian influenza and swine flu. The Minnesota Board of Animal Health mandates veterinarian-administered vaccines for these diseases, and farms must keep detailed records of vaccinations, especially during disease outbreaks that could affect trade and public health.

These examples highlight how state regulations vary based on disease risk, industry size, and local agricultural conditions. States with a higher risk of certain zoonotic or economically impactful diseases tend to have more stringent vaccination requirements and greater veterinarian oversight.

Australia: In Australia, the administration of vaccines in livestock production environments is governed by guidelines provided by the Australian Pesticides and Veterinary Medicines Authority (APVMA), which regulates the use of all veterinary medicines, including vaccines. These guidelines ensure that vaccines are used safely and effectively to protect both animal health and public safety.

Veterinarians typically oversee the use of vaccines, especially for diseases that are significant from a biosecurity or economic standpoint. They are involved in prescribing and advising on the use of vaccines, especially for more complex or high-risk diseases like foot-and-mouth disease, anthrax, or brucellosis. In cases where animal health is of critical importance to the region or where a notifiable disease is concerned, veterinary oversight is often mandatory. Veterinarians may also play a crucial role in advising on vaccination schedules, doses, and the proper handling of vaccines to ensure their efficacy.

While veterinarians play an important role in overseeing vaccinations, many routine vaccines are administered by farmers themselves. This is particularly common for widely-used vaccines for diseases like clostridial infections (e.g., blackleg and pulpy kidney), which are essential for protecting livestock like sheep and cattle. The APVMA permits farmers to administer these vaccines under the assumption that they have received proper training and understand the manufacturer's guidelines. Vaccines that farmers commonly administer include those for tetanus, botulism, and pinkeye, which are prevalent in Australia's diverse agricultural landscapes.

The APVMA requires that vaccines are stored, handled, and administered in strict accordance with the manufacturer's specifications to ensure their effectiveness. This includes proper refrigeration of live vaccines, avoiding exposure to extreme temperatures, and adhering to the correct dosage. Farmers must follow these protocols closely, and any deviations can result in ineffective vaccination or harm to the livestock. The APVMA also emphasizes the importance of record-keeping for vaccine administration, which helps in tracking animal health, disease prevention, and compliance with state and national regulations.

In addition to APVMA guidelines, state and territory authorities may have additional rules or programs for vaccinations based on local disease prevalence and livestock needs. For example, Queensland has specific guidelines for vaccinating livestock against tick-borne diseases, while Western Australia focuses heavily on diseases like Johne's disease, which are more prevalent in that region. Some states may require veterinarian involvement for specific diseases or during outbreaks to control the spread of infectious diseases.

Farmers administering vaccines must be adequately trained, as improper vaccine handling can reduce its efficacy or cause harm to animals. Training programs are often provided by veterinarians, livestock organizations, or local agricultural departments. These programs cover essential aspects of vaccination, such as proper injection techniques, recognizing adverse reactions, and understanding withdrawal periods before livestock products enter the human food chain.

While veterinarians in Australia typically oversee vaccination programs for high-risk diseases, routine vaccinations are frequently administered by farmers under the strict guidelines of the APVMA. State and territory regulations can further specify the involvement of veterinarians and the conditions under which vaccines must be administered, especially during disease outbreaks or in regions with specific health concerns.

Europe: In the European Union (EU), strict rules govern the administration of vaccines to livestock, as part of its broader animal health and welfare laws. These regulations are outlined under various legislative frameworks that focus on safeguarding both animal health and food safety. The EU has a harmonized approach to animal vaccination to prevent the spread of infectious diseases, which includes veterinary oversight for regulated diseases and guidance on the use of vaccines by farmers.

The EU's animal health laws require that veterinarians play a significant role in the administration of vaccines, especially for regulated diseases that pose significant public health or economic risks. Diseases such as foot-and-mouth disease, avian influenza, and bluetongue are strictly regulated, and vaccinations for these diseases can only be administered by licensed veterinarians. This ensures that there is proper control and monitoring of disease outbreaks, and that the vaccines are administered safely and effectively. Veterinarians are also responsible for reporting vaccination efforts to relevant authorities, ensuring compliance with EU disease control programs, and providing expert guidance on vaccination schedules and protocols.

In contrast to the strict controls for regulated diseases, EU law allows farmers to administer certain routine vaccines to livestock. These typically include vaccines for common diseases that are not considered a significant threat to public health or the wider agricultural sector, such as clostridial diseases in sheep or cattle. However, farmers must follow the instructions provided by the manufacturer and comply with national regulations for vaccine handling, storage, and administration. Farmers are also required to keep detailed records of all vaccinations administered, including the type of vaccine, date of administration, and the identification of the animals vaccinated.

While the overarching regulations are provided by the EU, each member state has its own implementation of these laws. This means that the role of veterinarians and the rules regarding farmer-administered vaccines can vary slightly from one country to another. In some countries, like Germany and France, veterinarians are more frequently involved in administering a wider range of vaccines, while in others, like the Netherlands or Denmark, farmers are granted more autonomy in administering routine vaccinations. These national adaptations are designed to reflect the specific agricultural practices and disease prevalence in each region.

A key aspect of the EU's vaccination rules is the requirement for detailed record-keeping, regardless of whether the vaccines are administered by a veterinarian or a farmer. These records are essential for traceability and play a critical role in disease surveillance and control across the EU. They also ensure that in the event of a disease outbreak, authorities can quickly assess vaccination coverage and the potential need for additional measures.

All vaccines used within the EU must be authorized by the European Medicines Agency (EMA) or national regulatory bodies. This ensures that only safe, effective, and high-quality vaccines are available for use. Unauthorized vaccines are prohibited, and their use can lead to severe legal consequences, including penalties for farmers or veterinarians.

The EU's comprehensive "Animal Health Law" (Regulation (EU) 2016/429) serves as the main regulatory framework for disease control, including vaccination efforts. This law sets out the roles and responsibilities of both veterinarians and farmers in preventing and controlling infectious diseases across the Union. It also establishes the conditions under which vaccination programs can be implemented and defines the criteria for when vaccination is compulsory or voluntary.

The EU places stringent controls on the administration of vaccines, particularly for regulated diseases, which require veterinary oversight. Farmers are allowed to administer certain vaccines, but they must comply with EU and national regulations, ensuring that vaccination efforts are safe, effective, and properly documented. These regulations are essential for maintaining animal health, public safety, and the integrity of the food supply chain across the European Union.

Vaccines

Livestock vaccination is a critical component of disease prevention and management in livestock production enterprises worldwide. When used correctly, vaccines can dramatically reduce losses by preventing outbreaks of infectious diseases that may otherwise devastate herds. Vaccination protocols, however, vary based on the type of vaccine, the region's disease prevalence, and the specific needs of the livestock operation [64].

Vaccines used in livestock production can be broadly categorized into live vaccines and inactivated (or killed) vaccines. Live vaccines generally provide long-lasting immunity with a single dose. They contain weakened versions of the pathogen, which stimulate the animal's immune system to recognize and fight off future infections. These vaccines are commonly used in herd health management where long-term immunity is needed. In contrast, inactivated vaccines require booster doses to maintain immunity. These vaccines contain killed pathogens, and while they are safer for some animals and situations, their effectiveness requires more frequent administration.

Another key component in the vaccination arsenal is anti-toxins. Anti-toxins are not vaccines but provide immediate, short-term protection against certain diseases by neutralizing toxins produced by bacteria. While they are used in emergency situations or immediate outbreaks, they do not offer long-term immunity and are generally used as a stop-gap measure before vaccination can take effect.

Clostridial diseases—caused by bacteria from the genus *Clostridium*—are widespread, found in soil, faeces, and even the gastrointestinal tracts of healthy animals. The bacteria can produce toxins, leading to often fatal diseases like tetanus, blackleg, malignant oedema, and enterotoxaemia (pulpy kidney). Vaccination is key in preventing these diseases, particularly for young or injured animals, as they are more susceptible [64].

For protection, most livestock producers administer 5-in-1 vaccines that protect against the five common clostridial diseases: tetanus, malignant oedema, enterotoxaemia, blackleg, and black disease. Young animals are usually vaccinated early, with booster shots required 4-6 weeks apart. In regions where botulism is a threat, separate vaccines or combinations like the 7-in-1 vaccine, which covers both clostridial diseases and leptospirosis, are used. Botulism prevention is critical in areas with phosphorus deficiencies, as animals may chew bones or decaying material that harbors the *Clostridium botulinum* bacteria.

Vaccination also plays a crucial role in preventing reproductive diseases like leptospirosis, vibriosis, and pestivirus, which cause infertility and abortions in cattle. Leptospirosis, caused by the bacteria *Leptospira*, is particularly concerning as it can also infect humans, making vaccination essential in high-risk regions. Vaccines for leptospirosis are available as standalone or in combination with clostridial vaccines, and both calves and breeding stock require two initial doses followed by annual boosters to maintain immunity [64].

Vibriosis, a venereal disease spread by bulls, can be controlled effectively through vaccination. Bulls are typically vaccinated twice, with booster shots required annually before the breeding season to prevent the disease's spread. Pestivirus is another major concern, especially for pregnant females. Vaccination of females prior to joining (mating) is necessary to prevent reproductive losses. Persistently infected animals can spread pestivirus to the entire herd, making herd-wide vaccination programs critical.

In feedlot environments, where animals are kept in close quarters and stress levels are high, respiratory diseases such as infectious bovine rhinotracheitis (IBR) are prevalent. Vaccines like Rhinogard™ provide protection against these diseases when administered intranasally upon arrival. Combining pestivirus vaccination with respiratory disease prevention is also common in feedlots, where maintaining strong immunity is essential to prevent outbreaks.

Different regions have specific vaccination needs based on prevalent diseases. In areas with known tick fever, such as Queensland, Australia, and parts of the Northern Territory, vaccination is crucial to prevent outbreaks. Tick fever is transmitted by cattle ticks and can cause significant economic losses [64]. Cattle moving into tick-endemic areas require vaccination, and permission from local veterinary authorities is often needed.

Anthrax, a notifiable disease in many countries, poses a serious risk in some regions, especially in summer when outbreaks are more common. Anthrax vaccination is mandated in areas like the anthrax belt in New South Wales, Australia, and requires strict quarantine and vaccination protocols to prevent the disease's spread [64].

Vaccines are also available for diseases that specifically affect young livestock, such as neonatal scours caused by E. coli and salmonella. These diseases often result from poor hygiene or close contact between animals. Vaccination of pregnant dams ensures that newborns receive antibodies through colostrum, providing them with early protection against these life-threatening infections.

The choice of which vaccines to use and when to administer them is influenced by local disease prevalence, environmental factors, and the production goals of the enterprise. In many parts of the world, farmers administer routine vaccines, while veterinarians oversee and administer vaccines for regulated or particularly dangerous diseases. Regardless of who administers the vaccines, it is essential to follow manufacturer guidelines, maintain good hygiene, and use handling and restraint facilities appropriately to minimize stress on animals and maximize the effectiveness of vaccines.

Vaccine Selection

Selecting the right vaccine for livestock is a critical aspect of animal health management and disease prevention. Vaccination helps protect animals from potentially devastating diseases, which can cause significant economic losses and affect animal welfare. However, it is important to carefully choose the appropriate vaccine based on the disease risk, the type of livestock, and the production environment.

Vaccines should be selected based on the diseases prevalent in the region and the specific needs of the herd or flock. Common vaccines, such as those for tetanus in horses or botulism in cattle, are often administered routinely. However, for more complex diseases, such as leptospirosis, which affects calves and breeding females, it is essential to consult with a veterinarian. A veterinarian can diagnose the disease and recommend whether vaccination is necessary, especially in areas where certain diseases are not common but pose serious risks if introduced.

In many cases, a veterinarian will recommend including vaccines as part of a comprehensive herd health program, which should also consider biosecurity measures and other disease prevention strategies. This comprehensive approach minimizes disease outbreaks, enhances animal productivity, and protects public health, particularly for zoonotic diseases that can affect both animals and humans.

For some diseases, there are multiple strains of causal organisms, and vaccines may cover several strains or disease complexes. This makes it critical to select the correct vaccine for the specific situation. For instance, in cattle, the choice between a 5-in-1 or 7-in-1 vaccine should be based on whether the herd is at risk of leptospirosis, a contagious bacterial disease that can cause stillbirths and abortions in late pregnancy. The 7-in-1 vaccine, which covers two strains of leptospirosis in addition to the five clostridial diseases (tetanus, blackleg, gas gangrene, Black's disease, and pulpy kidney), is more expensive than the 5-in-1, so it should only be used when necessary. For example, it may be more cost-effective to vaccinate dairy cattle and breeding beef females with the 7-in-1 due to their increased risk of leptospirosis, particularly in regions where the disease is prevalent.

Additionally, vaccines should be selected based on their registration for the specific type of stock. For example, there are vaccines such as three-in-one and six-in-one that are only registered for sheep and lambs, while other vaccines are registered for goats and kids as well. Using vaccines that are registered and recommended for the specific livestock type ensures efficacy and safety.

The administration and booster schedule for vaccines can vary by manufacturer, even for the same disease. For example, botulism vaccines for cattle may have different booster intervals, and it is important to follow the manufacturer's instructions to ensure proper immunity. Additionally, certain vaccines like those for infectious bronchitis or Newcastle disease in poultry also require adherence to specific dosage regimes to prevent outbreaks in flocks.

Since vaccines are expensive, it is important to purchase only what is needed. Over-purchasing vaccines, especially in small-scale operations, can result in wastage or the use of expired or compromised vaccines. For example, many poultry vaccines come in large-dose lots (e.g., 1000-dose packages), which may not be practical for small-scale producers. Storing opened or reconstituted vaccines for later use is not recommended because it can reduce the vaccine's effectiveness and introduce contaminants that could cause illness.

For cattle, 5-in-1 vaccines are the standard choice for protecting against clostridial diseases. Calves receive the first injection at marking and a booster four to six weeks later. The booster is critical for establishing long-term immunity. After the initial vaccinations, annual boosters are necessary to maintain protection. For female livestock, the booster should ideally be administered six weeks before giving birth to ensure passive immunity is passed to the offspring through colostrum.

Proper hygiene and handling during vaccination are essential for preventing infections at injection sites and ensuring the effectiveness of the vaccine. It is recommended to change needles after every 50 animals or fewer, and disinfect the needle between animals to reduce the risk of infection. Administering multiple vaccines simultaneously should be avoided to prevent overwhelming the animal's immune system and to reduce the likelihood of adverse reactions. If an individual administering vaccines accidentally injects themselves, they should seek immediate medical attention, particularly when dealing with vaccines containing oil-based adjuvants, which can cause serious reactions.

Selecting the right vaccine for livestock requires careful consideration of disease prevalence, the type of livestock, and the cost-effectiveness of the vaccine. While routine vaccinations like 5-in-1 or 7-in-1 for cattle provide broad protection against common diseases, consultation with veterinarians is essential for more complex diseases or where there are multiple vaccine options. Proper vaccine handling, administration, and adherence to booster schedules are crucial for ensuring long-term immunity and preventing disease outbreaks, ultimately promoting animal welfare and productivity.

Administering Vaccines

Storing vaccines correctly is crucial to ensuring their efficacy and protecting livestock from diseases. As vaccines are biological products, they degrade over time, even when stored properly. Therefore, the manufacturer assigns an expiry date that guarantees the vaccine's potency until that date, provided it is stored according to specific guidelines. The majority of vaccines must be refrigerated but never frozen. Freezing certain vaccines can not only reduce their potency but also lead to local reactions at the injection site in animals. It is essential that vaccines are stored in a cold environment, such as a refrigerator, where they are kept within a stable temperature range but not subject to freezing and thawing cycles, which can compromise the vaccine's effectiveness.

Every vaccine is subject to rigorous testing for safety and efficacy by relevant regulatory bodies, such as the Australian Pesticides and Veterinary Medicines Authority (APVMA). However, the vaccine's efficacy can be diminished if the user fails to follow the label instructions provided by the manufacturer. The label contains critical information about dosage, injection site, storage, and the recommended vaccination schedule. Users are required by law to adhere to these instructions unless directed otherwise by a veterinarian.

Failure to do so could render the vaccine ineffective, making the investment in vaccination futile. Proper handling and adherence to storage and usage guidelines ensure that the vaccine delivers the protection it promises.

Administering vaccines with sterile and appropriate equipment is another key factor in maintaining their efficacy. Most vaccines are administered via syringes and needles, but some may be delivered through water supplies or intranasally. Automatic syringes, commonly used with multi-dose containers, must be carefully calibrated to deliver the right dose to each animal. Needles should be sharp and short, particularly when delivering subcutaneous vaccines, as long needles may accidentally deposit the vaccine into muscle tissue, reducing its effectiveness.

To avoid introducing infections, the equipment must be sterilized between uses. This can be done using a pressure cooker or boiling the equipment for 40 minutes. Disinfectants or chemicals like methylated spirits should not be used, as they can deactivate the vaccine. After sterilization, syringes and needles should be stored in a clean, covered container to prevent contamination from dust or dirt.

Vaccines typically take 10-14 days to provide immunity. As such, vaccinations should be planned as part of a broader herd or flock health program, considering the type of livestock, seasonal changes, and the prevalence of specific diseases. It is important to vaccinate animals before major management events, such as castration, weaning, or shearing, to minimize stress and provide adequate immunity.

Proper vaccine handling is vital for maintaining its efficacy. Vaccines should be kept cool, ideally in an Esky, and protected from sunlight and heat. Keeping the equipment and vaccine away from dirt and dust is also essential to prevent contamination. When multiple vaccines are needed, they should not be mixed in the same syringe, as combined vaccines require careful balancing of components. Instead, separate syringes should be used, and the vaccines should be administered at different injection sites, at least 15 cm apart, and preferably on opposite sides of the animal's body.

For specific vaccinations, such as tick fever, it is advised not to administer other vaccines at the same time. If unavoidable, separate syringes and different injection sites should be used. Ideally, tick fever vaccinations should be done either two weeks before or four weeks after other vaccinations to minimize the risk of reactions, especially in adult animals.

It is critical to ensure that every animal in a herd receives the full recommended dose of vaccine. Automatic syringes should be regularly checked for correct dosage, and boosters should be administered as required to maintain immunity. Booster doses help strengthen the animal's immune system and prolong vaccine effectiveness.

Animal vaccines can pose risks to human health if accidentally injected. As a result, those administering vaccines must remain focused on the task and avoid performing other duties, such as restraining or moving animals, while handling the syringe. Proper safety protocols, including wearing protective gear and ensuring careful use of needles and syringes, can reduce the risk of accidental injury to humans.

Cattle Vaccination

Cattle vaccination is essential to maintaining the health, productivity, and profitability of livestock herds worldwide. Vaccinating cattle reduces the risk of disease outbreaks, which can lead to significant financial losses due to reduced productivity, illness, and even death. Prevention through vaccination is far more cost-effective than treating diseases after they spread. Moreover, vaccination helps ensure food safety by controlling diseases that can affect the quality of meat and dairy products. For example, preventing illnesses like tetanus in horses or botulism in cattle ensures animals remain healthy and viable for production [65].

Effectively administering vaccines requires an understanding of different methods and ensuring that appropriate techniques are used for the cattle's welfare. Vaccination methods generally include intramuscular (IM) and subcutaneous (SUBQ) injections, though some vaccines are administered orally or intranasally. Proper administration techniques ensure that vaccines deliver the best possible immune response and minimize complications, such as injection site infections or stress for the animals [65].

Vaccine Injection Methods for Cattle

Intramuscular (IM) Injection: Intramuscular injections are one of the most common methods used for cattle vaccination. The vaccine is injected directly into the muscle tissue, usually in the neck area, ensuring effective delivery of vaccines for diseases like clostridial infections. The recommended needle size varies by age and size of the animal. For example, 16-gauge needles are generally suitable for adult cattle, while 18-gauge needles are preferred for calves [65].

Subcutaneous (SUBQ) Injection: This method involves injecting the vaccine under the skin rather than into the muscle, which helps prevent potential carcass damage. Subcutaneous injections are frequently used to administer vaccines aimed at preventing diseases like bovine respiratory disease. By pinching the skin to create a pocket between the skin and muscle, farmers can safely inject vaccines with shorter needles, minimizing tissue damage.

Intranasal and Oral Administration: Intranasal administration is typically reserved for live vaccines used to prevent respiratory diseases like Bovine Respiratory Syncytial Virus (BRSV). This method requires restraining the cattle to ensure proper delivery into the nostrils. Oral administration is used for certain medications and vaccines that are delivered through bolus guns or drenches.

Figure 47: Cattle vaccination. Defense Visual Information Distribution Service, Public Domain, via Picryl.

Effective cattle vaccination relies on more than just choosing the right method. Ensuring aseptic techniques, selecting appropriate equipment, and administering accurate doses are key to achieving successful vaccination outcomes. Using sharp, sterile needles, changing them frequently, and cleaning the injection site can significantly reduce the risk of infections at the site of injection. Furthermore, accurate dosing not only ensures that each animal receives adequate protection but also helps avoid waste of expensive vaccines [65].

Proper handling and restraint of cattle during vaccination are critical to maintaining both animal welfare and the safety of farm workers. Before administering vaccines, it's essential to walk through the livestock handling area to identify potential bottlenecks and ensure that all workers are familiar with the vaccination procedure. Restraint methods, such as using chutes or head gates, prevent injury to both the animals and workers during the vaccination process.

Aseptic conditions during vaccination help prevent the introduction of contaminants at the injection site. Although achieving 100% sterility on a farm is challenging, basic hygiene practices, such as using clean needles and disinfecting the injection site, can significantly reduce the risk of infections. Changing needles between animals and using sterilized equipment ensures that cattle receive vaccines in a safe and effective manner [65].

Administering the correct dose of vaccines to each animal is crucial. Overdosing can lead to vaccine wastage, while underdosing may result in incomplete protection. Using properly calibrated syringes and regularly checking the equipment ensures the right amount of vaccine is administered. Additionally, maintaining detailed records of each animal's vaccination history helps manage booster schedules and track the herd's overall health [65].

To maximize the efficacy of vaccines, it's important to implement a tailored vaccination schedule that accounts for local disease prevalence and herd-specific conditions. Consulting with veterinarians ensures that farmers are using the right vaccines at the appropriate times. Additionally, monitoring cattle for adverse reactions post-vaccination, such as abscesses or irritation, helps detect any issues early and ensures prompt intervention [65].

Vaccines lose potency if they are not stored properly, so adhering to storage guidelines—whether that involves refrigeration or keeping them in a cool, dry place—is vital. Most vaccines require refrigeration, but some may need to be kept at room temperature. Ensuring proper storage and handling is simple but critical to maintaining vaccine effectiveness.

Biosecurity measures, such as quarantining new animals before introducing them into the herd and ensuring proper sanitation around the farm, help prevent disease transmission. These measures are especially important when using vaccines as part of a larger disease prevention program. Ensuring workers follow strict hygiene protocols and minimizing contact between potentially sick animals and the rest of the herd can greatly reduce the risk of disease spread.

Different groups within a herd, such as calves, pregnant cows, and bulls, have unique vaccination needs. Pregnant cows may require different vaccines to ensure the health of both the mother and calf, while calves need to be vaccinated at specific intervals to build immunity. Tailoring vaccination protocols to different life stages and conditions ensures that each group receives optimal protection [65].

The Importance of a Well-Structured Cattle Vaccination Plan

A well-thought-out and executed vaccination plan is vital to maintaining the health of cattle herds and the financial success of livestock operations. Healthy cattle are more productive, which directly impacts profits by reducing disease-related losses, veterinary costs, and downtime. Preventing disease through vaccination is far more cost-effective than treating outbreaks, and a proactive approach can help avoid the negative effects of illness on cattle performance, such as weight loss or reduced milk production.

A robust vaccination program protects individual animals and supports overall herd health by reducing the likelihood of disease outbreaks. Vaccinated cattle are more resilient and less likely to transmit contagious diseases. This is especially important for operations that interact with external factors like shared stock trailers, equipment, or neighbouring cattle. A well-vaccinated herd also improves biosecurity, making it easier to manage risks from external sources.

Several factors influence the creation of a cattle vaccination plan. Understanding the unique needs of your herd ensures that the plan is effective and efficient:

1. **Age and Health Status of the Herd:** Different age groups within the herd—calves, yearlings, adult cows, and pregnant cows—have varying immune responses and vaccination needs. Calves, for instance, require early protection against diseases like **Bovine Viral Diarrhea (BVD)** and **Bovine Respiratory Syncytial Virus (BRSV),** while adult cows may need boosters to maintain immunity.

2. **Prevalent Diseases in the Region:** Regional disease risks can significantly affect which vaccines should be prioritized. Consultation with veterinarians and

local agricultural authorities helps identify which diseases are common in the area and informs vaccine selection.

3. **Type of Cattle Operation:** Different types of operations—such as dairy, beef, or feedlot—face distinct challenges. For example, dairy cows may have different vaccination needs compared to beef cattle, particularly with respect to diseases like **leptospirosis**, which can affect milk production.

4. **Veterinary Resources:** While many ranchers can manage routine vaccinations, veterinary input is crucial for tailoring a plan to specific herd needs, particularly for new ranchers or those dealing with persistent health issues. Veterinarians can provide guidance on the appropriate timing and types of vaccines based on the herd's health history and regional disease risks.

A vaccination schedule typically includes core vaccines, which protect against widespread and highly contagious diseases, and optional vaccines, which may be necessary depending on specific risks or regional disease prevalence.

Core Vaccines:

- **Bovine Viral Diarrhea (BVD):** Prevents a virus that can lead to significant respiratory and reproductive issues.

- **Bovine Respiratory Syncytial Virus (BRSV):** A major cause of respiratory disease in cattle.

- **Leptospirosis:** A bacterial infection that can lead to kidney damage, abortion in pregnant cows, and reduced milk production.

- **Clostridium chauvei (Blackleg):** Causes rapid death in cattle, typically affecting young animals.

- **Clostridium septicum (Malignant Oedema):** Leads to severe swelling and often results in death.

- **Clostridium tetani (Tetanus):** A disease that causes muscle stiffness and can be fatal.

Optional Vaccines:

- **Bovine Herpesvirus Type 1 (BHV-1):** Causes infectious bovine rhinotracheitis (IBR), a respiratory disease.

- **Parainfluenza Virus (PI3):** A common contributor to respiratory disease complexes.

- **Pasteurella multocida:** A bacterium that can cause pneumonia in cattle, particularly in feedlots.

Vaccination Plan by Cattle Group

1. **Calves:** Early protection is key, with vaccinations such as BVD, BRSV, and clostridial vaccines administered shortly after birth. Vaccines are crucial during high-stress periods, like weaning, to prevent disease outbreaks.

2. **Yearlings:** At this stage, vaccines such as **BHV-1** and **PI3** are introduced alongside boosters for diseases like **BVD** and **BRSV**. This is essential for preventing respiratory illnesses during their development.

3. **Adult Cows:** Regular boosters for diseases like **BVD** and **leptospirosis** ensure ongoing protection, particularly during breeding seasons. Pregnant cows also require specific vaccines to protect the unborn calf from infections that could cause abortion or stillbirth.

4. **Pregnant Cows (Pre-Calving):** Specific vaccines, such as those for **rotavirus** and **coronavirus**, help protect newborn calves from digestive diseases, which are common causes of calf mortality in the first few weeks of life.

Implementing the Vaccination Plan

A successful vaccination plan relies on more than just selecting the right vaccines. Accurate dosage, correct injection techniques, and proper storage of vaccines are crucial. Following manufacturers' guidelines ensures that the vaccines remain effective and prevent adverse reactions. Keeping detailed records of vaccine administration helps manage booster schedules and track herd health.

Lamb Vaccination

Vaccination is critical in ensuring the health and productivity of lambs, beginning from their earliest stages of life. A comprehensive vaccination program protects lambs from several diseases, especially during vulnerable periods like marking and weaning. These procedures can weaken lambs' immune systems, increasing their risk of illness or death. Administering vaccines correctly, at the right time, significantly boosts lamb survival rates and ensures long-term herd health.

Before lambs are capable of developing their own immunity, they rely on maternal antibodies passed from the ewe's colostrum. When ewes are vaccinated four weeks before lambing, these antibodies are produced in the udder and transferred to the lamb when they consume the colostrum within the first 48 hours after birth. These antibodies provide protection for 6–10 weeks against diseases like tetanus and pulpy kidney. After this period, lambs must begin developing their immunity through their own vaccinations.

Lambs are particularly vulnerable during marking and weaning, both physically and immunologically. These processes create stress and, if not supported by proper vaccination, can lead to disease, reduced growth rates, or even death. Vaccination at these stages reduces risks, ensuring that lambs recover quickly and remain productive. For example, clostridial vaccines administered at marking help prevent a range of fatal bacterial diseases such as blackleg, malignant oedema, and tetanus.

Specific Vaccine Protocols for Lambs

Clostridial Diseases: Vaccination for clostridial diseases, such as pulpy kidney, black disease, and tetanus, is essential across all sheep breeds. The common vaccines, like Ultravac 5-in-1 and Glanvac 6-in-1, protect lambs from these diseases. Typically, lambs should receive two doses:

1. First dose at marking (usually when the lamb is around 3 weeks old).

2. Second dose at weaning, approximately four weeks later.

Caseous Lymphadenitis (CLA): For protection against cheesy gland (CLA), caused by the bacterium *Corynebacterium pseudotuberculosis*, lambs receive vaccines like Glanvac 6-in-1. This vaccine is administered along with clostridial vaccines. Like other vaccinations, lambs need two doses: one at marking and one at weaning. Annual boosters are necessary to maintain immunity, especially before stressful events like shearing or joining.

Erysipelas Arthritis: Lambs in areas prone to Erysipelas arthritis, a bacterial infection, should receive the Eryvac vaccine. The vaccination schedule involves two doses, one at marking and another at weaning. Vaccination against Erysipelas helps prevent polyarthritis, which can severely affect the mobility and productivity of lambs.

Scabby Mouth: Vaccination for scabby mouth, a viral infection causing lesions around the mouth and face, is recommended in areas where the disease is prevalent. Scabigard, a live vaccine administered at marking, provides 12 months of protection after a single dose. Monitoring for a vaccine "take" (a small scab indicating success) is essential to confirm effective vaccination.

The timing of vaccinations is crucial. Administering vaccines too early, while lambs still carry high levels of maternal antibodies, can interfere with the effectiveness of the vaccine. For this reason, vaccination protocols are carefully designed to ensure the first dose is given when lambs are at least three weeks old, followed by a booster four weeks later to strengthen immunity.

Annual booster shots are also vital to ensure long-term protection, especially for diseases like pulpy kidney, which can pose high risks during times of dietary changes, such as transitioning to lush pasture or grain feeding.

Proper vaccination technique is important for maximizing the effectiveness of the vaccine and minimizing adverse effects. Vaccines should be administered subcutaneously, on the side of the neck, avoiding the base of the ear and critical spinal structures. Using the correct needle gauge and length ensures the vaccine is delivered properly. For lambs, an 18-gauge needle of ¼ inch is typically recommended, inserted at a 45° angle.

Maintaining hygiene during vaccination is essential to prevent secondary infections. Clean needles should be used for each lamb, and vaccines should be stored and handled according to label instructions to maintain potency. Furthermore, human safety is important, especially when handling live vaccines like Gudair for Ovine Johne's Disease, as accidental human injection can lead to severe complications.

Vaccination plays a critical role in ensuring lamb health from birth through adulthood. Following a structured vaccination protocol, including booster doses and proper techniques, prevents serious diseases and promotes herd productivity. Combining vaccination with other management practices, like good biosecurity and handling hygiene, helps secure the health of both lambs and the broader sheep enterprise.

Using a Vaccinator

Vaccinators are essential tools in livestock vaccination programs, ensuring that vaccines are administered safely and effectively. Each vaccinator is designed with specific features to enhance accuracy, ease of use, and safety for both the animal and the administrator. Here's a detailed breakdown of some types of vaccinators and their benefits in livestock management:

1. Quickshot 1ml Injector

- **Administration Type:** 1ml injection

- **Product Compatibility:** Glanvac or Eryvac

- **Features:** The Quickshot injector is a simple device that delivers 1ml doses of vaccines. It is suitable for basic vaccination tasks where a protective shroud is not required. This device is easy to use, making it a common choice for small farms.

2. Glanvac 1ml Safeshot

- **Administration Type:** 1ml injection

- **Product Compatibility:** Glanvac or Eryvac

- **Features:** The Safeshot offers a higher level of safety with a protective needle shroud, which minimizes the risk of accidental needlestick injuries. This feature is crucial when working with livestock, especially for less experienced handlers, as it improves vaccination safety.

3. Glanvac Eryvac Dual Vaccinator 2ml

- **Administration Type:** Simultaneous 2 x 1ml injection

- **Product Compatibility:** Glanvac and Eryvac

- **Features:** This dual injector allows for the simultaneous administration of both Glanvac and Eryvac vaccines through a single injection. The needle is shrouded for added safety, reducing the risk of accidental self-injection. The ergonomic design of this device enhances usability, making it an efficient option for large-scale operations. It helps farmers save time and labor by administering two vaccines at once.

4. Gudair 1ml Sekurus Vaccinator

- **Administration Type:** 1ml injection

- **Product Compatibility:** Gudair, Glanvac, Eryvac

- **Features:** This vaccinator is designed with advanced safety features such as a two-stage safety locking mechanism and a shrouded needle. These features minimize the risk of accidental self-injection, which is particularly important when working with vaccines like Gudair, which are known for their viscous consistency. The Sekurus also has a top-fed draw-off system, which makes it easy to refill the syringe, ensuring accurate doses.

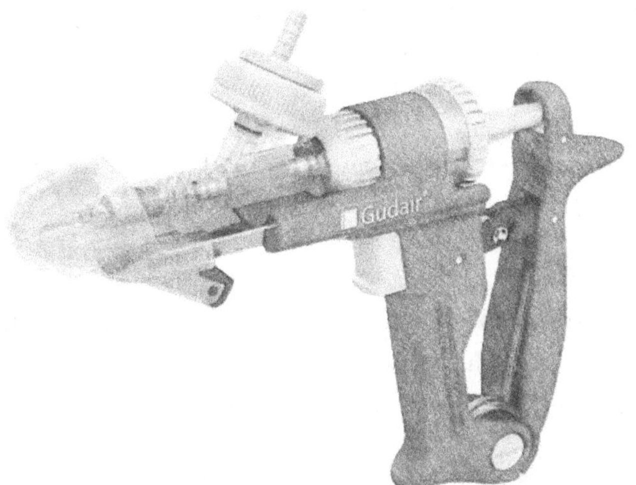

Figure 48: Gudair Sekurus 1mL Safety Vaccinator.

The Gudair Sekurus Vaccinator is specifically designed for the safe and accurate delivery of Gudair vaccines. It includes unique safety features, such as a safety release trigger and a V-lock mechanism, which minimize the risk of accidental self-injection. This is particularly important when handling potentially hazardous vaccines like Gudair.

The device is easy to use, service, and maintain, and it delivers a fixed 1ml dose with precision. The shrouded needle helps tent the skin for subcutaneous delivery, reducing the likelihood of adverse reactions at the injection site. This makes the Gudair Sekurus an ideal choice for large-scale sheep vaccinations, ensuring both operator safety and animal welfare.

To use the vaccinator effectively, follow these two stages for proper and safe administration:

Stage 1: Preparing the Vaccinator

1. **Squeeze the Safety Release Trigger** – Begin by gently squeezing the safety release trigger. This will engage the shroud guard slide mechanism. Applying too much pressure at this stage can cause the mechanism to lock up, making it difficult to use.

2. **Engage the Shroud** – As you begin injecting the sheep, apply gentle pressure on the shroud. The shroud will slide back along the metal guide, exposing the needle subcutaneously. This should be a smooth, controlled push.

3. **Avoid Excessive Force** – Don't increase the force through the vaccinator, as this can complicate the action by making it harder for the safety components to release.

4. **Let the V-Grip Collapse** – Do not manually squeeze the V-Grip. Allow it to collapse naturally as part of the vaccinator's mechanism.

Stage 2: Administering the Vaccine

1. **Full Retraction and Dose Delivery** – Once the shroud has fully retracted, the Safety V-Lock mechanism will be released, delivering a precise 1mL dose of vaccine. The vaccinator will automatically collapse, completing the injection.

2. **Proper Angle for Injection** – Maintain the vaccinator at a 45° angle to the sheep's skin. Push the vaccinator forward to create a tent in the skin for subcutaneous injection.

3. **Avoid Self-Injection** – Keep your body and hands away from the vaccinator during the process to avoid accidental self-injection. **Do not use your free hand to tent the skin**; instead, rely on the forward motion of the vaccinator to create the tenting.

This process ensures accurate dosing and reduces the risk of injury to both the operator and the animal while ensuring effective vaccine administration.

5. Scabigard Ezi-Grip

- **Administration Type:** Scratch 0.02ml injection

- **Product Compatibility:** Scabigard

- **Features:** This vaccinator is ergonomically designed for ease of use, especially when applying Scabigard, a live vaccine used to prevent scabby mouth in sheep. Its design helps ensure accurate dosing through a simple scratch method, making it easy to use even in large flocks.

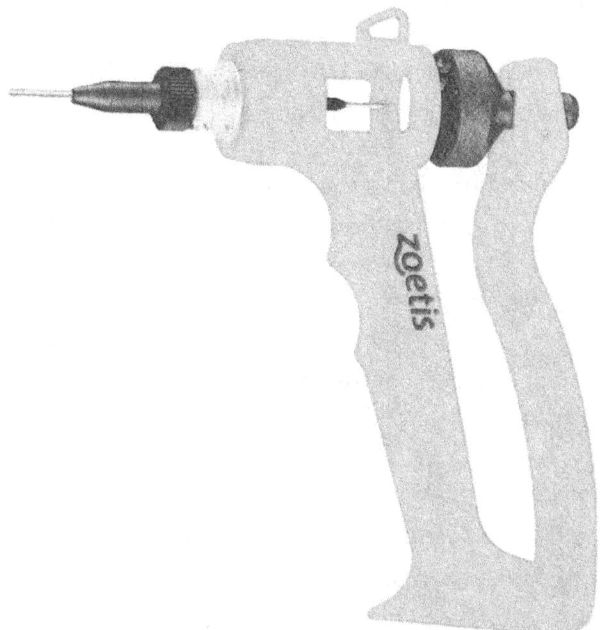

Figure 49: Zoetis Scabigard® EziGrip Vaccinator.

6. Scabigard Scratch Vaccinator

- **Administration Type:** Scratch 0.02ml

- **Product Compatibility:** Scabigard

- **Features:** This vaccinator has a low-profile design to minimize the risk of improper application. The 0.02ml dose is precisely administered, and handlers are advised to check for a "vaccine take" (scab) 10 days after vaccination to ensure successful application.

The **Scabigard® Applicator** is designed for easy administration of the Scabigard vaccine to sheep, offering a practical and safe way to apply the vaccine. The following provides a guide on how to use the applicator effectively:

Step-by-Step Instructions for Using the Scabigard® Applicator

Step 1: Preparation and Safety Measures

Before you begin using the Scabigard applicator, it is important to prepare properly:

- **Put on Personal Protective Equipment (PPE):** This includes gloves to prevent accidental contact with the vaccine, which is live and can cause skin infections in humans.

- **Prepare the Applicator**: Open the handle of the applicator and securely insert the Scabigard bottle onto the draw-off needle inside the applicator.

- **Ensure Proper Fit**: Close the gun handle, making sure the vaccine bottle is securely attached to the handle mechanism.

Step 2: Priming the Applicator

- **Priming**: Once the bottle is attached, hold the applicator tips facing downward. Pump the applicator around 10 times until a small drop of vaccine appears at the tips. This indicates that the applicator is primed and ready for use.

Step 3: Administering the Vaccine

- **Positioning**: Hold the applicator at a **45-degree angle** to the sheep's skin. Ensure that both prongs of the applicator are touching the skin to administer the vaccine.

- **Scratching the Skin**: Make a single scratch on the sheep's skin approximately 4–5 cm in length. Press firmly enough to scratch the surface of the skin, ensuring proper absorption of the vaccine, but avoid pressing too hard to prevent bleeding, which could dilute the vaccine and reduce its effectiveness.

- **Pumping**: Pump the applicator once before each scratch to release a small amount of vaccine onto the tips, ensuring that the vaccine is effectively applied.

Step 4: Important Precautions

- **Avoid Chemicals**: After applying the vaccine, do not apply dips, disinfectants, or other chemicals to the scratched area, as these could interfere with the vaccine's effectiveness.

- **Maintaining the Applicator**: Over time, the applicator tips may become dull from frequent use. It is important to either sharpen or replace the tips regularly to maintain their effectiveness in creating scratches.

- **Avoid Over-Scratching**: Do not scratch too deeply or cause bleeding, as this can dilute the vaccine at the application site, reducing its efficacy.

Safety Considerations

- **Human Contact Risk**: Scabigard is a live vaccine and can cause infections in humans if it comes into contact with skin. It's essential to cover any cuts or abrasions on your hands before handling the applicator. Additionally, always refit the safety cover after use to prevent accidental exposure.

- **Remove Safety Cover Before Priming**: Ensure that the safety cover is removed before priming the applicator, and refit it after you are done using the applicator to avoid accidental exposure to the vaccine.

Recording Withholding Periods and Other Details of Treatment

Maintaining proper records of livestock treatments, including vaccines, is a key responsibility for livestock owners worldwide. This responsibility ensures the safety of food products, the health of animals, and compliance with regulatory standards. Let's explore the importance of recording withholding periods and treatment details for livestock production enterprises, using examples from various regions.

Recording livestock treatments is critical for two main reasons: ensuring animal health and adhering to food safety standards. Proper documentation helps track the Withholding Period (WHP)—the time after treatment during which an animal or its products (such as milk or meat) cannot be sold for human consumption. This ensures that any veterinary drugs or chemicals have adequately cleared from the animal's system, protecting human health.

Additionally, the Export Slaughter Interval (ESI) must be observed for animals whose products will be exported. This interval often differs from the WHP, as international standards may vary. Livestock owners must strictly follow both WHP and ESI to avoid fines, product recalls, and other legal issues.

For instance, in Victoria, Australia, livestock owners must adhere to strict guidelines laid out by Agriculture Victoria and industry bodies like Meat & Livestock Australia (MLA). The Livestock Production Assurance (LPA) program requires all producers to document the date of treatment, the specific livestock treated, and the trade name, batch number, and dose of chemicals used. These records must be kept for at least three years or as long as the livestock remains on the property.

In Europe, similar regulations exist. The European Union's Animal Health Law requires farmers to keep detailed records of all treatments, including vaccines. This system is part of the broader traceability framework, which tracks animal movements and treatments from birth to slaughter, ensuring food safety and disease control.

In the United States, regulations set by the U.S. Department of Agriculture (USDA) also mandate the recording of vaccine use and veterinary treatments, particularly for animals that enter the human food chain. Failure to comply with these guidelines can result in significant penalties, including the potential rejection of livestock from meat processors or exclusion from export markets.

Practical Application: Using the National Livestock Identification System (NLIS)

The National Livestock Identification System (NLIS) in Australia, while primarily a tool for tracking animal movements and ensuring traceability, also helps producers document treatments like vaccinations. Through NLIS, producers can link individual animals to their treatment records, including WHP and ESI data. This system allows for better management decisions based on individual animal health histories, improving productivity and biosecurity.

Similarly, in the United Kingdom, the British Cattle Movement Service (BCMS) operates a traceability system for cattle. Livestock owners must report any treatments or movements of cattle within specific timelines, helping control disease outbreaks and maintaining public confidence in food safety.

Implementing a Vaccination Plan

A well-structured vaccination plan not only protects the herd from disease but also supports compliance with legal and market requirements. Vaccines should be chosen based on local disease risks, herd size, and farming objectives. For example, a dairy operation may prioritize vaccines for diseases like leptospirosis and clostridial infections, while a feedlot may focus more on respiratory diseases.

After administering vaccines, livestock owners must:

1. Record the type of vaccine used, including the manufacturer's details.

2. Document the dosage and batch number to track the specific treatment given to each animal.

3. Note the WHP and ESI for each vaccine to ensure compliance with legal standards before the livestock or its products enter the market.

Challenges and Solutions

Maintaining accurate records can be a challenge, especially for large operations. The use of technology can simplify the process. Electronic databases like NLIS in Australia or similar platforms in other regions allow producers to track treatments, movement, and performance data with minimal manual effort. These systems often come with backup options to prevent data loss, further streamlining the process.

For example, in New Zealand, the National Animal Identification and Tracing (NAIT) system offers an electronic platform that integrates with farm management software, allowing farmers to easily record treatments and monitor WHP and ESI compliance. This system helps mitigate the risks associated with manual record-keeping and ensures that farmers meet all necessary legal requirements.

Proper record-keeping of livestock treatments, including withholding periods and vaccination details, is a global requirement that ensures food safety, animal health, and compliance with national and international laws. Systems like NLIS, BCMS, and NAIT provide essential traceability and allow producers to manage their operations efficiently while adhering to legal and industry standards. Whether you are in Australia, Europe, the U.S., or elsewhere, keeping detailed records of livestock treatments is not only a legal obligation but also a practical way to protect your herd and business.

Withholding Period (WHP) Compliance

In the global livestock industry, identifying and isolating treated animals is essential for ensuring compliance with Withholding Periods (WHPs). This is crucial to avoid harmful chemical residues entering the food chain, which could pose health risks to consumers and jeopardize market access. WHPs are the minimum period required between the last administration of a veterinary or agricultural chemical and the time when animals or their products can be safely sold for human consumption. Failing to comply with WHPs or failing to isolate treated animals from non-treated ones could result in serious legal, financial, and reputational consequences for livestock producers.

One of the first steps in managing WHP compliance is identifying which animals have been treated with chemicals, such as antibiotics, pesticides, or hormone growth promotants (HGPs). This is often done using National Livestock Identification Systems (NLIS), which exist in many countries like Australia and New Zealand. These systems use ear tags or electronic identification (EID) to track individual animals from birth to slaughter, allowing producers to monitor treatment history, including when treatments were administered and when the WHP will expire. For instance, in Australia, the NLIS system is mandatory and forms part of the broader Livestock Production Assurance (LPA) program, which ensures that all livestock movements and treatments are tracked and recorded.

Similarly, in Europe, livestock producers must comply with regulations under the European Union's General Food Law, which mandates the identification and traceability of all animals that have received treatments. The British Cattle Movement Service (BCMS) in the UK, for example, maintains detailed records of cattle movements and treatments to ensure compliance with WHP requirements.

Residue contamination occurs when animals or their products (meat, milk, or eggs) contain traces of chemicals that exceed the Maximum Residue Limit (MRL). This can happen if WHPs are not observed, if incorrect doses of veterinary medicines are administered, or if animals are exposed to environmental contaminants such as heavy metals (lead or cadmium) or pesticides. Ensuring that livestock do not contain harmful residues is critical for maintaining consumer safety and access to international markets.

To prevent harmful residues, producers must follow strict guidelines, including:

- Following label directions for all chemical treatments or veterinary medicines.

- Observing grazing WHPs, which dictate how long animals must be kept off chemically treated pastures or feed crops.

- Isolating treated animals from non-treated stock to prevent cross-contamination. For example, in New Zealand, treated animals are often moved to separate paddocks or housed in isolation pens to ensure they are not processed for human consumption until the WHP has passed.

- Conducting regular farm risk assessments to identify potential sources of contamination, such as old machinery, dumpsites, or areas previously sprayed with hazardous chemicals.

Isolating treated animals is especially important in feedlots and intensive farming operations. In the United States, for example, the Food and Drug Administration (FDA) enforces strict residue monitoring programs to ensure that meat and dairy products are free from harmful levels of residues. Feedlots often use segregated pens to isolate animals that have been treated with veterinary drugs. Similarly, in Canada, the Canadian Food Inspection Agency (CFIA) mandates that treated animals be clearly identified and isolated during the WHP to prevent any risk of contamination in food products.

Isolation measures are also critical when animals are exposed to environmental contaminants such as lead from old machinery or batteries, which can be ingested by livestock. Farms in regions with a history of industrial activity or where persistent organic pollutants have been used in agriculture (e.g., old orchards treated with organochlorine pesticides) must take extra precautions to prevent exposure to these residues.

Once animals are treated, producers must maintain accurate records to ensure compliance with WHP and MRL standards. In Australia, livestock producers are legally required to keep records of all treatments under the LPA program. These records include details such as the treatment date, product name, batch number, dosage, and WHP or Export Slaughter Interval (ESI). These records must be updated whenever animals are treated, and producers must declare whether animals are within or outside the WHP on the National Vendor Declaration (NVD) form when selling livestock.

In South Africa, the Red Meat Industry Forum (RMIF) requires similar documentation under the Livestock Registration and Traceability System (LRTS), ensuring that treated animals are not prematurely slaughtered or sold.

Effective management of WHPs and the identification and isolation of treated animals are critical steps in preventing harmful residues in livestock production worldwide. By adhering to national and international regulations, following veterinary directions, and using systems like NLIS or similar traceability frameworks, livestock producers can safeguard public health, maintain market access, and protect the reputation of their farms. Whether in Australia, Europe, or North America, responsible management practices are essential to ensuring compliance with food safety standards and avoiding costly legal repercussions.

Isolating Sick Calves

Isolating sick calves is an essential practice in livestock management to prevent the spread of disease within calf-rearing environments. Proper isolation minimizes the risk of both direct and indirect transmission of pathogens to healthy calves, protecting herd health and maintaining productivity. Globally, farms implement these strategies to reduce disease outbreaks, enhance animal welfare, and adhere to biosecurity standards.

Cross Infection: Direct and Indirect Contact

The transmission of diseases among calves can happen through two primary routes: direct contact and indirect contact. Direct contact occurs when a sick calf directly interacts with healthy calves, transferring pathogens through bodily fluids like saliva, mucus, or faeces. Indirect contact happens when calves are exposed to contaminated materials, such as bedding, boots, clothing, or feeding equipment that have been in contact with a sick calf. For example, in New Zealand, where intensive calf-rearing is common, direct contact from group housing systems and shared feeding equipment increases the risk of spreading diseases such as Bovine Respiratory Disease (BRD) and scours (diarrhea caused by infections like rotavirus or E. coli).

In the United States, feedlots and dairy farms employ strict isolation and sanitation measures to prevent outbreaks of diseases like cryptosporidiosis, which can devastate entire groups of calves if transmitted indirectly through contaminated water or feeding equipment. In both cases, isolating sick calves in designated pens helps reduce this risk by physically separating infected animals from the rest of the herd.

Quarantine and Group Isolation

If illness is detected late, it's likely that the infected calf has already exposed its neighbouring penmates to the disease. In these cases, it's essential to quarantine the entire group instead of just the visibly sick calf. This approach reduces the chances of spreading the disease to other pens or groups. In European countries such as the Netherlands, where dairy farms are closely monitored for disease outbreaks, isolating entire groups of exposed calves and leaving neighbouring pens empty are common biosecurity practices. Solid barriers between pens are also installed to reduce the risk of transmission through airborne particles or accidental contact.

Moving Sick Calves: Reducing Risk

When transferring sick calves to isolation areas, farmers should plan and execute the process with care to minimize contamination risks. The movement should be done in a controlled manner, using separate pathways to avoid contact with healthy animals. This is particularly important in densely populated farming systems, such as those found in India, where diseases like foot-and-mouth disease and calf diarrhea can spread quickly in large herds. By creating separate isolation zones far from healthy livestock, farmers can protect the rest of the herd from exposure.

Hygiene and Disinfection

High standards of hygiene and disinfection are crucial in managing disease outbreaks. Farmworkers who handle sick calves must disinfect their hands, boots, and clothing after every interaction with infected animals. For example, in Australia, biosecurity guidelines emphasize the need for separate gear when attending to sick calves, such as using different overalls, boots, and feeding equipment to avoid cross-contamination. It is common practice to attend to sick calves last during feeding or other tasks, reducing the chances of spreading disease to healthy calves.

In Canada, farms dealing with calf diseases like Johne's disease ensure that disinfectants and proper sanitation measures are used not only on equipment but also on vehicles and tools that may have come into contact with infected animals. Strict biosecurity measures, such as using separate feeding bottles and keeping healthy calves far from sick ones, help contain outbreaks.

Examples from Around the World

- **United Kingdom**: Farms adhere to the **Farm Animal Welfare Committee (FAWC)** guidelines, which advocate for calf isolation facilities equipped with proper ventilation, bedding, and sanitation protocols to manage diseases such as **bovine tuberculosis.**

- **Brazil**: Large cattle ranches employ isolation protocols for calves with diseases like **bovine viral diarrhea (BVD)**. Isolation pens are often located far from the main herd, and workers follow stringent hygiene practices to prevent contamination.

- **Denmark**: Dairy farms have dedicated isolation areas for calves affected by **Mycoplasma bovis** and **calf pneumonia**. Calves are isolated early, and specific protocols are in place to clean and disinfect pens, equipment, and feeding tools.

Isolating sick calves is a globally recognized practice for preventing the spread of infectious diseases in livestock. From ensuring proper hygiene and sanitation to establishing solid barriers and designated isolation areas, farmers worldwide employ various strategies to protect their herds. By taking these precautions, farms can maintain herd health, reduce losses from disease outbreaks, and uphold food safety standards. Proper isolation, hygiene, and disease management contribute to improved animal welfare and more sustainable livestock production.

CHAPTER TEN

Completing the Treatment Process

Counting Animals Out, Preparing and Moving Along a Planned Route Without Damage to Person, Property or Environment

Safely Moving Cattle

Cattle, as prey animals, have natural instincts that influence their behaviour, especially when they are being moved. They are herd animals, meaning they seek comfort and safety in numbers. When an individual is separated from the herd, it often becomes stressed and may act unpredictably. This can pose risks to both the animal and the handler, so caution must be exercised when working with cattle.

Herd Instincts and Individual Isolation

When moving cattle, it's important to avoid separating individuals from the herd whenever possible. Isolated animals can become dangerous as they may feel vulnerable and stressed. Typically, if a cow breaks away from the herd, it will instinctively attempt to return to the group. Handlers should avoid getting between an isolated animal and the herd as this can increase the animal's stress and lead to dangerous behaviour such as charging or kicking.

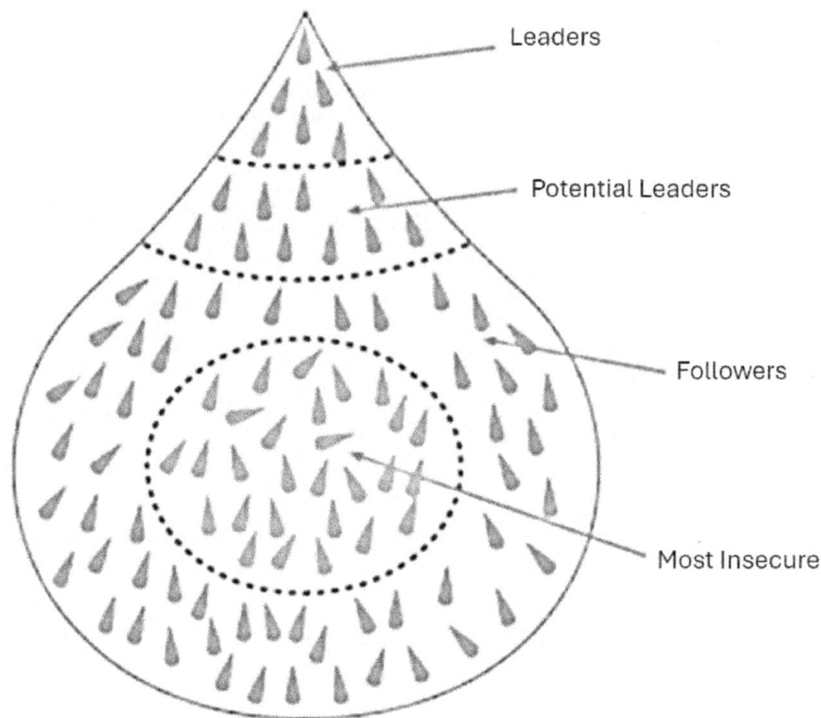

Figure 50: A mob of cattle is structured with leaders at the front guiding the group, potential leaders following closely behind, while the majority of the herd, including the most insecure animals, remain in the centre for protection and stability.

An example from ranching practices in Australia emphasizes the importance of respecting cattle's herd instincts. Australian cattle stations often cover vast areas, and cattle are moved over long distances. Handlers use calm, slow movements to allow the cattle to remain together and avoid stress. Separating an individual cow can cause unnecessary agitation, increasing the risk of accidents.

Startling Movements and Objects

Cattle are easily startled by sudden movements and unfamiliar objects. Their vision is designed to detect motion, so flapping materials or even wind-blown papers can cause them to balk or change direction unexpectedly. In New Zealand, farmers often take care to remove objects that might startle cattle in yards and races to ensure smooth movement through these spaces. Sudden changes in the environment can make cattle uneasy, leading to unpredictability during handling.

Flight Zone and Blind Spots

Cattle have a blind spot directly behind them, and entering this area too quickly can trigger a defensive reaction, such as kicking. Experienced handlers understand that approaching cattle from the side, rather than from behind, is safer because it avoids this blind spot. The concept of a flight zone, which is the animal's personal space, is key in moving cattle efficiently. The size of the flight zone varies depending on the animal's temperament and handling experience. In the United States, ranchers often train their cattle by regularly walking through the herd to reduce the flight zone, making cattle more accustomed to human presence and easier to handle.

When moving or drafting cattle through gateways, handlers should position themselves carefully, as cattle tend to move in circular patterns around them. By understanding how cattle move in relation to their handler, it is possible to control their movements with less stress. In Canada, ranchers moving cattle through races use a zig-zag motion behind the herd to maintain forward movement, ensuring that the animals can see the handler without feeling trapped or pressured.

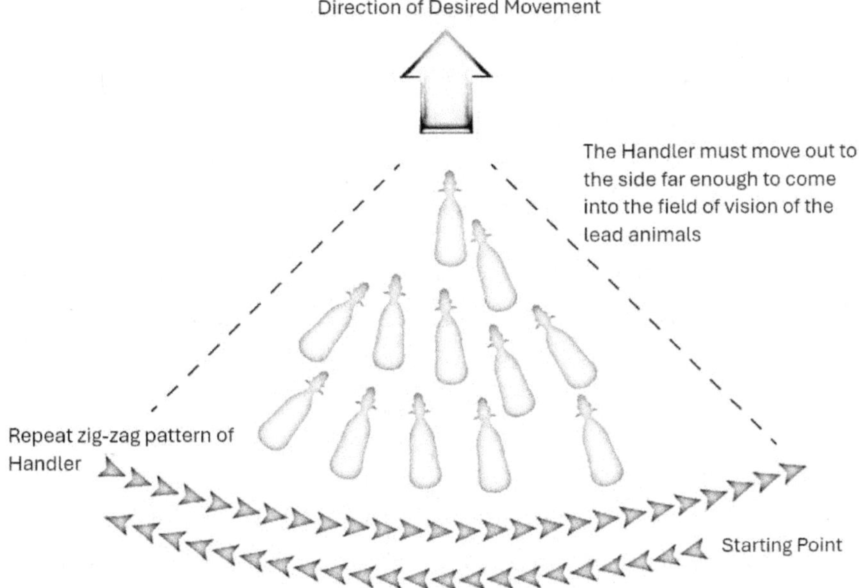

Figure 51: Ranchers move cattle through races by using a zig-zag motion behind the herd, allowing the animals to see them and maintain forward movement without feeling trapped or overly pressured.

Handling Cows with Calves

Special care is required when dealing with cows that have recently calved. Even the calmest cows can become aggressive and highly protective of their newborns. In the UK, farms dealing with dairy cows are particularly cautious during the calving season, ensuring that handlers do not approach newborn calves without appropriate precautions. Protective cows may charge at handlers who get too close, making it important to maintain a safe distance and allow the cow to feel in control of its environment.

Best Practices for Safe Cattle Handling

Effective cattle handling is built on four key principles: Position, Pressure, Movement, and Communication.

1. **Position**: Handlers should work from the side of the animal, ensuring they are visible to the cattle and avoiding the blind spot. This helps to maintain control and reduces the risk of startling the animal.

2. **Pressure**: Pressure should be applied in a controlled manner, and once movement is achieved, it must be released. Handlers can apply pressure by moving towards the cattle and then stepping back once the animals start moving.

3. **Movement**: The handler's movement can influence the cattle's speed and direction. Waving flags or using livestock talkers can encourage movement, but these tools should be viewed as extensions of the handler's body rather than instruments for hitting the animals.

4. **Communication**: Both verbal and non-verbal communication are essential when moving cattle. Calm, consistent commands help reduce stress, and clear communication between handlers ensures smooth operations during mustering or drafting.

Mustering and Yard Work

Mustering, the act of gathering cattle from pastures, should be done at a walking pace to prevent cattle from becoming agitated. Rushing cattle into the yards will result in stressed animals that need time to rest before further handling can occur. In South America, particularly in Brazil, large-scale cattle operations utilize these methods to ensure that cattle remain calm during mustering, reducing injuries and improving meat quality by avoiding stress-induced issues.

When cattle are moved through yards, the same principles apply. Handlers should position themselves where the cattle can see them and use parallel movement to encourage the animals to move forward through races. In Europe, particularly in Ireland, farmers use a combination of voice commands and light pressure to draft animals into crushes or pens, ensuring minimal stress and efficient movement.

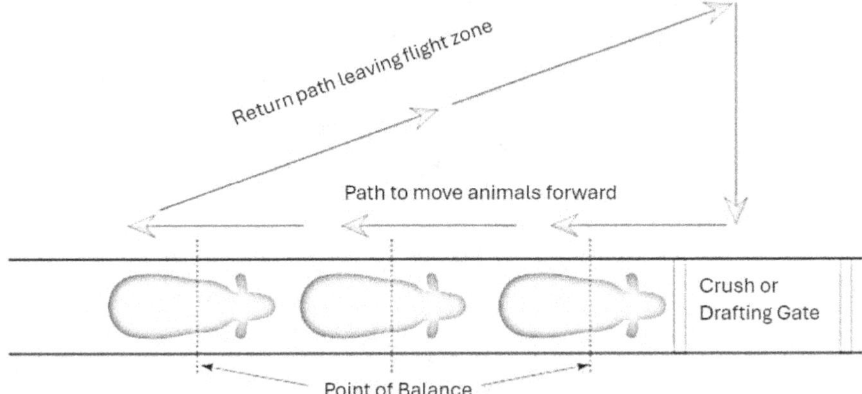

Figure 52: Parallel movement is used to draft animals, count them, and guide them along a race by walking in the opposite direction of their desired movement, encouraging them to move forward.

The safe movement of cattle hinges on understanding their natural behaviour, including herd instincts, flight zones, and reactions to stressors. By applying calm, controlled handling methods and maintaining effective communication, handlers around the world—from Australian outback stations to European dairy farms—can move cattle efficiently while minimizing stress and the risk of injury to both animals and humans. Proper planning, such as positioning gates for optimal flow and preparing equipment in advance, ensures smoother operations and better outcomes in both cattle health and productivity.

Monitoring Animal Health and Condition Post-Treatment and Reporting Abnormalities

Aftercare for sheep following drenching is crucial to ensure their well-being and the effectiveness of the treatment. After drenching, it is important to observe sheep closely for any abnormal signs such as coughing, convulsions, tremors, or unsteadiness, which may indicate toxicity. Toxicity risks are higher with certain drenches, especially those containing organophosphates, levamisole, or abamectin, particularly in lambs. If toxicity is suspected, it is essential to contact a veterinarian immediately for treatment.

One common issue during drenching is sheep spitting out the drench, which signals improper technique. The correct procedure involves holding the sheep's head level and placing the tip of the drench gun barrel over the sheep's tongue to ensure the dose is properly administered. If mistakes occur, restarting the process ensures the sheep receive the full dosage for effective treatment.

Proper pain management is an essential aspect of animal welfare, especially after medical procedures or injury. Training employees to recognize behavioural and physiological signs of pain—such as avoidance, vocalization, inflammation, or increased cortisol levels—helps ensure timely interventions. Restraining the animal correctly and administering pain relief, such as local anaesthetics or anti-inflammatory drugs, reduces discomfort and prevents further stress. Continuous monitoring post-treatment is also necessary to determine if additional pain management is required, and all treatments must be logged in individual health records for future reference.

When disease or injury occurs, the priority is to assess whether the animal can recover. If so, they should be segregated from the herd into a designated recovery area to avoid further complications. Providing prompt medical care, monitoring for any signs of worsening condition, and ensuring a clean, comfortable environment with access to food and water are crucial for rehabilitation. In cases where recovery is unlikely, timely decisions regarding euthanasia must be made in accordance with proper guidelines to avoid prolonging suffering.

Handling non-ambulatory animals (those unable to stand or walk) requires special care to minimize additional stress or injury. Such animals should be moved carefully using proper equipment like sleds or loaders and placed in an isolated, comfortable area. They must be monitored regularly, rolled from side to side to prevent muscle damage, and supported with slings or float tanks if necessary to help them regain mobility. If the animal shows no improvement, euthanasia may be considered following veterinary guidance.

These practices are essential for ensuring that livestock are treated humanely and effectively, especially after procedures like drenching or in cases of illness and injury. Proper aftercare, pain management, and monitoring help maintain the health and welfare of the animals, which, in turn, supports overall farm productivity.

Environmental Implications Associated with the Treatment

Food-producing animals, particularly major trade species such as cattle, sheep, pigs, and chickens, are subject to stringent regulations to ensure the safety of food and fibre products for human consumption. The use of chemicals, including veterinary drugs and treatments, on these animals is closely monitored to reduce risks to both human health and international trade. Key regulations ensure that treated animals are properly identified and monitored through tagging systems like the National Livestock Identification System (NLIS), isolation from other animals, and meticulous record-keeping throughout treatment and withholding periods.

Specific examples of regulations governing food-producing animals, particularly major trade species like cattle, sheep, pigs, and chickens, from around the world include:

1. European Union (EU)

- **Regulation (EC) No 178/2002**: Establishes general principles and requirements of food law within the EU, emphasizing traceability, food safety, and the prevention of contaminants in food products, including those derived from livestock.

- **Regulation (EC) No 470/2009**: Governs the use of veterinary medicines in food-producing animals and sets maximum residue limits (MRLs) for substances to ensure food safety.

- **Directive 2001/82/EC**: Lays down the guidelines for the veterinary medicinal products used in animals and defines withholding periods for medicines in food-producing animals to ensure that residues do not exceed MRLs.

2. United States

- **Food and Drug Administration (FDA)**: The FDA's **Center for Veterinary Medicine (CVM)** ensures that animal drugs, feeds, and veterinary products used on food-producing animals meet safety standards. Regulations include mandatory withholding periods after the administration of drugs to ensure safe residue levels in meat, milk, and eggs.

- **USDA National Residue Program (NRP)**: Monitors and tests meat, poultry, and egg products for harmful residues to ensure compliance with U.S. laws.

- **Federal Meat Inspection Act (FMIA)** and **Poultry Products Inspection Act (PPIA)**: These laws mandate the inspection of meat and poultry products to ensure they are free of harmful contaminants and safe for human consumption.

3. Australia

- **Australian Pesticides and Veterinary Medicines Authority (APVMA)**: The APVMA regulates the use of veterinary medicines in food-producing animals, ensuring compliance with maximum residue limits (MRLs). The **National Residue Survey (NRS)** monitors residues in meat and other livestock products to maintain safety.

- **National Livestock Identification System (NLIS)**: A traceability system for identifying and monitoring livestock to ensure food safety and quality standards are met.

- **Livestock Production Assurance (LPA)**: An on-farm food safety program that mandates the recording of chemical use, withholding periods, and movement declarations for livestock to prevent contamination.

4. New Zealand

- **Ministry for Primary Industries (MPI)**: Oversees the safety and quality of food products derived from animals. This includes the regulation of veterinary medicine use and ensuring that residues do not exceed acceptable levels in food products.

- **Animal Products Act 1999**: Requires all food-producing animals to be processed according to strict guidelines to ensure that products are free from harmful residues.

5. Canada

- **Canadian Food Inspection Agency (CFIA)**: Monitors and regulates the use of veterinary drugs in food-producing animals. The CFIA enforces compliance with MRLs set by **Health Canada** and ensures that food products derived from livestock are safe for consumption.

- **Safe Food for Canadians Regulations (SFCR)**: These regulations set safety standards for livestock production and the handling of veterinary drugs, ensuring that no harmful residues enter the food supply chain.

6. Japan

- **Food Safety Basic Act**: Establishes the fundamental principles of food safety in Japan, with specific regulations on the use of veterinary drugs in livestock to prevent harmful residues.

- **Japan's Positive List System**: Sets MRLs for chemicals and veterinary drugs in food-producing animals to ensure consumer safety, with strict controls on the use of agricultural and veterinary chemicals.

7. Brazil

- **Brazilian Ministry of Agriculture, Livestock, and Supply (MAPA)**: Regulates the use of veterinary medicines in livestock and ensures compliance with MRLs. MAPA enforces regulations for the export of meat products, adhering to both domestic and international food safety standards.

- **National Program for the Control of Residues and Contaminants (PN-CRC)**: This program monitors residues in animal products to ensure they comply with both domestic and international safety standards.

8. South Africa

- **Department of Agriculture, Forestry, and Fisheries (DAFF)**: Ensures that the use of veterinary drugs in food-producing animals is regulated and that residue levels are monitored to comply with safety standards for both domestic consumption and exports.

- **Meat Safety Act, 2000**: Governs the safety of meat products derived from livestock, ensuring compliance with withholding periods and monitoring for residues.

These regulations ensure the safety of food derived from animals by monitoring veterinary drug use, managing chemical residues, and preventing contamination. Each region's specific laws and guidelines are designed to meet both domestic and international safety standards.

In Australia, for instance, the NLIS is used to track livestock movements and treatments. This system includes electronic ear tagging to ensure that treated animals can be identified throughout their lifecycle. Farmers are required to maintain detailed records for at least two years, documenting the use of veterinary chemicals, treatment details, and withholding periods. These records help ensure that animals treated with veterinary drugs do not enter the food supply until it is safe to do so. Similarly, in the European Union and the United States, regulatory agencies enforce strict record-keeping to trace chemical use and ensure compliance with food safety standards.

Effective parasite control in trade-species animals includes both chemical treatments and non-chemical methods. For example, in tick-endemic regions, breeds that are more resistant to cattle ticks are favoured. In Australia, buffalo fly infestations in cattle are monitored, and treatment is only applied when fly numbers exceed specific thresholds. Strategic use of chemical drenches and pour-ons is crucial in overall parasite control programs, but it is equally important that the chemicals are used in line with label instructions, which include dosage, safety precautions, and withholding periods. Non-compliance with these guidelines can lead to unacceptable chemical residues in meat, milk, or wool, which may affect both domestic and international trade.

Globally, residue management is vital to prevent the contamination of animal products with chemicals such as pesticides or veterinary drugs. Residues can enter animal products either directly through treatment or indirectly through contaminated feed or water. For example, persistent organochlorine pesticides, which were once widely used but are now banned, can still be found in soils, affecting livestock that graze on contaminated land. International bodies like the Codex Alimentarius Commission set global standards for acceptable residue levels in food products, while national regulatory bodies such as Food Standards Australia New Zealand (FSANZ) and the U.S. Food and Drug Administration (FDA) ensure compliance with local residue limits.

In countries like Australia, programs such as the National Residue Survey (NRS) monitor residues in livestock products, helping ensure that agricultural exports comply with international standards. In regions with higher risks of contamination, targeted testing and traceability mechanisms help maintain the safety of food products. These efforts are vital for maintaining access to global markets, where even trace amounts of residues can lead to export restrictions.

Countries around the world employ various methods for monitoring chemical residues and contaminants in food products. For example, in Queensland, Australia,

Biosecurity Queensland actively samples agricultural products to detect chemical residues and contaminants. In the United States, the USDA and FDA monitor food safety through residue detection programs, while the European Union has strict residue limits that are enforced across member states. When residue levels exceed acceptable limits, authorities trace the contamination back to its source and take corrective action to prevent further contamination, ensuring food safety for both domestic and international consumers.

Cleaning Equipment and Worksite and Disposing of Waste, Including Animal Residues

Ensuring proper hygiene in handling livestock, maintaining clean equipment, and correctly disposing of waste is critical to preventing disease outbreaks, preserving animal health, and complying with biosecurity standards. This is particularly important for food-producing animals, such as cattle, sheep, pigs, and chickens, which are subject to strict global regulations to ensure the safety of food products for human consumption.

Equipment Hygiene and Storage: Across the world, livestock equipment, including storage containers, tools, and feeding devices, can harbor diseases, pests, or weeds if not properly cleaned and disinfected. Regular cleaning using suitable disinfectants on visually clean surfaces is crucial to ensure that tools and equipment do not carry harmful microorganisms or residues. For example, in Australia, regulations require that feeding equipment, drench guns, and other livestock handling tools are cleaned after each use to prevent contamination and disease spread. After use, drench guns should be thoroughly cleaned with warm soapy water and lubricated to maintain seal integrity and functionality.

In Europe, the EU Animal Health Law (Regulation 2016/429) emphasizes the importance of cleanliness and sanitation for equipment used in animal farming. The goal is to prevent cross-contamination between animals, especially between different farms or high-risk areas like quarantine zones. Equipment must be kept free from soil, manure, and other residues that could harbor pathogens .

Dedicated Equipment for High-Risk Areas: It is good practice to dedicate specific tools and clothing for areas or animals affected by diseases. This reduces the risk of spreading zoonotic diseases, which can be passed from animals to humans, like avian influenza. In Asia, specifically in countries like Japan and China, livestock farms maintain strict biosecurity protocols by segregating equipment used in different animal areas and limiting the transfer of gear from diseased to healthy zones .

Environmental Cleaning: The hygiene of animal environments, including farms and zoos, is crucial. In the United States, USDA guidelines recommend cleaning all areas where animals are kept using detergents and water, followed by appropriate disinfectants if needed. These recommendations apply across various animal-handling sectors, including livestock farming and petting zoos, where regular cleaning of surfaces such as fences, gates, and floors prevents contamination .

In Europe, farm facilities like stables and pens are required to undergo routine cleaning and disinfection to control the spread of animal diseases, such as foot-and-mouth disease. Disinfectants like quaternary ammonium compounds or peracetic acid are recommended, and farms must keep detailed cleaning logs to document their activities.

Handling Animal Residues: Disposing of animal residues properly is important for maintaining biosecurity and preventing disease outbreaks. In many parts of the world, including New Zealand and Brazil, farms are required to dispose of animal waste, such as manure and unused feed, in a manner that avoids contamination of clean areas. This is particularly significant in large-scale operations, where biosecurity protocols, such as in Brazil's beef industry, are designed to ensure waste is disposed of far from healthy stock to avoid contamination .

Personal Protective Equipment (PPE): Personal hygiene and the use of PPE during cleaning and waste handling are universal across farming sectors globally. In Australia, workers are required to wear gloves, masks, and boots when handling animal waste or cleaning equipment exposed to animals with known infections. This prevents the spread of zoonotic diseases and protects human health. Similarly, in the EU and Canada, strict guidelines dictate that staff must change protective outerwear if they move between different sections of the farm, especially when handling sick or quarantined animals.

Managing Chemicals and Residues: To manage chemical residues in food-producing animals, farmers in Canada and Australia follow strict protocols regarding the use of veterinary drugs and agricultural chemicals. These protocols are designed to ensure compliance with maximum residue limits (MRLs) in meat, milk, and eggs. These countries operate national surveillance programs, such as Australia's National Residue Survey, which monitors chemical residues to maintain food safety .

Global Examples

- New Zealand has strict regulations for managing livestock waste and equipment sanitation to preserve its reputation as a producer of clean and green agricultural products. Farm equipment is cleaned with biodegradable, non-toxic disinfectants to reduce environmental impact.

- In Denmark, one of Europe's largest pork producers, waste management and equipment sanitation are heavily regulated to prevent cross-contamination between herds and reduce antibiotic resistance by limiting the overuse of veterinary drugs .

Maintaining clean equipment and disposing of animal residues safely are key practices in livestock farming worldwide. These measures, combined with the use of dedicated tools for high-risk areas and following global biosecurity standards, help to prevent disease outbreaks and ensure the safety of food products derived from livestock. Each country's specific regulations, from Australia's stringent biosecurity protocols to the EU's detailed environmental cleaning practices, ensure compliance and protect both human and animal health.

Documentation

Treatment protocols are essential tools for managing the health of calves in a farm environment. They establish clear, step-by-step instructions for treating common diseases such as scours, pneumonia, and umbilical abscesses, and they ensure that every staff member follows the same procedures, even under stressful conditions. These protocols not only serve to provide consistency but also help maintain high standards of animal welfare, efficiency in treatment, and compliance with veterinary guidelines.

Globally, the use of documented treatment protocols in animal farming is a best practice. For example, in Canada, the Canadian Food Inspection Agency (CFIA) emphasizes the importance of clear treatment protocols in reducing calf mortality and ensuring compliance with health regulations. Similarly, in New Zealand, the Ministry for Primary Industries (MPI) encourages farmers to develop these protocols as part of their livestock health plans, especially when multiple staff members or contractors are involved.

Importance of Treatment Protocols

1. **Consistency in Treatment**: Treatment protocols allow farm workers to apply

consistent medical care. By documenting each step of the treatment process, workers know exactly what actions to take, minimizing the chances of mistakes. In Australia, treatment protocols are often designed for quick reference to ensure that, whether dealing with a sick calf or a larger mob, the process remains streamlined and effective.

2. **Training and Induction**: In farms across the United States, especially in larger operations, having treatment protocols in place is invaluable for training new staff. Staff can quickly learn how to handle common ailments without needing constant supervision. Farms that maintain a large staff often ensure that protocols are accessible to all, reinforcing the same level of care across shifts and among different workers.

3. **Record Keeping for Accountability**: Treatment protocols act as official records that can be referred to if complications arise or if there is a need to audit treatments. In Europe, specifically under the European Union Animal Health Law, detailed records of medical treatments are required to track disease outbreaks and assess the effectiveness of interventions. If there is an adverse reaction or a recurrence of disease, the protocol serves as a reference point for adjustments.

Creating Effective Treatment Protocols

- **In Consultation with a Vet**: In farms in Canada and New Zealand, farmers develop treatment protocols in consultation with veterinarians. The veterinarian helps ensure that the protocols are tailored to the specific needs of the herd, considering local disease prevalence and available treatments.

- **Simplicity**: Protocols should be simple and concise, ideally fitting onto a single page for quick reference. This is particularly important in high-stress situations when decisions need to be made quickly. For example, protocols for treating pneumonia in calves may include steps such as identifying symptoms, dosage amounts, and types of antibiotics or anti-inflammatory drugs.

Keeping Proper Documentation

Across many parts of the world, maintaining detailed records of all treatments administered to livestock is a legal requirement. In Australia, for instance, farmers are required to document the date of treatment, the animal identification or mob identification, the drug or chemical used (including its batch number and expiry date), dosage, and withholding periods (WHP). Such meticulous records are crucial for tracking animal health and ensuring food safety, particularly when animals are sold.

- **Withholding Periods (WHP) and Export Slaughter Intervals (ESI)**: When documenting treatments, it's essential to note the WHP and ESI for each chemical or drug used. This ensures that the animals are not sent for slaughter or sale before the drugs have cleared their system. In New Zealand, compliance with these periods is a major aspect of maintaining the country's stringent biosecurity standards.

- **Adverse Reactions and Contaminants**: Farmers are also expected to record any adverse reactions to treatments, which may indicate the need to adjust protocols. Similarly, animals exposed to physical contaminants, such as broken needles, should be marked and monitored closely to prevent complications.

Specific International Examples

1. **Australia**: The Livestock Production Assurance (LPA) Program mandates that Australian farmers keep records of all treatments and any potential physical contaminants. This helps safeguard the quality of food products, ensuring no harmful residues reach consumers.

2. **United States**: The National Animal Health Monitoring System (NAHMS) emphasizes proper documentation of treatments for disease control in food-producing animals. This system supports traceability and ensures animals treated with off-label drugs are handled responsibly.

3. **European Union**: Under EU Animal Health Regulations, strict documentation practices are required to ensure that treatments are administered according to veterinary guidelines. Records must include details such as dose rate, treatment duration, and withdrawal times to protect food safety and public health.

In conclusion, developing treatment protocols and keeping thorough documentation is critical for maintaining livestock health, training staff, and ensuring compliance with global food safety standards. From the European Union to New Zealand, these practices are integral to maintaining healthy herds and protecting the integrity of animal-derived food products.

Applying Enterprise Biosecurity Policies

Livestock disease outbreaks can cause severe economic harm to individual farms and the broader agricultural community, as seen in major incidents like the 2001 Foot and Mouth Disease outbreak in Great Britain and parts of Europe. This epidemic cost the UK economy an estimated £8 billion, severely impacting both local farmers and global trade. While diseases like Foot and Mouth grab global headlines, equally concerning are slower-developing conditions like bovine Johne's disease and drench-resistant worms, which may not be immediately devastating but can gradually erode livestock health, productivity, and market access.

Biosecurity is critical for preventing the introduction and spread of diseases in livestock, protecting not only individual farms but also the broader livestock industry and international trade. Countries like Australia and New Zealand emphasize strict biosecurity practices to maintain access to lucrative export markets. Australia's rigorous National Biosecurity Strategy ensures the country's agricultural exports meet the standards of trading partners, especially in the Asia-Pacific region, safeguarding multibillion-dollar trade agreements.

Strategies for Effective Biosecurity:

1. **Purchasing Animals with Known Health Status**: A **closed herd** is the best protection against disease introduction. However, when introducing new genetic material for breeding, such as semen or embryos, thorough disease testing is critical. In the U.S., for example, breeding programs often require animals to be free of diseases like **trichomoniasis** and **bovine viral diarrhea virus** (BVDV). Buying cattle directly from breeders who participate in health assurance programs, like CattleMAP in Australia for Johne's disease, minimizes the risk of introducing diseases into the herd.

2. **Quarantine Practices**: Newly introduced animals should be isolated for at least 4 weeks before being mixed with the main herd. This allows for monitoring of any emerging health issues. In regions like New Zealand, strict quarantine periods are enforced, particularly for imported livestock, ensuring that diseases like

Mycoplasma bovis, which significantly affected New Zealand's dairy industry, are detected early and managed effectively.

3. **Disease Prevention through Fencing and Vermin Control**: Preventing contact between neighbouring herds can minimize disease transmission. In areas like South Africa, double fencing or electric fencing is used to prevent nose-to-nose contact between infected and healthy animals, which can spread diseases such as bovine tuberculosis. Additionally, controlling vermin like rats, which can carry Leptospirosis, and wild birds, which can spread avian influenza, is essential. Effective pest control measures have been widely implemented in Canada, especially in regions with dense poultry farming, to prevent the spread of zoonotic diseases.

4. **Vaccination and Treatment**: Vaccination programs tailored to specific regions are a fundamental part of biosecurity. In countries like Brazil, known for its massive cattle industry, vaccinating against Foot and Mouth Disease is compulsory in certain regions to protect both domestic livestock and export markets. Additionally, routine treatments for internal and external parasites during quarantine periods, combined with regular monitoring, help ensure the health status of newly introduced animals matches that of the existing herd.

5. **Vehicle and Equipment Hygiene**: Dirty vehicles and equipment can introduce diseases to a farm. In European countries, particularly those that suffered from the 2001 Foot and Mouth Disease outbreak, farmers are now required to disinfect vehicles before and after transporting livestock. This practice prevents the spread of highly contagious diseases, which can decimate livestock populations.

6. **Managing Feed to Prevent Contamination**: Imported feed can introduce harmful weeds or diseases. In North America, farms are careful to ensure that imported hay or grain does not carry noxious weed seeds or diseases. Feeding stock in small, designated areas ensures that any contamination is confined and easier to manage, reducing the risk of farm-wide infestations.

A well-implemented biosecurity plan, tailored to the specific needs of the farm and region, can prevent catastrophic losses due to livestock disease. Farmers in countries like Australia, New Zealand, Canada, and Brazil have demonstrated the effectiveness of biosecurity practices in protecting their herds, ensuring long-term productivity, and maintaining access to global markets. By quarantining new animals, controlling vermin, and maintaining strict hygiene protocols, farmers can avoid the economic devastation that has been witnessed in past outbreaks like Foot and Mouth Disease. Effective biosecurity is not just a farm-level concern; it is essential to preserving the global livestock trade.

CHAPTER ELEVEN

Livestock Nutrition

Nutrition for Lotfed or Intensively Finished Animals

Ruminant livestock such as cattle, sheep, and goats, when placed in feedlot or intensive finishing systems, require carefully balanced rations to meet their nutritional needs and maximize growth potential. Effective feeding programs for these animals are crucial to ensuring optimal production, animal health, and the ability to meet market specifications. To achieve this, an understanding of ruminant digestive systems, particularly the role of the rumen microbes, is essential.

Nutritional Requirements: Ruminants rely heavily on the health and balance of microbes in their rumen, which break down feed into volatile fatty acids (VFAs), providing the animal with energy and nutrients necessary for body functions and growth. The microbial population—composed of bacteria, protozoa, and fungi—performs various functions, including starch utilization and fibre breakdown. A stable microbial population is necessary for proper digestion and animal health, making the feeding of both microbes and the animal an integrated process. Ensuring these microbes thrive is key to maximizing the feed's nutritive value.

Rations for Feedlot and Intensive Finishing: Rations in feedlots are designed to support rapid weight gain while maintaining animal health. These rations are primarily composed of energy sources, often up to 85%, with grain being the main component. High-energy rations are introduced gradually over 15 to 20 days, allowing the rumen microbes to adjust to the shift from roughage to grain. A sudden introduction of high-grain diets can result in conditions like rumen acidosis, where the rumen environment becomes overly acidic, leading to health issues.

The essential components of feedlot rations include:

- **Protein**: Often supplied by protein meals and sometimes supplemented with non-protein nitrogen to support microbial growth.

- **Energy**: Derived from cereal grains (processed to enhance digestibility), silage, molasses, and by-products such as citrus pulp.

- **Roughage**: Hay or other fibre sources to maintain the rumen's microbial balance and prevent acidosis.

- **Minerals**: Calcium, phosphorus, magnesium, and sulphur are vital for maintaining rumen function and overall animal health.

- **Buffers**: Rumen buffers, such as sodium bicarbonate, are added to high-grain diets to neutralize acid and support microbial function.

Additional Nutritional Enhancements

- **Rumen Modifiers**: Additives like monensin or virginiamycin help maintain the microbial balance in the rumen and improve feed efficiency.

- **Hormone Growth Promotants (HGPs)**: These promote faster growth by stimulating the animal's hormonal system. Their use is dependent on market requirements, as they can affect meat quality but are often managed through specific meat processing techniques to meet standards such as Meat Standards Australia (MSA).

Global Examples of Feedlot Nutrition Practices: In Australia, feedlots emphasize gradual ration changes to prevent digestive upset and optimize weight gain. The National Feedlot Accreditation Scheme (NFAS) ensures that feedlot operators meet high standards of animal nutrition, health, and welfare. Similarly, in the United States, the National Cattlemen's Beef Association (NCBA) provides guidelines on balancing rations to meet both growth and welfare standards, emphasizing the importance of managing the rumen environment when feeding high-energy diets.

In Brazil, which has one of the largest cattle herds in the world, intensive finishing systems focus on using local feed resources like sugarcane by-products to meet the animals' nutritional needs, while minimizing costs. The balance between high-energy feeds and roughage is critical in maintaining a healthy microbial population in the rumen, similar to practices observed in feedlots globally.

Health Monitoring and Deficiencies: Nutritional deficiencies are rare when well-balanced rations are provided, but care must be taken to ensure sufficient levels of key minerals like calcium and phosphorus. Daily monitoring of livestock health is essential, and any signs of nutritional deficiencies should be addressed by consulting a nutritionist or veterinarian. Maintaining a proper feed-to-water ratio is also critical to ensure optimal digestion and nutrient absorption.:

Withholding Periods and Residue Management: Another critical aspect of feedlot management is adherence to withholding periods (WHPs) and export slaughter intervals (ESIs), which ensure that animals treated with chemicals or supplements do not enter the food chain before residues have been metabolized. Livestock producers must maintain detailed records and use Commodity Vendor Declarations (CVDs) to certify that feedstuffs are free from harmful residues. This is vital for maintaining market access, particularly in export markets with stringent food safety regulations, such as the European Union and Japan.

In summary, feedlot nutrition requires a balanced approach that considers not only the animal's growth needs but also the health of the rumen microbes, while carefully managing feed composition and animal health. This approach ensures that animals achieve their full growth potential while meeting market requirements and maintaining overall health.

Feed Requirements of Low Body Condition Cows

Calculating Feed Requirements

To ensure the maintenance or improvement of low body condition cows, it is essential to calculate their feed requirements accurately. Figure 6 provides a method for estimating the daily feed required, particularly during challenging periods such as droughts when feed quality and availability may be low.

1. **Identify Liveweight**: Begin by determining the cow's liveweight, which is located on the left-hand vertical axis of the chart. This weight is typically measured in kilograms.

2. **Assess Feed Quality**: Next, evaluate the quality of the feed available using the feed metabolizable energy (ME) content, which is represented by the M/D (megajoules per kilogram of dry matter) located on the central line of the chart.

3. **Draw a Straight Line**: Using a ruler, draw a line between the animal's liveweight and the feed M/D value. This line will extend to the right-hand vertical line, which gives you the daily feed requirement in kilograms per cow. This value represents the amount of feed needed to maintain the cow's current liveweight.

4. **Adjust for Low Body Condition**: The calculation gives an estimate for the average cow's maintenance, but low body condition cows (bottom 25% of the herd) require slightly more feed. Offering them additional feed will help increase their liveweight over time. If possible, segregate these cows for specialized feeding to ensure they receive adequate nutrition.

Adjusting for Pregnancy and Lactation

Step 1: Pregnancy and Lactation Multiplier

The nutritional needs of cows vary significantly depending on their reproductive status. Cows in different stages of pregnancy and lactation require more energy and feed:

- **Dry and empty** cows require the base level of feed (× 1.0 multiplier).

- Cows in **early pregnancy (over 2 months)** need 1.2 times the base feed.

- Cows that are **8 months pregnant** should receive 1.4 times the feed to support both their maintenance and the growing foetus.

- **Lactating cows** need the most energy and should receive 1.6 times the base feed to support milk production and maternal maintenance.

Step 2: Adjusting for Dry Matter (DM) Content Different feedstuffs contain varying levels of water, which reduces the actual dry matter content. For example, hay typically contains around 10% water, while molasses contains up to 23%. To calculate the actual amount of feed needed ("as fed"), the dry matter content must be taken into account:

- Use the formula:

As fed = Amount at Step 1 (pregnancy and lactation) / Dry matter %

For example, if a cow requires 10 kg of feed per day on a dry matter basis and the hay contains 10% water, the "as fed" quantity would be:

$$\frac{10}{0.9} \approx 11.1 \text{ kg of hay per day.}$$

Additionally, when calculating feed for a group of cows, multiply the individual requirement by the number of animals. Adding 10% to the total amount accounts for wastage, ensuring cows receive adequate nutrition despite potential losses during feeding.

Practical Example:

If a cow weighs 450 kg and the feed has an M/D value of 9, following the chart in Figure 6 might indicate she needs approximately 8 kg of dry matter per day. If she is 8 months pregnant, you multiply this by 1.4, bringing the requirement to 11.2 kg of dry matter. If you're feeding hay (which contains 10% water), the "as fed" amount becomes:

$$\frac{11.2}{0.9} \approx 12.4 \text{ kg of hay per day.}$$

Nutrient Requirements of Beef Cattle

Feeding cattle effectively is essential for achieving desired production goals such as growth, maintenance, or reproduction. To ensure cattle receive proper nutrition, several critical factors must be taken into account, including the level of performance expected, the animal's nutrient needs, the quality of the available feed, and any nutrient deficiencies that may need correction.

Key Considerations for Feeding Cattle

1. **Performance Level:**The first step in feeding cattle is to define the level of performance required, whether it's for maintenance or growth. Maintenance means providing just enough nutrients to sustain the animal's current body condition, while growth requires additional nutrients to promote weight gain.

2. **Nutrient Requirements:**Cattle have specific nutrient needs, which depend on their life stage, size, and production goals. These requirements include energy, protein, vitamins, and minerals. For example, pregnant and lactating cows need more energy and protein than non-pregnant or dry cows. Bulls, heifers, and calves each have different nutrient demands based on their development stage and reproductive status.

3. **Feed Quality and Animal Performance:**Understanding the nutritive value of available feed, whether it's pasture, hay, or grains, is crucial. Feed quality varies by season, and the nutrient content must match the animal's requirements. For example, high-quality pasture provides more energy and protein compared to low-quality forage, so supplementary feeding may be needed to meet cattle's needs.

4. **Identifying and Correcting Nutrient Deficiencies:**After determining the nutrient needs of the cattle and assessing the quality of the feed, any deficiencies should be corrected. These deficiencies can be in energy, protein, minerals, or vitamins, and supplementation is often required to address them. For instance, cattle grazing on low-phosphorus soils may need phosphorus supplementation, especially in the wet season.

Nutrient Requirements Table Example: A typical nutrient requirement table shows the daily needs of cattle for key nutrients like metabolizable energy (ME), rumen degradable protein (RDP), undegraded dietary protein (UDP), calcium, and phosphorus. For example, a pregnant heifer in the last third of her pregnancy weighing 450 kg needs 74 MJ of energy, 614g of RDP, 23g of calcium, and 18g of phosphorus to maintain her condition and support foetal development.

Liveweight (kg)	Daily gain (kg)	Daily requirements			
		ME (MJ)	RDP/UDP[A] (g)	Calcium (g)	Phosphorus (g)
Pregnant heifers – last third of pregnancy					
350	0.4	62	512	20	15
400	0.4	68	564	22	16
450	0.4	74	614	23	18
Dry pregnant mature cows – last third of pregnancy					
350	0	45	373	12	12
350	0.4	58	496	20	15
400	0	50	412	13	13
400	0.4	64	532	22	16
450	0	55	452	15	15
450	0.4	70	579	23	18
500	0	59	492	17	17
500	0.4	75	623	25	20
550	0	64	532	18	18
550	0.4	80	665	26	21
Lactating first-lactation cows — with calf four months old					
350	0.1	96	798/145	27	19
400	0.1	103	849/128	28	20
450	0.1	108	898/112	29	22
Lactating mature cows — with calf four months old					
350	0.1	93	772/351	23	18
400	0.1	99	821/336	25	19
450	0.1	105	867/321	26	21
500	0.1	110	911/307	28	22
550	0.1	115	953/293	29	24
Bulls					
500	0.4	100	825	23	19
600	0.4	112	927	25	22
750	0.4	128	1063	26	25
800	0.4	134	1110	27	27

Figure 53: Nutrient Requirements Table example.

Similarly, lactating cows have higher energy and protein demands due to milk production. A 450 kg lactating cow with a 4-month-old calf needs 108 MJ of energy, 898g of RDP, and 29g of calcium to maintain her body condition and produce milk for the calf.

Primary Limiting Nutrient Principle: One of the most important concepts in cattle nutrition is the idea of the primary limiting nutrient. Animal performance can be restricted by the insufficient supply of a specific nutrient, even if other nutrients are available in abundance. For instance, if cattle graze on pastures deficient in phosphorus, the lack of phosphorus will limit their growth or reproduction, regardless of how much protein or energy they consume. Only by correcting the deficiency in the primary limiting nutrient can performance be improved.

An example of this is cattle grazing on low-phosphorus soils. In the wet season, when protein and energy are abundant, cattle may respond well to phosphorus supplementation. However, during the dry season, when protein and energy are limited, these nutrients become the primary limiting factors, and phosphorus supplementation may not have a noticeable effect.

Global Application of Nutrient Management: Around the world, cattle producers follow similar principles of nutrient management to optimize production. In regions like Australia, where phosphorus deficiency in soils is common, cattle are often supplemented with phosphorus, particularly in tropical areas during the wet season. In the United States and Canada, where feedlots are commonly used, balanced rations with adequate energy and protein are provided to ensure consistent growth rates.

In Brazil, a country with extensive pasture-based systems, cattle are frequently supplemented with energy and protein to counteract the low nutritional value of native pastures, especially during the dry season. Similarly, in Europe, precision feeding strategies are employed to ensure cattle meet specific growth targets while maintaining nutrient efficiency and minimizing waste.

Using Feeding Plans to Determine Adjustment in Response to the Monitoring of Livestock Condition and Pasture Growth

When seasonal conditions turn dry, either as part of annual feed gaps or due to prolonged periods of below-average rainfall, producers often face challenges with dwindling feed and water supplies. In such situations, supplementary feeding becomes a key strategy to maintain livestock health and productivity. However, developing a feeding program for livestock during these times requires careful planning and consideration of several factors.

Developing a Feeding Plan

Start Early with a Clear Strategy: It's crucial to have a well-defined feeding strategy in place, which includes deciding on the stock to sell and which to retain and feed. Producers must also consider other options, such as leasing or agistment, where livestock may be temporarily relocated to areas with better pasture conditions. Evaluating these options involves a cost-benefit analysis, including the cost of feed, labour, and market value of livestock. For instance, during severe drought conditions in Australia, many producers opt to sell older, less productive livestock to preserve resources for younger, more valuable animals.

Assessing the Property and Water Supply: Paddocks with poor water supply should be utilized first to avoid depleting resources in better paddocks. Areas prone to erosion or those with minimal ground cover should be avoided, or livestock should be confined to specific areas to reduce damage. A common practice is to use "sacrifice paddocks," where animals are confined to small areas to minimize soil erosion and over-grazing.

What to Feed? The type and amount of feed provided depend on several factors, including the livestock's physiological state, the availability of pasture, and production goals. Young animals and lactating females, for instance, require higher levels of protein and energy compared to dry or non-lactating animals.

- **Protein and Energy Supplements**: For high-energy and protein needs, lucerne hay, clover hay, or silage are often suitable. In situations where forage is unavailable, feeds such as lupins, faba beans, or dried distillers' grains can be fed in combination with roughage to balance the diet. In North America, grain-based feedstuffs like corn and soybeans are frequently used in drought conditions, while in Australia, lupins are a popular high-protein supplement.

Introducing Grain Gradually: Grain supplements, such as oats or barley, are often introduced as a high-energy source during drought conditions. However, it is important to introduce grains slowly to avoid acidosis—a metabolic disorder that can cause livestock deaths if grains are consumed too rapidly. Gradual adaptation is necessary, and supplements like limestone may be added to grain-heavy rations to balance calcium deficiencies.

How Much to Feed? If there is no pasture left and the livestock are fully reliant on supplementary feeding, full hand feeding becomes necessary. Tools like the NSW DPI's "Drought Feed Calculator" help calculate the exact amount of feed required for cattle and sheep. These tools also allow producers to compare the cost-effectiveness of different feed options.

The amount of feed provided can be adjusted based on the condition of the animals and the availability of dry standing feed. If animals are in good condition, slight weight loss can be managed, but younger or more vulnerable animals may need additional nutrition. Continual monitoring of livestock condition and pasture availability is essential, and rations should be adjusted as necessary.

Monitoring and Adjusting Feed: Once a feeding program is established, it's essential to monitor both pasture conditions and livestock health regularly. This involves adjusting the ration as necessary and ensuring that livestock have a gradual transition to green feed once the drought breaks. For example, after a prolonged dry spell in regions like South Africa, careful monitoring of livestock is needed as they adapt to fresh green pasture to avoid digestive issues caused by the sudden dietary change.

Supplementary feeding during drought conditions requires careful planning and monitoring. A well-structured feeding program can help mitigate the effects of reduced pasture availability and maintain livestock health, ensuring a sustainable operation even during challenging times. By using tools and strategies like the drought feed calculator, monitoring livestock conditions, and carefully selecting feed, producers can navigate periods of dry conditions more effectively.

The following presents a sample supplementary feeding plan for a herd of 100 beef cattle, aimed at maintaining condition during a dry season:

Feeding Plan Overview:

- **Objective**: Maintain body condition during drought and supplement for protein and energy deficiency.

- **Herd Size**: 100 cattle (split into two groups: dry cows and lactating cows).

- **Timeframe**: 60 days (or until a break in drought).

- **Pasture Availability**: Minimal dry standing feed, no green pick.

Step 1: Assess Cattle Performance and Grouping

- **Group 1**: 50 dry cows (maintenance feeding).

- **Group 2**: 50 lactating cows (higher nutritional requirements).

The dry cows will require less feed than the lactating cows, as their maintenance energy and protein needs are lower.

Step 2: Determine Nutritional Requirements

Using the **NSW DPI Maintenance Feed Table** as a guide:

- **Dry cows** (400 kg average live weight): Require 7 kg of hay/day with protein supplementation.

- **Lactating cows** (450 kg average live weight): Require 8 kg of hay/day plus additional high-energy and protein supplementation due to higher nutritional demands during lactation.

Step 3: Feed Type and Supplement Selection

- **Hay**: Mixed pasture hay at 8.5 MJ/kg DM for both groups.

- **Protein Supplement**: Cottonseed meal (43% crude protein) for lactating cows and Urea-based blocks for dry cows.

Step 4: Feed Calculation

For **dry cows**:

- 50 cows x 7 kg hay/day = 350 kg of hay/day.

- 0.5 kg/day of a Urea-based protein supplement to supply 75 g of protein/day.

For **lactating cows**:

- 50 cows x 8 kg hay/day = 400 kg of hay/day.

- 1 kg/day of cottonseed meal to meet protein demands.

Total hay per day = 750 kg. Total protein supplement per day:

- Urea-based block for dry cows = 25 kg/day.

- Cottonseed meal for lactating cows = 50 kg/day.

Step 5: Adjust for Dry Matter and Wastage

Considering 10% wastage:

- **Hay**: Increase total hay required to 825 kg/day (750 kg + 10% for wastage).

- **Protein Supplements**: Adjust supplement amounts to account for any wastage during feeding (no adjustment needed for urea blocks).

Step 6: Monitor and Adjust

- Conduct regular weight checks and adjust feed amounts based on animal condition and pasture recovery.

- If there is any dry standing feed, reduce hay supplementation slightly, ensuring animals have consistent access to protein.

Costs and Budget

- **Hay**: If hay costs $300/tonne, total hay cost for 60 days = 49,500 kg = $14,850.

- **Urea-based blocks**: If blocks cost $1,500/tonne, total cost for 60 days = 1,500 kg = $2,250.

- **Cottonseed meal**: If cottonseed meal costs $600/tonne, total cost for 60 days = 3,000 kg = $1,800.

Total estimated feed cost = **$18,900** for 60 days.

This plan serves as a guide and should be tailored to local conditions, animal performance, and availability of feed resources. Adjustments should be made based on regular monitoring of both livestock and pasture conditions.

Supplementary Feeding program

Planning and managing a supplementary feeding program is a crucial aspect of livestock management, especially during periods of drought or seasonal feed shortages. To ensure a cost-effective and successful feeding strategy, there are several key factors that producers must assess and plan for.

1. Define the Performance Level

The first step in any feeding program is determining the required performance level for the targeted livestock group. Producers must decide if the goal is maintenance (preventing weight loss and sustaining health) or production (promoting growth, milk production, or weight gain). For instance, cattle may need to gain 0.2 kg or 0.5 kg per day, depending on the production target. Each performance goal will require a different feeding strategy and nutrient balance.

2. Assess Nutritional Requirements

The nutritional needs of livestock vary depending on the desired performance. Maintenance typically requires lower levels of energy and protein, while production demands are higher. Nutrients such as protein, energy, and minerals must be adequately supplied to achieve growth or milk production targets. In scenarios like drought feeding, the key is to first identify which nutrient is the limiting factor in achieving the desired performance.

3. Evaluate Pasture Quality

An accurate assessment of the current nutritional value of pasture is critical in planning supplementary feeding. Near-Infrared Reflectance Spectroscopy (NIRS) is a valuable tool used globally to evaluate the energy and protein content in grazing cattle's diet. This technology helps determine the existing quality of pasture and helps pinpoint what additional nutrients are needed to meet livestock performance goals. For instance, in Australia, NIRS has been widely adopted by producers to monitor diet quality during dry seasons.

4. Determine Nutritional Shortfalls

Once the livestock's requirements are established and pasture quality assessed, the next step is identifying nutritional shortfalls. This gap between what the animals need and what the pasture provides will dictate the type and amount of supplements required. Typically, the most limiting nutrient—whether protein, energy, or minerals—is addressed first, as it will have the greatest impact on performance. For example, in the dry season, protein often becomes the limiting nutrient, leading to a reliance on protein-based supplements.

5. Select Cost-Effective Supplements

Producers must consider the cost-effectiveness of the supplements they choose. Factors like supplement availability, cost, and ease of feeding play a role in selecting the right feed. Protein meals, grains, molasses, or commercial supplements can be categorized based on the nutrients they provide, such as protein or energy. For example, in regions like the United States or Australia, combining molasses with urea can be a cost-effective solution for providing both energy and protein to cattle during the dry season.

Nutrient Deficiencies in Different Seasons

During the dry season, protein is often the first limiting nutrient as the quality of pasture declines. In contrast, during the wet season, phosphorus can become the primary limiting nutrient, especially on phosphorus-deficient soils. In some regions, like areas of 'basalt country' in Australia, salt and sulphur may also become limiting, requiring specific supplementation.

Managing Livestock After a Drought

After drought conditions improve and pasture begins to recover, feeding management remains critical. Livestock, particularly cattle, may spend considerable energy searching for fresh green feed, which can lead to energy deficits if the feed quantity is inadequate. Producers should continue supplementary feeding for two to three weeks after the drought ends to allow livestock to transition to green pasture gradually. Strip grazing can also help manage the transition by controlling intake and preventing digestive issues such as nitrate poisoning from weeds like capeweed.

Grazing Management and Livestock Numbers

When managing livestock grazing, one of the most important factors is determining how many animals a piece of land can sustain, which is often done using Adult Equivalents (AE). AE refers to the relative feed intake of different animals, using a standard reference animal, typically a 450 kg dry cow. This standard helps farmers compare animals of different classes and sizes more accurately when assessing their grazing pressure on land. For instance, a younger or smaller animal generally consumes more feed in proportion to its body weight than an older or larger animal, which is crucial when managing grazing effectively.

Determining Livestock Numbers Using AE

To determine livestock numbers, animals are often converted into their AE equivalent, which allows for standardizing different types and sizes of livestock. For example:

- A 400 kg heifer may be considered around 0.9 AE, while a larger animal such as a 600 kg bull may equate to 1.2 AE.

- Using AE also allows farmers to allocate proper grazing resources based on the animal's size and nutritional needs, ensuring balanced feed intake across the herd.

For instance, if a paddock can sustain 100 AEs, it can support either 100 adult cows or their equivalent in smaller, younger animals, ensuring no overgrazing or underfeeding occurs.

Grazing Management Plan

A well-developed **grazing management plan** helps optimize pasture usage by planning how and when pastures are grazed and rested. A good plan accounts for:

1. **Stocking Rate**: Determining how many livestock can graze a pasture without depleting it.

2. **Grazing Duration**: Setting the length of time animals spend on a specific pasture to avoid overgrazing.

3. **Resting Pastures**: Allowing pastures to regenerate after grazing periods.

Proper management prevents overgrazing, which can deplete vegetation, lead to erosion, and negatively affect both animal production and pasture health.

Principles of Grazing Management

- **Planned Grazing**: Grazing schedules should be determined based on pasture conditions, not left to happen by chance.

- **Stocking Rate and Pasture Utilization**: Ensure stocking rates allow cattle to graze 15% to 40% of the pasture, giving them enough choice while leaving sufficient vegetation to allow regrowth.

- **Forage Budgeting**: Regularly assess available feed and adjust stocking rates accordingly to meet production targets.

Compensatory Growth

A phenomenon often observed in livestock after periods of nutritional restriction, compensatory growth refers to a better-than-expected recovery in growth once high-quality feed is reintroduced. This occurs due to reduced maintenance energy requirements, increased efficiency in feed conversion, and higher protein deposition in the body. However, compensatory growth is difficult to predict and should be seen as a bonus rather than a strategy to rely on exclusively.

Rumen Modifiers and Their Role

Rumen modifiers are substances like Monensin or Bovatec that alter the rumen environment or the population of microorganisms. By improving feed conversion efficiency, these modifiers enhance livestock productivity. They work by increasing the production of propionate (a more efficient energy source for animals) while decreasing methane production. Additionally, they help increase the availability of protein to the animal, contributing to weight gain and growth.

However, these modifiers are more effective when animals are gaining weight at reasonable rates, typically above 0.5 kg/day, and may not provide benefits when cattle are merely maintaining weight or losing it.

Hormonal Growth Promotants (HGPs)

Hormonal Growth Promotants (HGPs) are another tool used in cattle production to increase growth rates and improve feed conversion efficiency. HGPs are generally based on male (androgen) or female (oestrogen) hormones that promote muscle growth or reduce muscle breakdown. However, their use is controversial and banned in some markets, making it important for producers to understand market requirements and declare their use properly.

Feed and Biosecurity

Farm biosecurity is the implementation of various practices aimed at preventing the introduction and spread of infectious diseases, pests, and weeds on farms. These measures are critical for safeguarding livestock health and ensuring the economic stability of farm operations. A well-executed biosecurity plan is essential to minimizing the risk of disease outbreaks, making the difference between a successful, profitable farm and one that suffers significant losses.

A robust farm biosecurity plan consists of two primary objectives:

1. Reducing the risk of introducing diseases, pests, and weeds onto the property.

2. Reducing the risk of spreading diseases, pests, and weeds within the property.

Consistent and vigilant implementation of these measures is crucial for maintaining the overall health and productivity of the farm.

Reducing the Risk of Introduction

Introduced Stock: When introducing new livestock, it is crucial to source animals from reliable and biosecurity-conscious suppliers. Livestock should be inspected prior to purchase to ensure they meet health standards, and all necessary documentation, such as the Livestock Production Assurance (LPA) National Vendor Declaration (NVD), should be secured. New livestock should be quarantined for at least seven days and monitored for any signs of illness. Additional precautions, such as drenching and vaccination, help ensure that the animals are free of parasites and diseases before being integrated with the rest of the herd.

Vehicle and People Movements: Vehicles and people can be significant vectors for disease and pest transmission. Farm visitors and contractors can unknowingly bring contaminants onto the property. To mitigate this risk, vehicle access should be limited, and visitors should use a single designated entry point. Contractors should be required to use clean equipment and protective clothing to avoid contamination.

Stockfeed: Stockfeed is another potential carrier of diseases, pests, and weeds. To reduce this risk, it is important to purchase stockfeed from reputable suppliers who operate under quality assurance programs. Always request a Commodity Vendor Declaration (CVD) to confirm that the feed is free from restricted animal material (RAM), which is prohibited for ruminants due to its connection with diseases like Bovine Spongiform Encephalopathy (BSE).

Boundaries and Feral Animal Control: Secure fencing is essential for keeping livestock contained and feral animals out, thus reducing the chances of disease transmission. Coordinating efforts with neighbouring farms to control wildlife such as pigs, deer, and other feral animals can further lower the risk of disease spread.

Reducing the Risk of Spread

Regular Monitoring of Livestock Health: Frequent health checks are vital for early detection of diseases. Any sick animals should be immediately isolated to prevent the spread of illness, and any unusual deaths should be investigated and reported to a veterinarian. Establishing a comprehensive herd health plan, including regular vaccinations and parasite control, is key to maintaining a healthy herd.

Managing Vehicles and People: Strict hygiene protocols should be followed to limit the spread of disease within the farm. Equipment and clothing should be cleaned and disinfected regularly, particularly after contact with sick animals. Visitors and workers should wear clean clothing and wash their hands after handling animals, especially when moving between different areas of the farm.

Safe Handling of Feed: Feed should be stored and handled carefully to prevent contamination. Feed should be kept in clean, dry areas, and troughs should be regularly cleaned to avoid contamination from pests or livestock waste. Keeping feed areas secure from wild and feral animals reduces the risk of disease transmission.

Water Management: Water sources can easily become contaminated by feral animals, chemicals, or diseases. Regular inspections and cleanings of water troughs are essential to maintaining water quality. Wildlife barriers can prevent access to livestock water sources, further protecting against contamination.

Farm biosecurity is a fundamental aspect of maintaining livestock health and protecting the economic viability of farm operations. By focusing on measures that prevent the introduction and spread of diseases, pests, and weeds, farms can maintain their productivity and profitability. Regular monitoring, consistent documentation, and diligent hygiene practices are the pillars of an effective biosecurity management plan.

Utilising Available Pastures

Pasture assessment is an essential part of managing livestock and ensuring the efficient use of available feed resources. The process involves evaluating both the quantity and the quality of the pasture to understand how much feed is available and its nutritive value for livestock.

Pasture quantity refers to the amount of feed available for livestock and is typically measured in kilograms of dry matter per hectare (kg DM/ha). The dry matter (DM) content of a pasture is the weight of the pasture once all the moisture has been removed. For example, young green pastures may contain only 20% dry matter (80% moisture), while more mature pastures may have 80–85% dry matter. Hay, for comparison, typically contains less than 20% moisture.

To accurately determine the dry matter content of pasture, a sample is collected, weighed, and dried using a dehydrator or microwave. Once all the moisture has been removed, the dry weight is divided by the initial wet weight to give the percentage of dry matter. This information helps calculate how much feed is truly available to livestock.

Pasture quality measures the concentration of key nutrients such as energy and protein. The most critical indicators of pasture quality are:

- **Digestibility (%):** Refers to the proportion of the pasture that can be broken down and absorbed by the animal.

- **Metabolisable Energy (ME, MJ):** The energy available to livestock after digestion, expressed in megajoules (MJ).

- **Protein Content (%):** The percentage of crude protein in the pasture, essential for animal growth and maintenance.

Pasture growth and quality are influenced by a variety of environmental factors, including water availability, temperature, and day length. These factors create seasonal variations in both the quantity and quality of the pasture, resulting in periods where feed is abundant and highly nutritious and others where it is scarce and lower in nutritional value.

For instance, during periods of adequate rainfall, pasture growth is vigorous, and nutrient levels are high. However, as the pasture ages, especially after flowering, quality declines rapidly. Frost can further reduce pasture quality, particularly in cooler regions.

Water availability, influenced by rainfall patterns, soil type, and ground cover, is the most critical factor for pasture growth. Soil type affects water retention; for example, sandy soils can hold less water but make nearly all of it available to plants, whereas clay soils have a higher water-holding capacity but also a higher wilting point.

Four Phases of Pasture Growth and Development:

1. **Phase 1:** This is the rapid growth stage, where pasture is lush, green, and highly nutritious. However, it is also susceptible to overgrazing.

2. **Phase 2:** Pasture quality is still high, but stems begin to grow, reducing the nutritive value slightly. This phase is considered optimal for grazing.

3. **Phase 3:** The pasture sets seed, and its quality declines rapidly. While the pasture is still abundant, it offers less nutritional value to livestock.

4. **Phase 4:** During this dormant phase, pasture quality is very low, especially after frosting, and is insufficient for supporting high animal productivity.

As pastures mature, the digestibility, metabolisable energy, and protein content decrease. This drop in quality significantly affects cattle performance, as lower digestibility reduces feed intake and increases retention time in the rumen, slowing down growth. The low metabolisable energy (ME) content limits the animal's ability to convert feed into body weight, and insufficient protein hinders efficient microbial fermentation in the rumen.

Tropically adapted pastures are more efficient in utilizing water but tend to have higher fibre content, making them less digestible than temperate pastures like ryegrass. For example, temperate grasses offer higher digestibility, which supports better animal growth compared to tropical species such as rhodes grass or black speargrass.

Legumes, unlike grasses, can fix nitrogen from the atmosphere and typically maintain higher nutritional quality throughout the season. For instance, young speargrass might contain 15-18% protein, while the legume shrub leucaena can have up to 28% protein in its shoots. Pastures containing a high proportion of legumes typically provide cattle with a better diet quality than grass-only pastures.

The stocking rate—the number of animals per unit of land—directly impacts both liveweight gain and pasture condition. While increasing the number of livestock can boost production per hectare, it often reduces liveweight gain per animal. Over time, heavy stocking rates lead to the decline of desirable pasture species and the invasion of less palatable and less productive species.

By adjusting stocking rates, farm managers can strike a balance between livestock production and pasture health. Generally, utilizing only 15% to 40% of pasture promotes both cattle growth and pasture sustainability, with higher utilization rates appropriate for more fertile and wetter regions and lower rates for drier areas.

Understanding and assessing the quantity and quality of available pasture is fundamental for effective grazing management. By monitoring factors such as dry matter content, digestibility, energy, and protein levels, farmers can make informed decisions about stocking rates and the timing of grazing to maximize livestock productivity and pasture health. The integration of legumes, proper water management, and adherence to the principles of rotational grazing can further enhance both pasture quality and animal performance.

Implementing Grazing Management Plans for Sustainable Stocking Capacity

Effective grazing management is critical for ensuring the sustainability of pastures and maintaining livestock health. While grazing industries, such as beef production, are generally seen as less harmful to the environment compared to cropping systems, they still impact soil, plants, air, water, and natural resources like fuel. The challenge lies in balancing livestock needs with environmental conservation, as this will shape the future of the industry and influence profitability, market access, and ecological sustainability.

Historically, European settlers in Australia applied their traditional farming methods, which were not suited to the region's climate and soils. Over time, it became evident that these methods were unsustainable. Today's grazing management reflects a shift towards working in harmony with the land and climate, ensuring that practices are sustainable and adaptable to environmental limitations.

Today's beef production is far more complex, with the need to balance market demands for environmentally friendly, chemical-free food with efficient farm operations. Beef producers need to consider several factors to ensure sustainability:

- **Sustainable pasture management** to maintain both production and biodiversity.

- **Soil and water management**, including measures to prevent erosion and manage water efficiently.

- **Efficient use of resources** such as fuel and minimizing greenhouse gas emissions.

- **Responsible stock management** to ensure the welfare of animals and prevent overgrazing.

- **Property management planning**, including the flexibility to adjust grazing patterns and adapt to changing markets and environmental conditions.

Pasture health is paramount for sustainable livestock production. The way animals graze can have a significant impact on pasture condition. During times of drought or low pasture availability, producers often face the difficult choice between preserving livestock or protecting pasture. Overgrazing, particularly during drought, can lead to long-term damage to pastures, reducing their productivity and requiring expensive restoration efforts.

To prevent overgrazing, the growth phases of pasture should be closely monitored. Pasture has three phases:

- **Phase I**: Short, young pasture with limited energy reserves, where growth is slow.

- **Phase II**: Optimal growth phase, where pasture grows rapidly and has high nutrient quality. Grazing during this phase maximizes production.

- **Phase III**: Plants mature and set seed, leading to a decline in quality, though the bulk remains available for grazing.

Managing grazing pressure to keep pasture in Phase II as long as possible is key to optimizing both pasture and livestock productivity. However, resting pastures to allow for regeneration and seed production is equally important to ensure long-term sustainability.

Stocking rates—the number of animals per unit of land—must be adjusted according to the season, pasture availability, and livestock needs. Overgrazing can deplete desirable plant species and promote the growth of less nutritious annuals, leading to a drop in overall pasture quality. Conversely, too light a stocking rate may allow lower-quality, less palatable species to dominate, reducing the overall productivity of the pasture.

Stocking density, or the number of animals in a specific paddock, is a powerful tool in pasture management. High stocking density can help control undesirable species and manage pasture quality, while low density may improve individual animal performance but could allow lower-quality species to thrive. Flexibility in managing stocking rates based on seasonal conditions and pasture growth rates is critical for sustainable operations.

Rotational grazing—where pastures are rested periodically to allow for regrowth—is a proven strategy for improving pasture quality. Spelling pastures during key growth phases helps maintain a healthy balance of plant species and ensures that perennial grasses are not overgrazed. Properly managed, rotational grazing can enhance pasture resilience, reduce the need for external inputs like fertilizers, and promote a more stable grazing system.

Diversifying farm enterprises by integrating grazing and cropping operations can further promote sustainability. Perennial pastures, when included in crop rotations, help maintain soil health, increase biodiversity, and break pest and disease cycles. Additionally, legumes such as lucerne can improve soil fertility by fixing nitrogen, reducing the need for synthetic fertilizers.

Retaining natural vegetation and establishing wildlife habitats on grazing properties can also have significant ecological benefits. For example, insectivorous birds that live in undisturbed bushland adjacent to pastures can help control pests, reducing the need for chemical interventions.

Australia's highly variable climate presents a significant challenge for beef producers. Droughts, in particular, can force producers to destock early to preserve pastures and ensure long-term sustainability. Property management plans should account for these environmental fluctuations, providing flexibility to adapt stocking rates, manage water resources, and reduce soil erosion.

Producers also face increasing pressure from markets that demand sustainably produced beef. Meeting these expectations requires not only sound grazing management practices but also monitoring and record-keeping to demonstrate environmental and animal welfare compliance.

Implementing a grazing management plan is essential for maintaining sustainable stocking capacity and ensuring the long-term health of pastures. Flexibility, regular monitoring, and proactive adjustments to stocking rates based on environmental conditions are key to success. By integrating responsible pasture, water, and animal management practices, producers can optimize livestock productivity while safeguarding the environment for future generations.

Soil Management: Ensuring Longevity and Productivity

Avoiding Soil Compaction: Soil compaction, often referred to as "soil pugging," is a common consequence of cattle farming. Compaction occurs when the hooves of cattle press down on the soil, leading to physical damage. While the most severe examples of pugging are easily noticed, even light compaction caused by cattle treading can significantly impact the soil. Research shows that commercial cattle stocking rates can result in yield reductions of at least 20% due to compaction. The severity of compaction depends largely on soil type, with clay-heavy soils being more vulnerable, especially when wet. When possible, farmers should move cattle to lighter, sandier soils during heavy rain to minimize damage. Compacted soils resist root penetration, hinder water infiltration, and reduce aeration, ultimately harming both plant growth and soil structure. On the other hand, in more arid regions with sandy soil, hoof action can help bury seeds, facilitating pasture establishment.

If cattle frequently create tracks to access water troughs, these areas can channel surface water, potentially leading to soil erosion. Rotational grazing is one of the most effective methods to limit soil compaction and manage treading damage.

Maintaining Ground Cover: Ground cover plays a crucial role in preventing soil erosion, and the necessary amount of ground cover varies depending on soil type, slope, and location. For instance, research in New South Wales shows that at least 70% ground cover is needed on red soils with a slope of around 10% to prevent erosion. This general guideline helps prevent excessive runoff, which not only leads to soil loss but also reduces the amount of water that infiltrates the soil. Without sufficient ground cover, water loss due to runoff can significantly hinder pasture growth, reducing its potential.

Overgrazing is another common issue that exacerbates soil erosion. Producers often face periods of erratic rainfall and limited feed supply. In these times, planning for supplementary feeding or destocking is crucial. Allowing animals to graze excessively when ground cover is low can result in severe damage, not just to the pasture but to the soil as well. If ground cover falls to dangerous levels, cattle should be moved, sold, or confined to prevent them from depleting the pasture entirely.

Fencing Layout: Proper fencing layout is essential for managing grazing efficiently and protecting soil health. Factors such as soil type, topography, pasture type, and livestock size should influence fence placement. Traditional square paddocks can make management more challenging when dealing with areas of varying soil types or terrain within the same paddock. For example, different sections of a paddock may have varying pasture growth rates due to changes in soil or slope, and they may need to be managed separately. Ideally, areas with uniform characteristics should be fenced together and managed as a unit.

Smaller paddocks offer greater flexibility in managing grazing and preventing overgrazing. In mixed farming, temporary electric fencing can be a cost-effective solution to subdivide pastures without heavy investments in permanent infrastructure.

Water Management: Water availability and quality are vital components of grazing management. Watering points influence paddock layout and impact grazing distribution. If water sources are too far apart, cattle may overgraze areas close to the water and undergraze other areas, leading to uneven pasture use. This also forces cattle to expend more energy walking to water, slowing their weight gain.

Where possible, water should be provided in troughs or ground tanks rather than giving cattle free access to watercourses. Allowing cattle to use natural streams can lead to soil erosion along stream banks, water contamination, and downstream pollution, including the potential for algal blooms. Research indicates that even without fencing off watercourses, cattle will often prefer to drink from troughs if they are placed on their regular route to water.

Minimizing Fertilizer Runoff: When applying fertilizer, it's essential to avoid contamination of watercourses. Fertilizers should not be spread over or near water sources, especially in areas prone to runoff during heavy rains. Improper use of fertilizers can lead to nutrient-rich runoff, which can pollute water sources and contribute to algal blooms, negatively impacting both livestock and the environment.

Sustainable Management: Sustainable soil management goes hand in hand with proper grazing practices, water management, and environmental conservation. Minimizing soil compaction, maintaining sufficient ground cover, carefully planning fencing layouts, and managing water sources effectively all contribute to the long-term sustainability of the land. Farmers must balance the needs of their livestock with the health of the soil, ensuring that their practices today do not compromise the productivity of their farms tomorrow. In this way, proper soil management not only supports the immediate economic goals of beef production but also preserves the environmental integrity of the land for future generations.

Feed Budgeting

Feed budgeting is a critical tool that enables cattle producers to manage their feed resources effectively. It allows for better planning, particularly during periods of limited pasture availability, such as drought or winter. By calculating how much feed is required to sustain cattle and supplementing it when necessary, producers can minimize the risks associated with overgrazing and maintain animal health and productivity. Feed budgets can also be used to compare different feed options and estimate the cost of supplementary feed, ultimately assisting in more informed decision-making.

The tactical feed budget is designed for use when there is still some pasture available. It helps producers make decisions based on current pasture conditions and the nutritional needs of their cattle. This method can be applied throughout the season as pasture growth and feed demand fluctuate. Several key data points are required to complete a tactical feed budget:

1. **Number and Class of Cattle**: Knowing the number and type of cattle is essential for determining their collective feed requirements. Different classes, such as mature cows or growing young stock, have different energy and nutrient demands.

2. **Cattle Liveweight**: The average liveweight of the cattle should be used for the budget period. For mature cows, the liveweight may remain constant, but for younger cattle, a projected weight increase should be considered. For instance, if young stock are expected to grow from 300 kg to 350 kg over the budget period, an average liveweight of 325 kg would be used.

3. **Current Feed on Offer (FOO)**: This measures the amount of pasture available for grazing, expressed in kilograms of dry matter per hectare (kg DM/ha). This value varies depending on the growth stage of the pasture and its species. Pasture with a high moisture content, such as young green pasture, has a lower dry matter content than more mature, dried-off pasture.

4. **Pasture Quality**: The energy content of pasture is a key component in determining how much additional feed might be necessary. Pasture quality decreases as the season progresses. For example, green pasture in the growing season might have an energy value of 10 MJ ME/kgDM, while pasture that has dried out by mid-summer may only provide 6 MJ ME/kgDM.

5. **Grazing Area and Time Frame**: The area in hectares where cattle are grazing and the length of the budget period (in days) must be factored in to determine how much pasture is available during that time.

6. **Required Performance**: Cattle performance goals, such as maintenance or growth, affect the amount of feed required. Maintenance-level feeding supports 0 kg/day growth, while young stock may need to grow 0.5 kg or 1 kg per day, requiring more energy.

7. **Energy Requirements**: The energy needs of the cattle are based on their liveweight and the performance target. This can be found in energy requirement tables that list the energy needs for different cattle classes.

8. **Minimum Pasture Cover**: This is the lowest level of pasture (in kg DM/ha) that cattle should graze. For example, in drought conditions, it is recommended not to graze below 1,000 kg DM/ha to protect pasture health and prevent soil damage.

9. **Pasture Growth Rates**: Estimating the rate at which pasture grows during the budget period, measured in kg DM/ha/day, is necessary for calculating how much new pasture will be available. Underestimating growth rates is safer than overestimating. Some tools, such as the Pastures from Space program by CSIRO, provide regional growth rate data.

When pasture availability does not meet the energy needs of the cattle, supplementary feed must be provided. The amount of supplementary feed required is calculated by determining the pasture deficit—the difference between the pasture available and what is needed—and then selecting an appropriate feed source.

For instance, if grazing down to 1,000 kg DM/ha results in a deficit of 52,800 kg DM, and pellets with an energy value of 12 MJ ME/kgDM are selected, the feed budget would calculate the total amount of pellets required to meet the cattle's energy needs. The dry matter content of the pellets (90%) must also be considered, as this affects how much of the feed is usable. In this case, 39.11 tonnes of pellets would be required to fill the gap left by the pasture deficit.

Feed budgeting is an invaluable tool for cattle producers, enabling them to make informed decisions about managing pasture and supplementary feed. It ensures cattle have adequate nutrition throughout the year, especially during periods of low pasture availability, while protecting pasture health and preventing overgrazing. By incorporating all the essential factors, from cattle class and liveweight to pasture growth rates and feed quality, producers can optimize their feed resources and reduce unnecessary costs.

Pearson's Square for Feed Balancing

Pearson's Square is a practical method used by livestock producers to balance the diets of cattle, particularly in situations where pasture is not available, and the animals rely entirely on supplemental feed. This tool helps in formulating feed rations that meet the nutritional needs of livestock, focusing on both energy and protein levels. Incorrect feeding can lead to deficiencies or excesses, which can negatively impact livestock health and productivity, making Pearson's Square an essential tool for managing cattle nutrition.

In drought or periods of pasture scarcity, cattle may need feed sources that are balanced in terms of energy and protein. Using Pearson's Square, two different feeds (e.g., grains and hay) can be combined to meet these needs. For example, in the case of a 500 kg cow with a one-month-old calf, the daily energy requirement might be 105 MJ of Metabolizable Energy (ME) with a crude protein (CP) level of 10%. Additionally, the cow can consume up to 10.7 kg of dry matter (DM) per day.

Suppose the feed sources available include barley, which is high in energy but lower in protein, and hay, which may provide roughage and some protein. Pearson's Square can help determine the appropriate amount of each feed type needed to meet both the energy and protein requirements. The result is a balanced feed ration where the cow-calf unit consumes the right amounts of both feed types, ensuring that they meet their daily nutritional needs.

For the example cow-calf unit, the diet might require:

- 3.0 kg of barley (2.7 kg DM) to provide energy,

- 9.2 kg of hay (7.8 kg DM) to complement the protein requirement.

This combined diet of barley and hay would provide 105 MJ ME per day and 10% crude protein, supporting both the cow's maintenance and the calf's growth.

Key Considerations in Feed Balancing

Digestibility and Consumption: It's critical to ensure that the total amount of feed offered can be consumed by the cow-calf unit. In this example, the diet provides 10.5 kg DM, which is within the cow's maximum daily intake of 10.7 kg DM. Regular monitoring of cattle body condition is crucial to ensure that the feed amounts are appropriate. Adjustments might be necessary if animals begin to lose condition, indicating that they need more energy or protein.

Mineral and Vitamin Supplementation

When cattle are entirely removed from pasture and rely on a complete feed ration, mineral and vitamin deficiencies can arise. Supplementation becomes necessary to prevent deficiencies, especially in diets with high grain content.

Calcium Deficiency: Diets with a high percentage of grain, particularly more than 50%, are often deficient in calcium. Adding agricultural limestone at a rate of 2% (2 parts limestone to 100 parts grain) can prevent calcium deficiency. If roughage makes up at least 50% of the diet, calcium supplementation may not be needed.

Sodium Deficiency: High grain diets can also lead to sodium deficiency. Adding salt at 1% (1 part salt to 100 parts grain) helps address this issue unless the stock's water contains sufficient salt.

Buffers Against Acidosis: Grain poisoning, or acidosis, is a concern when feeding high grain diets. Adding buffers like sodium bentonite or sodium bicarbonate at 2% (2 parts buffer to 100 parts grain) can help prevent acidosis, particularly in the first 30 days of grain feeding. After 30 days, the buffer amount can be reduced to 1%.

Vitamin Supplementation

Vitamin A: Cattle that are off green pasture for three months or more may develop a Vitamin A deficiency, as this vitamin is found in green pasture, green hay, or yellow maize. A deficiency can lead to health problems in cattle. To prevent this, an injection of Vitamins A, D, and E provides sufficient protection for about three months.

Vitamin E: Vitamin E deficiency can also develop, especially in drought rations. Supplementation through injections of Vitamins A, D, and E can correct this and ensure cattle remain healthy in situations where their feed does not provide adequate vitamin levels.

Monitoring Grazing

Effective pasture management is critical for maintaining healthy, productive grazing systems. Pastures that contain a high density of perennial grasses, ample clover content, and few weeds are considered of higher quality compared to those with mostly annual grasses and minimal clover. The ability to manage and monitor these characteristics over time determines pasture productivity and its suitability for livestock grazing.

Pasture health should be monitored continuously, especially for signs of improvement or degradation. Ideally, an increase in perennial plants and clover alongside a reduction in weed prevalence are indicators of improved pasture quality. After a two-year period, it's important to assess whether the pasture requires further inputs such as fertilization, reseeding, or over-sowing. A qualified land management advisor or agronomist can assist in deciding on appropriate actions to improve pasture health [66].

Regular soil and plant testing should be conducted to ensure that nutrient levels are adequate for optimal plant growth. For example, soil pH should be monitored to track any increases in acidity, which may necessitate the application of lime to neutralize the soil. Soil amendments such as lime, dolomite, and trace elements can address nutrient deficiencies. Correcting soil acidity and maintaining soil organic matter will encourage the activity of soil microbes and earthworms, enhancing nutrient cycling and improving soil structure.

It is essential to understand the growth cycles of pasture species to align grazing practices accordingly. Grazing management involves allowing the pasture to grow to an optimal height before being grazed, and then resting the pasture to encourage regrowth. Native perennial pasture species may serve as a low-input alternative to introduced grasses, requiring less maintenance while still offering sustainable forage for livestock.

Weed management is a vital aspect of pasture management. Establishing vigorous perennial grasses and clover stands helps suppress weed growth through competition. Additionally, careful sourcing of hay and grain, along with diligent inspection of incoming stock, can prevent the introduction of new weeds. Control methods like spray topping can effectively manage annual weeds, while spray grazing is useful for broadleaf weeds.

Grazing management practices must be tailored to the type of livestock and the specific conditions of the pasture. As a rule of thumb, grazing should begin when pasture is between 6-10 cm in height and continue until it is reduced to 2-4 cm. Over the summer months, a two-week grazing period followed by a 6-10 week rest period is ideal. It is critical to maintain sufficient ground cover to prevent erosion and soil loss, especially during periods of low pasture productivity. When feed quality or quantity declines, supplementary feeding is essential to maintain livestock condition [66].

The key to effective pasture management is matching livestock requirements with available feed in terms of both quality and quantity. Pasture quality is often more important than pasture quantity in determining animal growth. For example, livestock will increase feed intake as pasture quality improves, leading to greater energy intake and liveweight gain. Therefore, managing pasture quality—ensuring the presence of nutrient-rich, green leaves rather than fibrous stems—will maximize livestock productivity [66].

Rotational grazing systems help minimize selective grazing, where livestock pick out the most nutritious parts of the pasture, leaving behind less palatable plants. This can lead to a decline in pasture quality over time, as the less desirable plants dominate. By rotating livestock between paddocks, pasture recovery is encouraged, ensuring both high productivity and species diversity in the long term.

A key aspect of grazing management is aligning livestock reproductive cycles with seasonal pasture growth. Timing lambing or calving to match peak pasture availability ensures that feed is plentiful when nutritional needs are highest, minimizing the need for supplementary feed. This approach optimizes pasture utilization and reduces operational costs.

Effective pasture management hinges on regular monitoring, soil health management, appropriate grazing strategies, and weed control. By matching pasture availability and quality with livestock nutritional needs, producers can maintain sustainable and productive grazing systems. Sustainable practices such as rotational grazing and aligning livestock cycles with pasture growth patterns will help maximize both pasture productivity and livestock performance over time.

Monitoring Tools

Monitoring tools have been developed to help farmers and advisors in Southern Australia assess and maintain the sustainable carrying capacity of paddocks. These tools, based on results from the Grassland Productivity Program (GPP), aim to help farmers make informed decisions about stocking rates and pasture management to ensure long-term sustainability. A key feature of these tools is the integration of groundcover thresholds—indicating when stocking rates should be adjusted to prevent overgrazing and maintain environmental health.

The tools are most suitable for use in winter-dominant rainfall areas in Southern Australia that grow improved pasture species like subterranean clover, phalaris, perennial ryegrass, and cocksfoot. They also consider common annual grasses such as ryegrass and barley grass. While the tools provide a framework for calculating potential carrying capacity, this is not a recommendation for running paddocks at maximum capacity. Instead, it offers a benchmark, prompting questions about whether soil fertility or other factors are limiting pasture productivity.

The tools are particularly useful for moderately fertilized pastures with perennial grasses and annual clovers, but they are not designed for use on native pastures in low-fertility soils.

Dry Sheep Equivalent (DSE): The DSE is a standard unit used to compare the feed requirements of different livestock and to calculate the carrying capacity of a farm or paddock. One DSE represents the amount of feed needed by a 2-year-old, 45 kg Merino sheep (dry and non-pregnant) to maintain its weight, equivalent to 7.6 megajoules (MJ) of metabolisable energy per day. This unit helps standardize comparisons across livestock types.

Potential Carrying Capacity: This term refers to the number of stock (expressed as DSE per hectare) that can be supported on a paddock through most years. During droughts, carrying capacity will decrease as pasture growth is significantly reduced. The potential carrying capacity provides farmers with a target for what can be sustainably supported, helping them balance profitability and environmental impact. It is calculated using historical data from the GPP and reflects what is possible under good management practices.

Groundcover and Sustainability

Groundcover is the amount of plant material covering the soil, expressed as a percentage. Maintaining sufficient groundcover is essential for preventing soil erosion. In Southern Australia, a minimum of 70% groundcover is recommended on flat or slightly sloping land, while erosion-prone soils require 80-90% groundcover. If groundcover falls below these thresholds, stocking rates should be reduced to protect the environment.

Steps for Monitoring Carrying Capacity:

1. **Calculate the Paddock Area:** Exclude areas such as remnant vegetation or cropped sections from the total paddock area.

2. **Determine the Critical Time for Groundcover:** The lowest point of groundcover may vary depending on climate and grazing management. In cooler regions, late autumn may be the critical time, while in hot, dry areas, summer-autumn may be the highest risk period.

3. **Estimate Groundcover:** Assess groundcover levels at the critical time of year and ensure they meet recommended thresholds for different soil types and slopes.

4. **Calculate Current Carrying Capacity:** Use livestock records to determine the stocking rate (in DSE per hectare) over a typical year. This calculation can be done using grazing days (number of animals x DSE x number of days) or through software tools designed for such assessments.

5. **Assess the Length of the Growing Season:** The length of the growing season influences pasture growth and, subsequently, carrying capacity. Estimating the

growing season for your region helps fine-tune potential stocking rates.

6. **Check Soil Fertility (Phosphorus Levels):** Soil fertility, particularly phosphorus (P) levels, directly impacts pasture growth and carrying capacity. Monitoring and maintaining appropriate P levels are key to maximizing pasture productivity.

7. **Compare Current and Potential Carrying Capacity:** Once the potential carrying capacity is calculated, compare it to your current capacity to assess whether adjustments are needed for sustainability.

Balancing stocking rates to avoid overgrazing is crucial for long-term profitability and environmental sustainability. Farms operating below potential carrying capacity with healthy groundcover can explore ways to increase stocking rates through better grazing management and soil fertility improvements. Conversely, farms exceeding potential carrying capacity with low groundcover should take immediate steps to reduce stock numbers or improve pasture management.

Grazing Management for Horses

Horses are naturally designed as "browsers," meaning they graze selectively, moving from one area to another, often covering significant distances. In the wild, they travel up to five kilometres daily, avoiding areas soiled by manure to minimize parasite risks. This natural grazing pattern becomes problematic when horses are confined to pastures, as they do not return to graze manure-contaminated areas, resulting in underutilized, nutrient-rich patches of pasture and overgrazed, poor-quality patches, often referred to as "horse-sick" pastures.

Horses rely on smell, touch, and taste when selecting forage, with smell being the most important factor in determining palatability. Horses typically avoid areas contaminated by manure, leading to uneven grazing patterns. These behavioural tendencies contribute to the formation of horse-sick pastures, characterized by patches of tall, ungrazed grass around manure piles and short, overgrazed sections. This selective grazing not only reduces pasture quality but also opens the door for weed invasion.

Horses graze for long periods, typically spending 16 to 18 hours a day grazing. Their diet consists of forages, such as grasses and clover, supplemented as needed depending on the quality and quantity of the pasture. A mature horse can consume forage equivalent to 1.5-2.0% of its body weight daily.

Manure plays a major role in creating horse-sick pastures. Horses' grazing patterns and the deposition of manure in specific locations cause large patches of pasture to grow tall and rank due to increased nutrients and a lack of grazing. This leads to further rejection of those areas by horses, perpetuating the cycle. Soil damage from hoof action, especially when horses gallop, further worsens the condition by breaking up the pasture sward and exposing bare ground. This can lead to soil compaction, favouring weed growth and reducing pasture productivity.

A typical horse can produce up to 7 kg of dry manure per day, significantly affecting pasture composition. Over time, horse-sick pastures can reduce the carrying capacity of a property by as much as 50%, diminishing the productivity of the land.

Managing Horse Pastures

Manure Management: Effective manure management is crucial to prevent horse-sick pastures. The most effective method is daily manure collection, which helps maintain pasture health and reduces the risk of parasitic infections. Other methods, such as pasture harrowing—spreading manure to break up clumps and expose parasite larvae to the elements—can be helpful but should be done carefully. Harrowing is most effective in hot weather after rainfall to dry out the larvae.

Rotational Grazing: Rotational grazing is another effective strategy to manage horse pastures. This involves grazing horses in one section of a pasture for a set period (typically around two weeks), followed by a rest period (six weeks), allowing the pasture to recover and regrow. This approach can help reduce the development of horse-sick pastures and improve overall pasture quality. Rotational grazing also encourages more uniform grazing and reduces the risk of overgrazing in certain areas.

Cross-Grazing with Other Livestock: Introducing other livestock, such as sheep or cattle, into the grazing rotation can help manage pasture health. Sheep and cattle do not graze in the same selective manner as horses, and their grazing behaviour can help manage and reduce the areas of tall, under-grazed grass around manure piles. Cross-grazing can also help improve pasture hygiene by breaking parasite life cycles.

Pasture Maintenance: Other pasture management practices, such as slashing or mowing tall patches of grass, can help improve the palatability of the pasture for horses. By removing tall, poor-quality forage, the pasture can regrow more evenly, leading to better grazing conditions.

Feed Options

Cattle

Feeding options for beef cattle during challenging seasons, such as a long dry summer followed by a late autumn break, require careful planning to ensure adequate nutrition and maintain cattle health. When pasture and conserved fodder are limited, producers can consider several strategies, such as selling cattle, agisting cattle off the farm, applying nitrogen fertilisers, or purchasing additional fodder such as grain or hay.

The first step in addressing feed shortages is to classify cattle into priority groups based on their nutritional needs. High-priority groups, such as cows rearing calves or young replacement heifers, require close attention, while low-priority groups, such as steers or dry cows, may be candidates for sale or agistment. Agisting cattle can reduce the on-farm feeding burden, but this option should be carefully costed against the expenses of purchasing additional feed.

A feed budget is crucial for determining the deficit between the cattle's nutritional requirements and the available fodder or pasture. This calculation enables producers to estimate how much extra feed, primarily energy, will be needed to sustain cattle through the winter. It is essential to compare feed alternatives based on their energy content and cost-effectiveness.

Pasture growth can be the most cost-effective source of nutrition during winter. Managing grazing to extend the rotation period can help build a "feed wedge" of available pasture ahead of the herd. This practice not only supports pasture recovery but also ensures that pastures have enough height to respond well to nitrogen fertiliser, which can enhance pasture growth during cooler months.

Applying nitrogen fertiliser is a cost-effective strategy for increasing pasture growth. The response time for nitrogen fertilisation is typically four to six weeks, so early application is key to maximizing its benefits. Nitrogen should be applied to responsive paddocks to stimulate pasture growth, especially in areas with established perennial pastures.

In times of fodder shortages, grain is often a cheaper alternative to hay in terms of energy value. For instance, 1 kg of grain with an energy content of 12 megajoules (MJ) is approximately equivalent to 1.5 kg of hay at 8.5 MJ. However, when introducing grain to cattle, it should be done gradually to prevent grain poisoning, which can occur if cattle are introduced to grain too quickly.

Grain can be fed on the ground, in troughs, or behind electric fencing to reduce waste. Oats are one of the easiest grains to feed because they do not require crushing or cracking, and their husks help prevent rapid grain poisoning. Combining hay and grain, especially during the transition period, can provide balanced nutrition while helping to stretch valuable hay reserves.

For producers with the option to finish cattle for slaughter, feedlots may provide a solution. Commercial feedlots offer contract services, allowing cattle to be finished for market under controlled conditions. However, careful budgeting is necessary to ensure the profitability of feedlotting. Producers should calculate the total cost of feeding, including feed and other expenses, and compare it to the expected market price of finished cattle to determine if the investment will yield a return.

The daily energy requirements for beef cattle vary depending on factors such as age, weight, and physiological state. For example, a 450 kg cow with a calf requires about 90 MJ of energy per day, while a dry cow in late pregnancy needs around 65 MJ. Forage quality also influences the amount of feed needed. High-quality pasture or hay can significantly reduce the need for supplementary feeding. For instance, 4 kg of pasture dry matter per day can reduce the amount of hay needed for a cow and calf by about half.

Feeding Options for Pigs

Dry Sows: The nutritional needs of dry (pregnant) sows should be managed carefully based on factors such as their size, housing, and body condition. Overfeeding pregnant sows can lead to excessive weight gain, which can reduce feed intake during lactation. A balanced diet of around 2-2.5 kg of grain-based feed per day is usually sufficient for most dry sows. Larger sows require more feed to meet their maintenance needs. For replacement gilts, additional protein can enhance reproductive performance and longevity by supporting skeletal development during their growth phase.

Lactating Sows: Lactating (wet) sows have higher nutritional demands, with over 80% of their energy intake being used for milk production. A sow producing milk for a litter of 10 piglets typically requires 85 MJ of energy daily and 55 grams of lysine, which can be met with about 6 kg of feed containing 14 MJ DE per kg and 0.55 g/MJ of available lysine. First-litter sows usually need a different diet or supplement due to their lower appetite compared to older sows. Feed intake is gradually increased from 2-2.5 kg per day after farrowing to ad libitum levels by four to seven days post-farrowing, depending on the herd's practices.

Increasing Intake: In hot conditions, lactating sows tend to reduce feed intake, which can delay oestrus after weaning and impact milk production. To mitigate this, farmers can feed more frequently during cooler hours, use high-energy, dense diets, provide wet feed (with proper hygiene), and ensure water is cool and available with an adequate flow rate. Improving feed palatability, such as adding flavourings or checking feeder designs for accessibility, can also help encourage higher feed intake.

Creep Feeding for Piglets: From around 10 days of age, suckling pigs need fresh diets with no less than 16 MJ DE/kg, which are highly palatable and easy to digest. This diet should continue for the first week post-weaning to ease the transition and maintain growth rates.

Weaner Pigs: Weaners require a diet that caters to their limited digestive capacity and growth needs. Their diet should be high in energy (14.5-15 MJ) and fed without restriction until they reach 20-25 kg live weight. Since feed intake limits growth during this phase, the diet should be digestible, fresh, and palatable, with regular trough checks for soiling.

Grower Pigs: Grower pigs, typically from 20 kg to 45-60 kg, undergo rapid lean growth, so their diet should provide 14-14.5 MJ DE/kg with an available lysine-to-energy ratio of 0.65 g/MJ. Like weaners, growers should be fed without restriction to maximize growth.

Finisher Pigs: Finisher diets focus on optimizing growth, feed efficiency, and carcass quality. Diets are typically adjusted according to factors like market requirements, genetic potential, and environmental temperatures. Finishers may require a restricted feed intake in some cases, with a maximum daily energy intake of 30-34 MJ to avoid excessive fat deposition. Males often require a higher energy and protein diet compared to females, especially during the finisher phase.

Split-Sex Feeding: Split-sex feeding is a useful practice for improving feed-use efficiency by tailoring diets to the distinct needs of male and female pigs. Female pigs, particularly those over 50 kg, tend to deposit more fat and require less energy than males, making separate diets for each sex more efficient in terms of feed conversion and carcass quality.

Supplementary Feeding in Grazing Systems

Supplementary feeding is a common practice in grazing systems to ensure that livestock's nutritional needs are met, particularly during periods when natural feed availability is low, such as in drought conditions. The decision to implement supplementary feeding depends on various factors, including business objectives, seasonal conditions, and available resources. It may be regularly integrated into the production cycle to match feed supply with demand or be reserved for emergencies.

Animal Health Considerations: Animal health should always be the top priority when determining the need for supplementary feeding. During droughts, if it becomes apparent through feed budgeting that supplementary feeding is necessary but not feasible due to financial or labour limitations, farmers should consider selling or agisting livestock before they fall below a condition score of 2. When supplementary feeding is required for specific production goals, such as weight gain or reproduction, the nutritional needs of ruminants should be carefully assessed, and appropriate rations formulated.

Feed Budgeting: Feed budgeting is a critical process that helps farmers predict whether the available feed is sufficient to meet livestock requirements. This process involves estimating pasture availability and quality, calculating livestock feed demand, and comparing the two. Based on these predictions, corrective actions can be taken, such as adjusting stocking rates or preparing to supplement with additional feed. The use of tools like feed demand calculators helps farmers make informed decisions and avoid both overgrazing and undergrazing, optimizing pasture productivity.

Production Objectives: The type of supplementary feed selected should align with the production objectives of the enterprise. For example, if the goal is to maximize weight gain, high-energy supplements like grain may be the best option. Conversely, if the objective is to support pregnant animals and prevent diseases linked to nutritional deficiencies, such as grass tetany or pregnancy toxaemia, providing roughage, such as hay, may be more appropriate. Each approach has its labour and economic implications, which must be considered.

Nutrition in Supplementary Feeding: To maintain proper rumen function and overall animal health, supplementary feeding should provide a balanced diet, including protein, energy, roughage, and essential minerals. For instance, grains can provide energy, while roughage, such as hay or silage, supports rumen function. Additionally, minerals such as calcium and phosphorus are essential and are often delivered through mineral licks to ensure animals do not consume excessive amounts.

Utilization of Dry Pasture: During times when dry pasture or stubble is the primary forage, supplementary feeding with protein-rich supplements (like lupins) can improve pasture utilization. However, feeding high amounts of supplementary feed (more than 100 grams per head per day) can lead to substitution, where animals prefer the supplement over pasture, reducing the efficiency of pasture use.

Sheep Requirements and Feed Budgeting: Feed budgeting is especially important for sheep, as their requirements change based on factors such as body condition, fleece production, and reproductive status. Ewes have varying energy demands, especially during late pregnancy and lactation. For example, during the last 50 days before lambing, their energy requirements increase rapidly. Supplementary feeding calculators and tables of energy requirements help farmers tailor their feed programs to the needs of their flock.

Feeding Methods and Frequency: The method of feeding, such as the use of troughs or ground feeding, and the frequency of feeding, are important for ensuring equitable feed distribution among animals. For example, dry sheep can be fed once or twice a week, while ewes in late pregnancy may need to be fed every second day. Introducing new feeds should be done gradually to avoid digestive issues such as acidosis, and roughage should be offered before grain to ensure proper rumen function.

Types of Supplementary Feed for Sheep

Cereal Grains: Cereal grains are a fundamental component of sheep rations or supplements due to their high energy content (10–13 MJ/kgDM) and availability. Grains such as wheat, barley, oats, and triticale provide varying levels of crude protein (5–15%), making them versatile energy sources.

Introducing cereal grains must be done cautiously to avoid digestive issues like acidosis. The high starch content in grains, especially wheat and triticale, poses a risk of ruminal acidosis, which can lead to serious health issues for sheep. Gradual introduction over 10 to 20 days allows the rumen to adjust to the starch-rich diet. Oats are the safest grain due to their lower starch levels, making them a better option for transitioning to grain feeding. It's also essential to vaccinate sheep against pulpy kidney disease when feeding high-starch diets.

An example introductory schedule involves starting sheep on 50 grams of grain per day and gradually increasing the amount. For example, after 14 days, the ration could reach up to 450 grams for a lactating ewe .

Seconds Grain (Screenings): Seconds grain consists of smaller or partially formed grains, often higher in crude protein and lower in starch compared to fully formed grains. This makes it a valuable inclusion in mixed rations. However, since the nutrient content of seconds grain can be variable, it should be tested for energy and protein levels. Despite being a useful supplement, it must be introduced gradually to avoid acidosis, as it still contains some starch.

Lupins: Lupins are highly regarded for their protein content and energy levels. With minimal starch and high fibre, lupins pose a much lower risk of acidosis compared to cereal grains. However, the sudden introduction of large amounts of lupins can lead to ammonia toxicity, so a gradual introduction is still necessary. Lupins are often considered cost-effective despite their higher price, as they reduce labour by allowing infrequent feeding intervals—sometimes as long as three weeks between feedings. The best practice for feeding lupins involves scattering them across the paddock, minimizing aggression among sheep and ensuring equitable feed distribution.

Beans, Peas, and Vetches: These legumes provide both energy and protein but come with a high level of starch, requiring careful introduction to avoid acidosis. These feeds are best introduced slowly and should be fed at least twice a week. Sheep might take time to adjust to beans, peas, and vetches if unfamiliar with them.

Canola: Canola seeds offer high energy and protein content but are less efficient for sheep if not processed. Whole canola seeds often pass through the animal undigested due to their small size and tough seed coat. Crushing or milling the seed improves its digestibility. However, caution must be exercised as canola oil, when released in the rumen, can coat fibre and reduce digestion efficiency. Canola seconds, with lower oil content, are often better utilized.

Pellets: Pellets offer convenience as they provide a balanced diet in each serving, ensuring uniform feed distribution. Pellets can vary significantly in energy content, so it's important to check their nutritional value and ensure they're appropriate for the sheep's requirements. Some pellets have high energy content, posing a similar acidosis risk as grain, while others are lower in energy and safer. Pellets should be stored properly to prevent spoilage and pest infestations.

Roughage: Sheep need roughage (at least 10% in their diet, 15% for lactating ewes) to maintain healthy rumen function and prevent acidosis. Common sources of roughage include hay, silage, and straw. Roughage alone may not provide sufficient nutrients for high-performance sheep, so it is often supplemented with grains or protein meals. Roughage quality varies greatly, so testing for energy and protein is recommended when planning feeding programs.

Silage: Silage is generally more nutritious than hay because it is harvested at an earlier stage of plant growth. Silage made from pasture or cereal crops can have an energy content of 8.5–11 MJ/kgDM and protein levels between 7% and 25%. Grass-based silage is suffi-cient for maintenance but may not support production goals without supplementation. Cereal-legume silage is more nutrient-dense and suitable for supporting growth.

Hay: Hay provides essential roughage in sheep diets. Cereal hay is commonly used, with its nutritional value depending on the time of cutting. Early-cut cereal hay has moderate digestibility and higher protein levels, while late-cut hay tends to have lower nutritional value. Mixed cereal-legume hay is more nutrient-dense and palatable to sheep. Lucerne hay, with its high protein content, is ideal for lactating ewes and growing lambs.

Straw: Straw is a low-cost roughage option, especially in drought conditions. However, it is low in protein and energy and less than 50% digestible. Straw can complement grain-based diets but should not be fed as the sole source of nutrition, as sheep would lose weight on a straw-only diet. Barley straw is the most energy-rich, followed by oats and wheat straw.

Grazing Strategies

Grazing strategies are critical to managing both animal health and pasture productivity. The choice of grazing method can significantly influence the sustainability and profitability of livestock enterprises. Farmers often use a variety of techniques, such as set stocking and rotational grazing, to meet different objectives depending on the season, pasture conditions, and livestock requirements. This adaptable approach is known as tactical grazing.

Set Stocking: Set stocking refers to the practice of keeping livestock in one paddock for extended periods, with minimal or no movement. The key characteristics of set stocking include:

- **Continuous Grazing:** The paddock is not rested, and livestock graze year-round, with stocking rates typically calculated to accommodate this continuous use.

- **Supplementary Feeding:** In times of feed shortages, supplementary feeding (e.g., hay or grain) may be needed to balance the livestock's feed requirements. This system is simple, requiring little infrastructure or labour, but risks overgrazing and potentially damaging the pasture over time if stocking rates are not well managed.

Rotational Grazing: In rotational grazing, livestock are moved between different paddocks in a planned sequence. This system allows for periods of rest between grazing, promoting pasture recovery and growth. Rotational grazing offers several advantages:

- **Increased Stocking Rates:** Since the pastures are rested, more animals can be grazed overall compared to set stocking.

- **Improved Pasture Management:** The practice supports the persistence of palatable, perennial pasture species, as plants are grazed at their optimal growth stage, allowing for proper regrowth before the next grazing cycle. However, rotational grazing requires more infrastructure, such as fencing and water sources, and can demand more labour compared to set stocking.

Tactical Grazing: Tactical grazing combines both set stocking and rotational grazing strategies in a flexible manner, allowing farmers to adjust grazing practices based on the needs of the pasture and livestock throughout the year. This approach can help optimize pasture use, manage feed supply, and support different livestock classes' growth and reproductive needs. Tactical grazing is well-suited to farms with existing infrastructure that supports multiple grazing methods.

Grazing to Manage Pasture Species: The choice of grazing method can also impact pasture species composition. Grazing management is particularly crucial during pasture reproduction and establishment phases. For example, targeted grazing can be used to suppress undesirable species by preventing seed set, while desirable species can be encouraged by resting the paddocks during critical growth periods.

Supplementary Feeding: Supplementary feeding is a valuable tool that can complement any grazing strategy. It helps balance feed supply during periods of feed shortage or when pasture quality declines. Supplementary feeds, such as hay, grain, and protein meals, can address specific nutritional deficiencies and enhance livestock productivity. However, the cost of supplementary feeding must be weighed against its benefits, and adjustments to the livestock management calendar or enterprise mix may reduce these costs.

Grazing Native Pastures: Grazing strategies for native pastures include continuous grazing, rotational grazing, cell grazing, and spell grazing. Each method has its benefits and challenges:

- **Continuous Grazing:** This is simple and low-labour, but can lead to overgrazing in preferred areas if stocking rates are too high.

- **Rotational and Cell Grazing:** These approaches allow for better pasture management, giving plants time to recover between grazing events, but require more infrastructure and labour.

- **Spell Grazing:** Involves resting pastures at key times, such as during wet seasons, to allow seed set and root replenishment. This promotes long-term pasture health but can concentrate livestock in fewer paddocks, increasing the risk of overgrazing.

Crops like cereals and canola can be grazed without sacrificing yield, offering an additional feed source, especially during the autumn-winter feed gap. Grazing crops provides multiple benefits, such as reducing supplementary feed costs and promoting animal weight gain due to the higher nutritive value of crops compared to pastures during the winter. However, careful management is required to prevent damage to crop yields, and livestock should be removed before the crops enter critical growth stages, such as stem elongation.

Successful grazing strategies, whether rotational, set stocking, or a combination of both, depend on effective pasture management, ensuring optimal utilization, and avoiding overgrazing. By aligning grazing practices with pasture growth patterns and livestock needs, farmers can achieve sustainable production and maintain the health of both their animals and the land.

Feeding Grain to Cattle

Feeding grain to cattle is a common practice in several scenarios, such as during droughts, as a supplement to grazing, or in feedlot systems. Each situation requires careful management to ensure cattle health and optimal feed efficiency.

Situations for Grain Feeding

1. **Drought Ration**: Grain can be the primary food source when pasture is scarce. In this scenario, cattle are fed approximately 1% of their live weight in grain. For example, a 300 kg cow would require around 3 kg of grain per day.

2. **Supplement to Grazing**: When pasture is available but insufficient, grain is used to supplement the grazing diet. The amount of grain ranges between 0.5% and 1.5% of the animal's live weight per day.

3. **Lotfeeding**: In intensive systems like feedlots, grain can make up the majority of the cattle's diet. Cattle in these systems can consume up to 2% of their live weight in grain daily, with total feed intake reaching up to 2.5% of their body weight, as the grain is mixed with other feed components.

Grain must be introduced gradually into the diet of cattle to prevent digestive disturbances, such as grain sickness or acidosis, which results from a rapid change in diet. Cattle unaccustomed to grain can suffer from a build-up of lactic acid in the rumen, causing severe illness or death.

Steps for Introduction:

- Start with a small quantity of grain, about 0.5 kg per head per day.

- Gradually increase the amount by 0.5 kg every second day until the desired ration is achieved.

- Monitor cattle closely for signs of digestive distress, such as scouring or reduced feed intake, and adjust the grain increment accordingly.

Shy-feeders or horned cattle may require separate feeding to ensure even consumption across the herd.

To minimize the risk of acidosis, fibrous grains like oats are safer to feed than low-fibre grains such as wheat. Buffers, such as sodium bicarbonate or sodium bentonite, can be added at 1% of the grain mix during the introductory phase to neutralize acid in the rumen. These additives can typically be phased out after a month once the cattle have adapted to grain feeding.

Roughage, such as hay, is essential for maintaining rumen function and preventing acidosis. In drought feeding, roughage may not be necessary after the introductory phase if there is some residual forage available in the paddock. In feedlot systems, roughage is crucial and should be chopped and mixed with grain to ensure a balanced intake.

During the introduction of grain, cattle should be fed daily. Once they are accustomed to the ration, feeding every second day can be an effective practice, particularly in drought or supplementary feeding scenarios. Some farmers may extend this to feeding every third day, but this should only be done after cattle have adjusted to the grain.

Processing grain (rolling, milling, or crushing) can improve its digestibility. However, processed grain poses a higher risk of acidosis, so coarse rolling is preferable. Whole grain is safer when fed separately from roughage, while processed grain is suitable when mixed with roughage in feedlot systems.

Feeding Methods

- **Trough Feeding**: Troughs minimize wastage but require sufficient space for all cattle to feed simultaneously. Each animal needs about 500 mm of trough space.

- **Ground Feeding**: In drought situations, feeding grain in small, spread-out heaps on the ground can prevent bullying and milling, allowing all cattle to access the grain. However, processed grain should not be fed on the ground as it leads to significant wastage.

Feeding grain to cattle, when done correctly, can provide essential nutrients, especially during periods of pasture scarcity. However, careful management is required to avoid digestive issues, and proper introduction and monitoring are essential to ensure the health and productivity of the herd.

Supplementary Feeding to Horses

In the face of drought, horse owners are legally responsible for ensuring their animals' welfare, meaning horses must not be allowed to starve or become distressed. Ignoring the deteriorating conditions could lead to significant problems as pasture feed diminishes, and once horses start losing condition, it becomes more costly and challenging to recover. The key message here is to plan early and establish clear action deadlines to address feed shortages.

Agistment, or relocating horses to another property with available pasture, can be a more economical solution compared to feeding during a drought. This practice reduces grazing pressure on the home property, improving the feed situation for any remaining horses. Typically, horses with high maintenance requirements, such as foals, lactating mares, and stallions, should stay on the home property, while dry mares and idle stock are ideal candidates for agistment. However, it's essential to inquire early about agistment options, as available paddocks become scarce as droughts progress, leading to increased costs.

Supplementary feeding is an expensive and labour-intensive process, especially in a drought. The objective is to feed horses enough to maintain a condition score of at least 2. Some categories of horses, such as lactating mares and weanlings being prepared for sale, might require a temporary increase in rations. It is essential to carefully calculate budgets since profit margins are often slim during drought. Fortunately, the quality of dry pastures and failed crops can sometimes be better than anticipated, helping reduce feeding costs.

In extreme situations where some horses are unsaleable and other options are exhausted, humane destruction might be considered. While this approach might seem drastic, particularly for companion animals like horses, it can sometimes be the most responsible choice for large-scale breeders. Local government departments or veterinarians can provide guidance on the most humane methods of destruction.

Breeding and weaning strategies can be adjusted during a drought to reduce nutritional demands. For example, delaying breeding or not breeding at all for a season can significantly cut down on the cost of feeding. Lactating mares, for instance, require up to 70% more feed than dry mares. Early weaning is another strategy that can reduce feed costs and simplify management, allowing mares to be managed as dry stock, which lowers their maintenance needs.

In a drought scenario, it's crucial to determine the weight and condition of your horses and their specific nutritional needs. The goals might vary—maintaining idle horses, supporting pregnant or lactating mares, or allowing for growth in weanlings. The feeding strategy involves six steps:

1. Determining horse weight and condition score.

2. Calculating the nutritional requirements based on different classes of horses.

3. Understanding the nutritional value of various feeds.

4. Determining feed intake capacity.

5. Calculating nutrient requirements and feed costs.

6. Implementing management strategies to optimize feeding.

It's essential to know the weight and condition of your horses to match feeding to their needs. While not everyone has access to scales, girth measurements can provide reliable estimates of weight, and condition scores can help monitor horses' body condition throughout the drought. Maintaining horses at a condition score of at least 2 prevents costly efforts to regain lost weight later.

The nutritional needs of horses vary depending on their class—maintenance horses, pregnant mares, lactating mares, young stock, and stallions all have distinct requirements. For instance, breeding mares in late pregnancy or lactation need significantly more energy and protein. Reducing feed costs can involve early weaning, selecting only the best mares for breeding, or even skipping a breeding season entirely.

Horses can consume up to 3% of their body weight in feed daily, but it's essential to balance this intake with their nutritional needs. For example, a 500 kg horse could consume around 12.5 kg of feed daily. To calculate the most cost-effective feeding options during drought, you need to consider the energy content of feeds, measured in megajoules (MJ), and calculate costs based on MJ of energy provided per unit of feed. Budgeting accurately helps to ensure horses receive adequate nutrition while minimizing costs.

Feed options during drought vary widely in their energy and protein content. Grains, hay, and silage are common choices, with grains generally providing higher energy per kilogram. In contrast, roughage such as hay is essential for maintaining gut health and digestion but provides less energy. Owners should also monitor protein intake, as certain feeds, like grass hay, might be low in protein and require supplementation with high-protein feeds such as lupins or lucerne hay.

In addition to energy and protein, minerals and vitamins must be considered. Calcium and sodium are often deficient in grain-based diets, and these can be supplemented through salt licks or additives like limestone. Vitamins, such as A and E, may also become deficient in drought conditions, particularly in young stock that have been without green pasture for extended periods.

Finally, ensuring adequate water supply is critical during a drought. Horses require around 25 to 30 litres of water daily, and this can rise significantly in hot weather. Monitoring water quality is also crucial, as high salt levels or contamination from algae can pose health risks.

Monitoring Condition and Live Weight Response to Feeding

Monitoring and reporting the condition and live weight response to feeding is a crucial part of livestock management, especially in cattle and dairy operations. This process requires attentive observation and analysis to ensure that animals are receiving adequate nutrition and that their health and productivity are maintained. Here's a detailed explanation of the process:

Effective observation goes beyond simply looking at the herd. Observers must notice and interpret the subtle signals that animals exhibit. These signals provide key information about their wellbeing, such as their physical condition, behaviour, and environment. The observer must ask critical questions: *Is everything as it should be? Is there a potential risk to the animals' health or comfort?*

Observations should be focused, open-minded, and compared against established standards or benchmarks. This involves systematically observing the herd, looking for deviations in body condition, coat appearance, cleanliness, and behaviour. For example, differences in body condition might indicate whether certain animals are receiving enough feed, while abnormalities like swollen feet or awkward gait can signal underlying health issues such as lameness.

Observation routines should be regular and cover all animals, including cows, heifers, and bulls. Ideally, dairy cows should be observed at least three times daily, while other livestock may need less frequent checks. This allows early detection of problems and enables timely intervention. When issues like deformities or health concerns are identified, such as swollen feet from hard floors, determining the cause and effect is essential to address the root issue.

Good record-keeping supports this process by allowing managers to track changes over time and assess the impact of different feeding strategies. Data collection can include weight, condition scores, milk yields, and health records, all of which contribute to more informed decision-making.

Assessing whether the herd is uniform in terms of size, body condition, and appearance can help identify underlying issues. Differences among animals may reveal problems in feed distribution, health, or general welfare. For instance, a significant portion of the herd being either too thin or too fat indicates a long-term imbalance between feed intake and utilization, requiring adjustments in feeding strategies or management practices.

Cow behaviour often reflects their physical and environmental needs. Observing behaviours such as eating, lying down, or abnormal movements helps determine if there is an issue with their feed or living conditions. For instance, cows that are consistently seen standing rather than lying down may not have enough comfortable bedding, or cows that are slow to eat may have health or feed quality issues. Identifying and understanding these behaviours can prevent larger problems, such as illness or underperformance.

Cows often give out subtle signals, such as reluctance to move, sudden changes in eating habits, or an increase in licking or grooming. Observing these signals and linking them to potential causes—whether nutritional, environmental, or health-related—allows for prompt intervention. These cow signals can help prevent the escalation of minor issues into more significant problems, such as a decline in milk yield or severe health issues.

Certain groups of livestock, such as high-yielding milkers, heifers, or calves, are more vulnerable and require closer monitoring. These groups are often the first to exhibit signs of nutritional deficiencies or environmental stress. Monitoring these indicator animals can provide early warning signs of broader herd issues. Similarly, risk-prone locations like feedlots, calving pens, or long walkways need careful attention to prevent injuries or stress among the animals.

Seasonal changes and specific life stages, such as calving or weaning, also require enhanced observation. For example, the hot summer months may stress cows, increasing their water needs, while calves are particularly vulnerable during weaning. Anticipating these times of greater risk allows managers to take preventative measures, such as providing additional shade or adjusting feed.

Finally, all key observations should be recorded, and any necessary actions documented. Writing down observations provides clarity and ensures that all farm staff are informed of any changes or required actions. Sharing this information with veterinarians or other professionals also ensures that expert guidance is available when needed.

Chapter Twelve

Maintaining Livestock Water Supplies

Monitoring and maintaining water quality and availability for livestock is a critical aspect of farm management, especially in regions experiencing variable climates and water scarcity. Ensuring the reliability of water supplies involves careful observation and periodic testing to prevent contamination, deterioration, or interruption in supply.

Water quality is crucial for the health and productivity of livestock. The key aspects of water quality include factors such as salinity, pH levels, toxic elements, and contaminants like algae. The pH levels of water should ideally be between 6.5 and 8.5, as extremes outside this range can cause digestive upsets in animals, leading to reduced water intake, poor appetite, and diminished productivity. If water is too acidic or too alkaline, treatments like adding lime or alum can correct the imbalance, although these should be applied with care [67].

Salinity is another key factor influencing water quality. High salinity levels can lead to increased water consumption as livestock attempt to balance their internal salt levels. If salinity levels exceed tolerable limits, they can cause significant health issues, such as scouring in weaner sheep and reduced growth rates. Measuring the Total Dissolved Solids (TDS) in water and monitoring its electrical conductivity (EC) are standard practices to evaluate salinity. Correcting salinity-related issues may require the use of alternative water sources or desalination techniques.

Toxic elements such as arsenic, mercury, lead, and selenium, though naturally occurring in some water sources, can reach harmful concentrations, especially in underground water supplies. Regular testing for these elements helps prevent long-term health issues in livestock. If toxic elements are detected, corrective measures, such as changing water sources or implementing filtration systems, may be necessary.

Water contamination from chemical residues, pollutants, and biological sources is a constant risk in farming, especially where pesticides and herbicides are used. Contaminants can be introduced into water sources through runoff, posing risks not only to livestock health but also to human consumers of animal products. Contaminants may accumulate in animals over time, rendering their produce unfit for consumption. Therefore, testing water for chemical pollutants is crucial, and preventive measures, such as limiting pesticide use near water sources and constructing barriers to prevent runoff, can be effective strategies.

Algal blooms, particularly those caused by blue-green algae, are another common water quality issue. Algal blooms thrive in stagnant water with high nutrient levels, and some species can produce toxins harmful to both animals and humans. Regular inspection of water bodies for signs of algal growth is essential, particularly in warmer seasons when blooms are more likely to occur. Approved algicides, like Coptrol Aquatic Algicide®, may be used to treat affected water bodies, but strict adherence to regulations regarding their use is necessary to prevent contamination of natural water systems.

Water consumption in livestock varies based on several factors, including the environmental temperature, the animals' diet, and their physiological state. For instance, animals consume more water during hot weather to regulate their body temperature through evaporative cooling. Stock grazing on dry or fibrous pasture also require more water to aid digestion. In such conditions, it is crucial to ensure that water supplies are not only available but of adequate quality to meet increased demand.

Moreover, young animals, pregnant or lactating females, and animals in poor condition are less tolerant of poor-quality water. Their physiological needs require higher water quality to maintain health and productivity. Monitoring these vulnerable groups closely ensures that any issues with water quality or supply are addressed before they escalate into more severe problems.

Ensuring clean and accessible watering points is another important aspect of water management. Troughs and tanks should be regularly cleaned to prevent algae growth and the buildup of contaminants. Contaminated or dirty water can lead to livestock rejecting the water source, causing dehydration and reduced productivity. Low water levels in dams or tanks, especially during dry periods, can force animals to wade through mud, further contaminating the water and risking the health of the herd.

The reliability of water supplies is also critical. In some areas, evaporation and seepage can significantly reduce available water, particularly during hot weather. A well-maintained water budget—accounting for factors such as water usage, evaporation losses, and available storage capacity—helps ensure sufficient water supplies during critical periods, preventing forced destocking or interruptions to farm operations.

Regular inspection of water supplies, testing for quality, and ensuring reliable and clean watering points are essential practices in livestock management. Understanding the environmental, chemical, and physical factors affecting water quality allows farmers to take corrective actions when necessary, ensuring the health and productivity of their livestock. By developing a comprehensive water management plan, including regular testing and monitoring, farmers can safeguard their operations against the risks posed by poor water quality and supply interruptions.

Water Supply for Sheep and Beef Cattle in Stock Containment Areas

Water supply management in stock containment areas (SCA) is critical for maintaining the health and productivity of sheep and beef cattle, especially during emergencies like droughts. These areas are designed to house livestock in controlled conditions, where they rely completely on the manager for food, water, and shelter. Proper water management ensures not only survival but also helps reduce stress and maintain optimal animal welfare [68].

The quality of water provided in stock containment areas is paramount. It must be fresh, clean, and free from harmful substances like salt, algae, and heavy metals. Contaminants such as blue-green algae, which can be toxic, and pollutants like chemical residues from surrounding environments, can harm livestock if present in water sources. Regular testing of water supplies, particularly in areas where the water source is questionable, is vital to ensure it meets safety standards.

Salinity is one of the most significant factors affecting water quality for livestock. Excessive salt levels in drinking water can negatively affect growth, body condition, and overall health. Livestock managers should adhere to maximum salinity limits, which vary depending on the class of stock. For instance, while mature sheep can tolerate water with up to 9,300 μS/cm (5,600 mg/L) of Total Dissolved Solids (TDS), weaner lambs should not consume water exceeding 6,000 μS/cm (3,600 mg/L). It's essential to introduce stock to water with high salt content gradually to prevent health issues like dehydration or scouring [68].

Other harmful ions and elements, such as magnesium, arsenic, and lead, also pose risks when present in water at elevated levels. These contaminants can lead to various health problems, such as diarrhea, scouring, or more severe conditions like liver damage and infertility. Regular testing for these elements is critical, particularly when using non-traditional water sources.

The amount of water required by livestock in containment areas fluctuates based on several factors, including the season, livestock type, and physiological conditions like pregnancy or lactation. For instance, lactating cattle may need as much as 120 litres of water per day during peak summer, while dry cattle typically consume around 80 litres. Sheep in containment areas have lower water needs, with dry sheep requiring about 6 litres daily in average conditions but up to 10 litres in peak demand during hot summer days.

Planning for worst-case scenarios is essential, as water demands can increase dramatically during extreme temperatures or drought conditions. Livestock in high-temperature environments can consume 40-80% more water than usual, making it crucial to ensure that sufficient water supplies are available. Shaded or deep water sources are recommended, as animals prefer cooler water and are less likely to drink warm or stagnant water, which can lead to dehydration and stress [68].

Ensuring the cleanliness and functionality of watering points is another vital aspect of managing water supplies in SCAs. Troughs should be cleaned regularly to prevent contamination from dust, straw, and other organic materials. Water that is stagnant or polluted can lead to reduced water intake, which may result in dehydration and poor animal health. Ideally, troughs should be cleaned and refilled every 1-2 days to maintain water palatability and freshness [68].

In terms of infrastructure, the placement and design of water systems should minimize waste and evaporation. For instance, open water storages are prone to evaporation losses, which can significantly reduce available water during hot months. Seepage and inefficient water distribution systems can also result in water shortages, further stressing livestock.

Planning for these potential issues through proper design and regular maintenance of water storage facilities is essential.

Water supply management in stock containment areas is integral to maintaining livestock health and welfare during challenging conditions such as droughts. By ensuring access to clean, fresh, and adequately distributed water, managers can help mitigate the risks posed by environmental stresses, improve livestock wellbeing, and maintain productivity. Regular monitoring, testing, and infrastructure maintenance are key strategies for ensuring that water supplies remain sufficient and safe for all stock classes in containment areas.

Water Quality

Water supply management is essential for ensuring the health and productivity of cattle, and automatic waterers are the most efficient method for providing consistent water access in paddocks or yards. When automatic systems are not available, well-sized water containers must be provided. These should be appropriately scaled to meet the varying needs of livestock based on factors such as age, weight, and weather conditions. The water supply should cater to the animals' needs during different seasons, production levels, and dry matter intake from feed.

Water Sources

Water for livestock typically comes from one of five primary sources:

- **Rainwater tanks**: These can be effective for smaller operations but are subject to limitations due to rainfall variability.

- **Scheme water**: Connected water systems are not common for large-scale commercial farms but can be a reliable source if available.

- **Surface water**: Dams, creeks, and rivers serve as valuable water sources, especially for larger properties.

- **Groundwater**: Bores, siphons, and springs provide access to underground water, often in areas where surface water is scarce.

- **Strategic community water supplies**: Managed by local Shires, these are usually reserved for emergency use during periods of drought or other water shortages.

Larger farms often rely on farm dams, while smaller operations might make more use of rainwater or scheme water. Regardless of the source, water management, particularly its quality and accessibility, remains a priority.

Chemical treatment should be limited to farm dams and tanks, as treating natural watercourses, springs, or groundwater can have detrimental environmental impacts. Chemicals like copper sulphate, calcium hypochlorite, and ferric alum are used to control algae, which can block water outlets and harm livestock by tainting or poisoning the water. However, these chemicals should be used with care, as they can also harm aquatic life like fish and crustaceans. Additionally, removing nutrient inflow into water sources can help prevent algae buildup, reducing the need for treatment.

Pollution by animals or debris, such as diseases carried in by birds, can contaminate water supplies, potentially leading to lowered production or even livestock fatalities. Regular cleaning of tanks, troughs, and other water sources is essential to maintaining water hygiene. To address water odour issues, aeration—where water is splashed onto a board before entering the tank—can help disperse dissolved gases that cause unpleasant smells.

Cloudy water, while not usually harmful to livestock, can be problematic for irrigation and homestead use. Various methods like adding alum or lime can help clear muddy water, though experimentation is often necessary to find the most effective treatment. For more severe cases, creating a system of two settling tanks can allow for better water management by providing time for particles to settle in one tank while water is drawn from the other.

Water Infrastructure

Checking the components of water supply systems for wear, deterioration, or malfunctions is an essential task, especially as livestock and agricultural operations rely heavily on consistent and efficient water access. This process involves assessing every part of the system, from the source to the distribution points, ensuring that it operates reliably under various conditions, particularly during high-demand periods like summer.

The health and well-being of livestock are directly impacted by the quality and availability of water. Inadequate or malfunctioning water systems can lead to dehydration, stress, and lower production levels. For example, during spring, livestock get a significant portion of their water from green pastures, but as the season progresses and pastures dry out, water demand increases significantly. This makes it crucial to assess water systems before the summer heat exacerbates water shortages or quality issues.

A comprehensive water system assessment involves inspecting pipes, pumps, tanks, and troughs for signs of wear or malfunction. Over time, pipes may become clogged or damaged, pumps may lose efficiency, and valves can become stuck or worn out. These issues not only impact the flow rate but can also cause a waste of water, which is particularly costly during periods of low rainfall or drought.

For example, valves are critical components that control water flow to livestock troughs. If valves are of poor quality or become worn, they can fail to provide adequate water flow, leading to livestock not receiving enough water. In some cases, stuck valves can lead to significant water loss, quickly draining tanks, especially if the system relies on gravity-fed water.

Another key factor is ensuring that water flow rates are sufficient for the livestock's needs. For example, sheep tend to move in mobs, and if the trough fills too slowly, only dominant animals will drink, leaving more timid animals without access to water. Flow rates of 1 to 1.5 litres per second are recommended for smaller mobs, while larger groups may require flow rates of up to 2 litres per second. Increasing the diameter of inlet pipes can help improve flow rates, ensuring that troughs refill quickly and consistently.

Installing water storage tanks near troughs can help manage water supply during peak demand periods. Tanks should be sized to hold at least three days' worth of water in case of a system failure. This is particularly useful in systems where water must be piped over long distances or where flow rates are naturally slow. In addition to storage, water points should be strategically placed to minimize the distance that livestock must travel to access water. Ideally, stock should not need to travel more than 1.5 km to access water, ensuring that all animals have sufficient access to hydration.

Rainwater collection systems can be highly effective in providing a clean and low-salinity water source for livestock. Proper system design, including roof catchment areas and downpipe sizing, ensures that rainwater collection is maximized. However, these systems require regular maintenance, such as cleaning gutters and ensuring that tanks are covered to prevent contamination.

Figure 54: Farm dam in the Adelaide Hills near Strathalbyn. Doug butler, CC BY-SA 4.0, via Wikimedia Commons.

Overall, regularly inspecting and maintaining the components of a water supply system ensures a reliable, clean, and efficient water source for livestock, especially during periods of high demand or environmental stress. Regular monitoring, proactive repairs, and upgrades are crucial to prevent failures and maintain animal health and productivity.

Hillside Dams

Hillside dams are an efficient solution for water storage, especially on sloping land where natural runoff can be captured and stored. Their effectiveness largely depends on the construction methods used, particularly the sealing of the base and inner walls with low-permeability clay to prevent leakage. Additionally, improving catchment areas around the dam can significantly enhance water collection, as runoff from natural landscapes like crop land or pasture might not always be sufficient to fill the dam.

Hillside dams are best suited to gently to moderately sloping land in agricultural and pastoral regions, especially where additional water supply is needed. They are often placed near but away from natural watercourses, such as rivers or streams, to avoid contamination and ensure that any overflow can safely drain into nearby waterways. The success of the dam depends on a sufficient catchment area, whether natural or enhanced through methods like roaded catchments, to ensure the dam can be reliably filled.

There are several designs for hillside dams, each suited to specific landscape and water storage needs:

1. **Rectangular or Square Excavated Tanks**: These are the most common, constructed with three or four walls, typically sealed with clay. A three-walled tank is open on the uphill side, with soil used to construct the remaining walls. The four-walled design optimizes capacity and reduces siltation.

2. **Double Dams**: This design, featuring two smaller dams connected, helps reduce evaporation by keeping the smaller dam topped up using water from the larger dam. This system is particularly useful in areas with high evaporation rates, as it minimizes the surface area exposed to the sun.

3. **Ring and Turkey Nest Tanks**: These tanks are built on flatter ground, typically where shallow saline water tables are present. The walls are made from earth "borrowed" from the tank itself, and the structure is completely enclosed to retain water. Turkey nest tanks, in particular, are filled by pumping water from external sources like creeks or other dams. They are positioned high in the landscape to allow gravity-fed water distribution.

The effectiveness of a hillside dam hinges on several key factors:

- **Site Selection**: Careful assessment of the slope, topsoil depth, and catchment area is crucial. Test pits or drilling to at least 1 meter below the proposed excavation depth are essential to evaluate the soil's suitability for water retention.

- **Clay Content**: The clay content of the site must be at least 25% to ensure stability and minimize leakage. This clay is vital for creating the seal that prevents water from seeping through the dam walls and floor.

- **Catchment Area**: To ensure sufficient runoff into the dam, the catchment area must be large enough to supply adequate water. For example, 10 hectares of catchment area can typically supply 1000 cubic meters of storage. In some cases, roaded or improved catchments are created to increase runoff during rainfall.

- **Depth and Shape**: The dam must be deep enough to account for evaporation losses, with common depths ranging from 5 to 6 meters. The dam's shape, often rectangular, square, or circular, helps maximize water retention while minimizing evaporation. Circular dams are preferred on flatter ground as they offer the smallest surface area relative to volume, reducing the amount of water lost through evaporation.

Managing excess water is another critical component of hillside dam design. Overflow must be safely diverted away from the dam walls to avoid erosion. The crest of the dam is set at the maximum water level, and a mechanical spillway or a piped inlet with debris protection is installed to handle overflow. In some cases, a silt pit is included to trap debris before water enters the dam, helping to maintain water quality.

As part of the overall design, steps must be taken to prevent debris and silt from entering the dam, as well as to regulate the inflow from high-runoff events. Structures like silt pits or inlet pipes are often fitted with trash racks or strainers to prevent debris from entering and clogging the dam or pipes.

Managing Salinity

Pumping Groundwater for Salinity Control

Groundwater pumping is an effective technique used by land managers to lower local watertables, helping mitigate salinity issues in specific areas. While it is not common in broadacre agricultural systems due to high costs and challenges with safe disposal, it can be a practical solution for protecting valuable assets such as townsites, roads, or railway lines. However, this method is typically recommended as part of a larger, integrated water and salinity management program.

The principle behind groundwater pumping is based on lowering the watertable through a process known as creating a "cone of depression." This is achieved by pumping water from a bore (or well), which draws the watertable down in a localized area around the bore. The cone of depression prevents the saline watertable from rising to the surface and damaging the soil within that area. This reduction in the watertable helps maintain the usability of the surface soil and protects plant life from the harmful effects of salinity.

The success of groundwater pumping varies depending on the specific characteristics of the groundwater system. Pumping tends to be more effective in areas with permeable aquifers, which allow water to flow freely and discharge at the surface. In contrast, systems with low-permeability groundwater or clay-rich soils are less responsive to pumping, leading to a reduced area of influence.

In optimal situations, pumping can impact areas ranging from 10 to 20 hectares for lower-yielding systems and up to 400 hectares for high-yielding systems using multiple wells. However, in less favourable conditions (e.g., clay-rich soils or multi-layered aquifers), the effectiveness is significantly reduced.

Before embarking on groundwater pumping, several critical factors must be taken into account. Farmers and land managers must notify the Commissioner of Soil and Land Conservation at least 90 days prior to beginning the process. Pumping is only effective when an unconfined aquifer with a deep, highly conductive profile is present. This often requires drilling several test bores to locate a suitable area for pumping.

The design of the pump system is crucial for its success. The area requiring protection needs to be carefully identified to ensure that the cone of depression from the pump covers the target area. Multiple production bores may be necessary to achieve sufficient drawdown across a larger area.

Groundwater pumping discharges saline water, which needs to be safely disposed of. This water is often highly saline and must be directed to an evaporation basin or pond to avoid contaminating other water bodies or land areas. Furthermore, once the watertable has been depressed, the salt present in the surface and shallow subsurface soils must be leached out of the root zone to restore plant growth.

Another major challenge associated with groundwater pumping is the cost. Pumping systems, especially those powered by electricity or fuel, have high installation, operational, and maintenance costs. However, more cost-effective options, such as wind or solar-powered pumps, can be considered for areas where power sources are limited.

Hydrological assessments are essential to identify the most suitable locations for boreholes and pumps. The aim is to find areas with the most productive aquifers that will yield sufficient water. In many cases, high-yielding aquifers are found in valleys where alluvial sediments have been deposited by ancient rivers. Geophysical surveys can help locate these aquifers, particularly in areas where the aquifer is thickest.

Additionally, understanding the characteristics of the aquifer is crucial for designing a successful pumping system. Factors such as aquifer depth, water quality, disposal options, and long-term yield must be carefully considered. Pumping systems should be tailored to the specific conditions of the aquifer, with boreholes located in the areas that will have the most significant impact on lowering the watertable.

Siphon systems can be an effective alternative to groundwater pumping, particularly in areas where the land has a natural slope. Siphons work by passively moving water from areas of higher hydraulic head to areas of lower head through a pipeline. This method is less costly than pumping and is often used in hillside seeps where salinity is a concern.

Siphons convert widespread discharge into a single point of discharge, making it easier to manage and reduce environmental damage.

Groundwater pumping for salinity control can be a valuable tool for land managers looking to protect high-value assets or sensitive areas from the harmful effects of salinity. However, it requires careful planning, significant investment, and an integrated approach to water and salinity management. By understanding the complexities of groundwater systems and utilizing appropriate pumping technologies or alternative methods like siphons, land managers can effectively control salinity and improve soil and crop conditions.

Relief Wells for Water Resources and Managing Salinity and Waterlogging

Relief wells are an effective tool for managing groundwater, particularly in areas suffering from waterlogging and salinity caused by leaky artesian systems. These wells are designed to release confined groundwater to the surface by tapping into the artesian pressure stored within deep aquifers. They can serve dual purposes: providing a continuous water source and helping to alleviate salinity and waterlogging in vulnerable areas.

Relief wells rely on the natural artesian pressure within confined or semi-confined aquifers to force water up through a borehole to the surface. This pressure is created by groundwater being restricted in its flow, often due to geological barriers like dykes or fracture zones. When these wells are drilled, the confined groundwater is released, creating a continuous discharge. This discharge reduces the hydraulic head (water pressure) in the area, effectively preventing the saline water from reaching the surface soil, which can cause waterlogging and salinity.

Relief wells are typically situated in low-lying areas that are prone to waterlogging or salinity, where they can mitigate the extent of these issues by allowing controlled groundwater discharge. The success of a relief well depends on the artesian pressure and the permeability of the aquifer it taps into. In areas where this pressure is strong—often ranging from 1 to 8 meters above ground level—these wells can produce a steady flow of water that may also be suitable for livestock.

One of the primary advantages of relief wells is their cost-effectiveness. They are relatively inexpensive to operate and maintain compared to other methods, such as pumps. Additionally, relief wells are space-efficient, as they take up little ground area while providing significant environmental benefits, such as reducing salinity and waterlogging.

However, relief wells only work where artesian pressure is present. They cannot completely eliminate waterlogging or salinity but can significantly reduce its severity. They also cannot lower the groundwater table below the level of the well's casing above ground. In areas with high artesian pressure, the flow can be increased by fitting pumps to the well, but in low-pressure zones, this may not be necessary.

To construct a relief well, a borehole is drilled into the ground until it reaches a confined aquifer with sufficient artesian pressure. For wells with low pressure (less than 3 meters), a perforated casing is inserted into the borehole to allow groundwater to flow into the well. A gravel pack is placed around the casing to maximize water entry, and a cement plug is installed to prevent water from leaking up the outside of the casing.

For areas with higher artesian pressure (greater than 3 meters), a larger casing is required to withstand the increased pressure. This casing is cemented into place above the pressure zone, and the well can be capped if necessary. Relief wells are usually drilled to depths of 10 to 25 meters, depending on local conditions and the depth of the aquifer.

Relief wells are particularly effective for managing saline seeps, areas where groundwater seeps to the surface, bringing salt with it. These seeps can cause significant environmental damage by increasing soil salinity and reducing vegetation growth. By relieving the pressure and allowing the groundwater to discharge in a controlled manner, relief wells can reduce the spread of salinity and improve the quality of the surrounding land.

In dryland salinity areas, relief wells can speed up the discharge of saline water, reducing the impact of evaporation and minimizing the area affected by salt accumulation. They are most effective when combined with other water management techniques, such as siphons or pumps, to enhance water movement and prevent the buildup of salinity.

Relief wells also have applications beyond salinity and waterlogging management. In rural areas, they can be used to supply water for livestock or even aquaculture ponds. The continuous flow from a relief well can help maintain water levels in earth ponds, improving water quality and stability. Additionally, in riparian environments (near rivers), relief wells can reduce the salt concentration of groundwater entering the ecosystem, benefiting both flora and fauna by creating more stable conditions for plant and animal life.

Relief wells offer a practical and cost-effective solution for managing waterlogging and salinity in areas with artesian aquifers. By allowing groundwater to discharge in a controlled manner, these wells help alleviate the pressure that leads to saline seeps and waterlogged soils. While they are not a comprehensive solution to salinity, they play a crucial role in integrated water management systems, improving both land and water quality in affected areas.

Water Reticulation Systems

Allowing livestock direct access to dams and natural watercourses can significantly harm both the environment and water quality. This practice leads to soil erosion, contamination of water sources, and damage to ecosystems, as well as potential legal issues depending on local environmental regulations.

When livestock have free access to watercourses or dams, their movements can cause soil erosion along the banks. As animals congregate around the water, their hooves disturb the soil, creating pathways for runoff. This increases sedimentation in the water, which can clog water systems and reduce water quality by introducing pollutants, including faecal matter and organic waste. This contamination not only makes the water less suitable for livestock consumption but also poses risks to downstream ecosystems and even human water supplies.

Another major issue with livestock in natural watercourses is the introduction of pathogens and nutrients such as nitrogen and phosphorus from animal waste. High concentrations of these nutrients can lead to eutrophication, where excess nutrients cause algal blooms that deplete oxygen levels in the water, harming aquatic life. Additionally, pathogens such as E. coli can thrive in these conditions, posing a health risk to animals and humans.

To mitigate these problems, the safest and most environmentally sound practice is to provide water for livestock through a reticulated water system, which involves using pipes to deliver water from a protected source to water troughs in paddocks. By doing this, farmers can prevent animals from accessing and damaging natural water bodies, ensuring that water quality remains high. Reticulated water systems also have the added benefit of allowing better control over water distribution, which can be more efficient for the management of livestock, particularly in larger operations.

In many areas, regulations may prohibit or restrict livestock access to natural watercourses to protect ecosystems and ensure compliance with environmental laws. These regulations are in place to prevent environmental degradation and preserve the quality of public water resources.

Protecting dams and watercourses by fencing them off and reticulating water to troughs not only safeguards water quality but also promotes sustainable land and livestock management. It is a proactive approach to balancing agricultural needs with environmental conservation, reducing the likelihood of contamination, and promoting healthier ecosystems.

Allowing livestock direct access to open dams, seeps, or waterways can cause significant environmental and operational problems. When animals freely enter these natural water sources, the surrounding soil and vegetation suffer from degradation, often leading to erosion and the destruction of critical plant life. Additionally, livestock may suffer injuries or even death, particularly in situations where water levels are low and the ground becomes muddy and unstable. Furthermore, livestock waste, such as urine and faeces, can contaminate water, leading to the spread of disease and a reduction in water quality, which affects both livestock and the broader ecosystem.

Instead of allowing animals to access natural water sources, using water troughs is highly recommended. Troughs offer several advantages: they are mobile, allowing for flexibility in their placement, and can be easily monitored for cleanliness and proper function. For example, for sheep, a trough should provide at least one meter of length per 100 sheep, with a flow rate of about 1 litter per second for every 300 sheep. Cattle, being larger animals, require greater trough space and water volume. Troughs also reduce the risk of livestock trampling natural water sources, preserving the environment and preventing contamination.

Water reticulation systems are essential for managing water in larger operations, such as feedlots. These systems provide reliable and controlled water distribution, minimizing the environmental impact and ensuring that water reaches the animals in the quantities needed. There are two main types of water reticulation systems: gravity flow and pressurized systems. Gravity flow relies on placing water storage higher than the delivery point, eliminating the need for pumps and infrastructure, and ensuring a simple, failure-resistant system. In contrast, pressurized systems use pumps to transport water, allowing for greater flexibility in areas without natural elevation.

When designing a water reticulation system, several factors must be considered, including the required volume of water, the distance over which water must travel, and the necessary flow rates and pressure. A properly designed system should meet peak demand during periods of high water use, typically during hot weather or when livestock numbers are at their highest.

Additionally, feedlots and other large-scale livestock operations must comply with various guidelines and regulations, including those related to water supply security and environmental protection. The design and maintenance of water systems must ensure both legal compliance and operational efficiency. For instance, the National Beef Cattle Feedlot Environmental Code of Practice (Australia) mandates that feedlots maintain a secure water supply, both in terms of legal rights and physical capabilities, ensuring that operations can be sustained under normal conditions.

Regular maintenance of water reticulation systems is crucial for preventing leaks, blockages, and equipment failures, all of which can reduce water availability and increase operational costs. Leaks, for example, can lead to waterlogged conditions, creating breeding grounds for pests and affecting livestock health and productivity. Furthermore, exposed elements like pumps and pipes must be protected from damage caused by cattle and machinery.

Water Troughs

Water troughs play a crucial role in ensuring that livestock, especially cattle, have consistent access to clean, adequate, and good-quality water. Properly designed and maintained trough systems are essential for the survival, welfare, and optimal performance of cattle, while also minimizing any potential negative environmental impacts associated with livestock management.

The primary design objectives of water trough systems are to provide cattle with clean, cool, and palatable water in sufficient volumes to meet their hydration needs. These systems should be constructed in a way that allows all cattle easy and frequent access to water. Additionally, the troughs must be built to withstand damage from both the cattle and any cleaning equipment, ensuring durability. Regular and easy cleaning of the trough's interior and exterior is also a key consideration. The system should prevent manure accumulation under or around the trough, avoiding conditions that promote fly breeding or vermin. Finally, the trough system must be designed to avoid water drainage problems or create wet areas, which could lead to pen maintenance issues.

Water troughs in feedlots must comply with several regulatory standards, including for example in Australia, the Australian Animal Standards and Guidelines for Cattle (DAFF, 2013), the National Guidelines for Beef Cattle Feedlots in Australia (MLA, 2012), and the National Beef Cattle Feedlot Environmental Code of Practice. These regulations ensure that feedlots maintain secure and legal access to water for cattle and address environmental concerns associated with water supply systems. Here are some examples of such standards in other countries:

1. United States - Beef Quality Assurance (BQA) Program

In the U.S., the Beef Quality Assurance (BQA) program, managed by the National Cattlemen's Beef Association (NCBA), sets voluntary guidelines for feedlot management, including water troughs. The BQA emphasizes providing clean, sufficient water and proper maintenance of water troughs to ensure the health and welfare of cattle. While not legally mandated, these standards help cattle producers demonstrate their commitment to sustainable and responsible farming practices.

The Environmental Protection Agency (EPA) also regulates water management in feedlots under the Concentrated Animal Feeding Operations (CAFO) rule, which requires large feedlots to manage water systems in ways that minimize runoff and prevent contamination of nearby water sources.

2. Canada - Code of Practice for the Care and Handling of Beef Cattle

In Canada, the National Farm Animal Care Council (NFACC) sets the Code of Practice for the Care and Handling of Beef Cattle. This code specifies that feedlot operators must provide cattle with access to clean water at all times and outlines best practices for maintaining water quality in troughs. It also requires water systems to be monitored regularly and to ensure that troughs are accessible, clean, and free of contaminants.

3. European Union - EU Animal Welfare Standards

In the European Union, animal welfare standards are governed by EU Directive 98/58/EC concerning the protection of animals kept for farming purposes. This directive requires farmers to ensure that cattle have access to a sufficient supply of fresh water. While it does not directly regulate water troughs, member states enforce water supply standards under broader animal welfare regulations.

For example, in the UK, the Department for Environment, Food & Rural Affairs (DEFRA) mandates that all cattle must have continuous access to fresh water. These regulations ensure that water troughs are maintained and accessible to livestock in feedlots.

4. New Zealand - Animal Welfare (Cattle and Sheep) Code of Welfare 2016

New Zealand's Animal Welfare Act 1999 and the Code of Welfare for Cattle and Sheep 2016 regulate the supply of water in feedlots. The code mandates that cattle must have access to fresh drinking water at all times. Water troughs must be designed to provide sufficient access and quantity, and regular inspections are required to ensure the water quality is maintained.

5. South Africa - Animal Protection Act

In South Africa, the Animal Protection Act, 1962 and the associated guidelines ensure that water troughs in feedlots are maintained to provide livestock with adequate and clean water. The act requires farmers to prevent unnecessary suffering by providing a sufficient supply of fresh water, ensuring the design and maintenance of troughs are consistent with animal welfare principles.

Technical Requirements

Capacity and Flow Rate: A key factor in trough design is ensuring that the water delivery system can meet the daily and peak water demands of cattle. This involves considering the flow rate of the water reticulation system and accounting for temperature and cleaning processes. Low-volume troughs, designed as long, narrow, and shallow, are preferable over larger, high-volume options because they reduce water wastage and facilitate easier cleaning.

Trough Shape and Size: Most water troughs are rectangular, as this shape aligns well with feedlot fence lines and allows for easier cleaning. The troughs' internal cross-section is typically either U-shaped or trapezoidal. The recommended linear trough space per animal is a minimum of 25 mm during normal weather and up to 75 mm during hot conditions to ensure adequate water access during peak heat.

Materials and Durability

Water troughs must be constructed using materials that are durable, long-lasting, and resistant to wear and tear from cattle and cleaning processes. Reinforced concrete is the most common material, as it stabilizes water temperature and withstands damage. Poly-ethylene and fiberglass are alternatives, though they require UV resistance for longevity. Surface coatings, such as epoxy resin, can protect troughs from wear and make cleaning easier, especially as the interior surfaces deteriorate over time due to water hardness and cattle interactions.

Cattle Access and Protection

Water troughs should prevent cattle from stepping into them, using exclusion bars or protective frames made from steel or timber. The float valves and supply pipes need to be protected from damage by cattle or machinery to prevent costly repairs and minimize issues like overflow, which can lead to bogging and pen floor damage. Additionally, cattle tend to prefer cooler water, so shading the trough or burying water pipes can help maintain lower water temperatures in hot weather, improving cattle hydration.

Maintenance and Cleaning

The area around the troughs sees heavy traffic from livestock, making regular cleaning and proper design essential to prevent muddy conditions and minimize wear. Concrete aprons around the trough, typically 3 meters wide, provide stable ground and allow for machinery access during cleaning. These aprons must have non-slip surfaces and be designed to handle the load of cleaning equipment without sustaining damage. Proper grading of the aprons ensures efficient drainage and prevents standing water, which could lead to sanitation issues.

Flow Control and Drainage

Water troughs require reliable flow control systems to maintain constant water levels and prevent overflow. Sleeved or seated valves with ball floats are commonly used to regulate water flow. Trough placement in feedlots also needs careful consideration—troughs should be placed near feed troughs to encourage frequent drinking but far enough to avoid contamination from feed. Proper drainage systems, such as concrete spoon drains or sewerage lines, ensure that water from cleaning or overflow is safely and efficiently removed from the pens.

A well-designed and maintained water trough system ensures cattle have continuous access to fresh water, promotes animal welfare, and minimizes environmental damage. By adhering to regulatory requirements, ensuring adequate trough capacity, and using durable materials, feedlot managers can create efficient, long-lasting water systems that support the overall productivity of their livestock operations.

Pumping Systems

Windmills

Windmills, a traditional technology that has remained effective over the years, have seen little change since their early designs. They continue to be useful in rural and remote areas, especially for water pumping. Their main advantages are longevity and ease of repair. With regular servicing, windmills can operate for decades, and most employees on pastoral stations have the skills to fix them, reducing the need for specialized technicians. However, a significant drawback is their classification as elevated work platforms, which introduces Occupational Safety and Health (OSH) considerations during maintenance, especially as workers must operate at heights.

One limitation of windmills is "wind droughts"—periods of low wind velocity that reduce their efficiency. This is particularly problematic in regions like northern Western Australia, where wind speeds drop between August and November, coinciding with times when cattle are under the most stress due to high temperatures and increased water needs. Auxiliary fuel-powered pumps often become necessary during these periods, adding substantial costs related to fuel, refuelling logistics, and pump maintenance. Additionally, in hilly or tree-covered areas, wind speed near the ground may be insufficient to drive the mill efficiently, requiring larger fans and taller towers, which increase installation costs.

In coastal areas prone to cyclones, windmills are especially vulnerable, and their destruction can create water supply crises, costing around $10,000 per unit to replace. To mitigate cyclone damage, operators can install larger water storage tanks, feather off the windmill blades, or lower the windmill structure to the ground before the storm. However, such measures require pre-installed equipment, preparation, and skilled personnel.

Fuel Powered Pumps

Diesel and petrol-powered pumps are widely used, often complementing windmills during wind droughts or in scenarios where a higher water flow rate is needed, such as deep water tables or long-distance water transport. They are especially common near cattle yards and homesteads. While fuel-powered pumps are reliable, they come with significant costs related to motor maintenance, fuel, and the logistics of delivering fuel to remote locations. In an effort to reduce refuelling needs, some operators install larger fuel tanks, but this practice can backfire if motors run for extended periods without servicing, leading to motor damage or even complete failure if the pumps run dry.

Water Pumps

Water pumps play a crucial role in agriculture, often being as essential to production as tractors. They are used to move water from sources like rivers, dams, or bores to storage tanks or irrigation systems. In many regions, especially where grid electricity is not available or too costly, diesel-powered pumps are common. However, electric pumps are generally preferable due to lower maintenance requirements, ease of automation, and compatibility with renewable energy sources like solar power. These pumps can run solely on solar power or in combination with grid or battery storage systems.

Choosing the right pump involves considering factors such as the daily water needs, the distance water must travel, and the vertical lift required. Pumps must be able to move sufficient water during daylight hours (for solar-powered systems) and efficiently handle the demands of the farm. There are two main types of pumps: dynamic and positive displacement.

- **Dynamic pumps**, like centrifugal pumps, use rotating blades to move water by changing the flow direction. These pumps are ideal for large tasks like irrigation due to their efficiency and flexibility.

- **Positive-displacement pumps**, like helical rotor and diaphragm pumps, work by altering the volume in a closed system to change pressure. These pumps are often used in specific agricultural applications requiring higher pressure or dealing with thicker liquids.

Solar Powered Pumps

Solar-powered pumps are becoming increasingly popular for agricultural water supply, especially in areas with low water demand variability. While the upfront cost of solar installations can be comparable to windmills, the cost of solar power has remained stable, and improvements in component quality and installation techniques have increased the reliability of these systems. Solar pumps are ideal for remote locations as they reduce the need for frequent maintenance visits and eliminate fuel logistics.

However, solar pumps are not without their challenges. Panels must be installed in areas with minimal shading, and regular cleaning is necessary to maintain efficiency, as dust accumulation can significantly reduce performance. Additionally, submersible pumps powered by solar systems have had mixed results, with some failing after a couple of years and others lasting much longer.

Reticulation Systems

Reticulation systems offer a way to transport water from a source to various points across a property. This is particularly useful in regions where water points are not ideally located for grazing patterns. By piping water to better locations, farmers can maximize pasture use and reduce pressure on fences. Reticulation can be expensive, especially when covering long distances, but it may be the only way to utilize productive grazing lands with poor water resources, such as saline flats.

Proper planning is essential when laying reticulation pipes, including considering future developments. While a smaller diameter pipe may be cheaper initially, it could increase long-term costs due to higher pumping resistance or the need for larger pipes in the future. Reticulation systems must also account for factors like pipe join integrity, airlocks, and the potential for gas accumulation in artesian supplies, all of which can be managed through careful design and the use of appropriate technology.

Maintenance

Maintaining water systems and equipment in accordance with manufacturer specifications, livestock requirements, and a comprehensive maintenance plan is critical to ensuring long-term efficiency and minimizing operational disruptions. Effective maintenance not only guarantees a reliable water supply for livestock but also ensures that water systems remain in good condition, reducing potential breakdowns and associated repair costs.

Dams play a crucial role in farm water supply systems and require consistent monitoring and maintenance to prevent leaks, erosion, and degradation of their structure. If dams develop leaks, they waste water and can contribute to soil erosion, undermining the dam's stability. It is essential to patch any weak areas with clay and topsoil to ensure integrity. Fencing off dams to exclude livestock and wildlife like rabbits is recommended to prevent further damage. The fence should be placed at least 10 meters away from the dam's high-water line, ensuring minimal disturbances.

In cases where livestock must access dam water, limiting access to a single, controlled point can help reduce erosion along the dam's edges. Stabilizing the entry path with rocks can further minimize soil displacement. Furthermore, dams should be designed with multiple functions in mind, including fire-fighting accessibility, for which vehicle access gates are crucial. Encouraging revegetation around dams with native species can improve water quality, reduce evaporation, and create wildlife habitats, offering a multifaceted approach to maintaining ecological balance and water management.

Maintaining a dam involves consistent checks to identify early signs of wear or damage, which could escalate into more serious issues. The dam's bank should be covered with a layer of topsoil and vegetated to prevent erosion. Any signs of structural issues, such as cracks or slumping, should be addressed promptly. Settling is normal for dams, but uneven settling may indicate poor compaction during construction and could require professional attention. Wet spots or tunnels in the dam wall signal seepage, and these areas should be dug out and repacked with clay-based material to prevent further leakage.

The spillway is an essential feature of a dam, designed to manage overflow during periods of heavy rain. Keeping the spillway clear of debris and maintaining a strong vegetative cover is necessary to prevent erosion and ensure that water flows safely back into the drainage line. If trickle flows are eroding the spillway during wet seasons, a pipe may be installed to divert these smaller flows, preserving the integrity of the spillway for flood events.

The use of surface or submersible pumps depends on the water source and the needs of the livestock or irrigation systems. Surface pumps are mounted above water sources like dams or rivers and work by drawing water upwards and pushing it to storage or usage points. However, they have limitations in terms of the vertical distance they can draw water, usually not exceeding 7.6 meters. Submersible pumps, installed below the water level, are more efficient for deeper sources as they push water directly to its destination. Both types require regular maintenance to ensure efficient operation and prevent issues like cavitation, which can damage the pump if not properly addressed.

Dam systems should be regularly inspected, especially after heavy rainfall, to identify signs of erosion, cracking, or unusual water flows. This proactive approach prevents costly repairs and ensures the safety and functionality of the dam. Regular monitoring also includes keeping an eye on water turbidity, which could indicate internal dam erosion.

References

1.Saeed, H.A., et al., *Role of Animal Behaviour and Welfare in Livestock Production and Management.* Biological and Clinical Sciences Research Journal, 2023. **2023**(1): p. 442.

2.Sinclair, M., C. Fryer, and C.J.C. Phillips, *The Benefits of Improving Animal Welfare From the Perspective of Livestock Stakeholders Across Asia.* Animals, 2019. **9**(4): p. 123.

3.Narayan, E., et al., *A Retrospective Literature Evaluation of the Integration of Stress Physiology Indices, Animal Welfare and Climate Change Assessment of Livestock.* Animals, 2021. **11**(5): p. 1287.

4.Alemayehu, G., et al., *Animal Welfare Knowledge, Attitudes, and Practices Among Livestock Holders in Ethiopia.* Frontiers in Veterinary Science, 2022. **9**.

5.Animal Health Australia, *Livestock welfare.* 2023, Animal Health Australia.

6.Herrero, M., et al., *Biomass Use, Production, Feed Efficiencies, and Greenhouse Gas Emissions From Global Livestock Systems.* Proceedings of the National Academy of Sciences, 2013. **110**(52): p. 20888-20893.

7.Buddle, E.A., H.J. Bray, and R.A. Ankeny, *"I Feel Sorry for Them": Australian Meat Consumers' Perceptions About Sheep and Beef Cattle Transportation.* Animals, 2018. **8**(10): p. 171.

8.Santeramo, F.G., et al., *Considerations on the Environmental and Social Sustainability of Animal-Based Policies.* Sustainability, 2019. **11**(8): p. 2316.

9.Erian, I., M. Sinclair, and C.J.C. Phillips, *Knowledge of Stakeholders in the Livestock Industries of East and Southeast Asia About Welfare During Transport and Slaughter and Its Relation to Their Attitudes to Improving Animal Welfare.* Animals, 2019. **9**(3): p. 99.

10.Menchetti, L., et al., *Application of a Protocol to Assess Camel Welfare: Scoring System of Collected Measures, Aggregated Assessment Indices, and Criteria to Classify a Pen.* Animals, 2021. **11**(2): p. 494.

11.Koidou, M., et al., *Temporal Variations of Herbage Production and Nutritive Value of Three Grasslands at Different Elevation Zones Regarding Grazing Needs and Welfare of Ruminants.* Archives Animal Breeding, 2019. **62**(1): p. 215-226.

12.Tafa, S.T., T.S. Mengistu, and T.E. Beriso, *A Pilot Survey on Factors Limiting Veterinary Service Delivery Systems at Animal Health Facilities in Hawassa, Shashemene, and Negele Arsi, Ethiopia.* Ethiopian Veterinary Journal, 2023. **27**(1): p. 172-185.

13.Nuvey, F.S., et al., *Management of Diseases in a Ruminant Livestock Production System: A Participatory Appraisal of the Performance of Veterinary Services Delivery, and Utilization in Ghana.* 2023.

14.Willis, R.S., et al., *Australian Livestock Export Industry Workers' Attitudes Toward Animal Welfare.* Animals, 2021. **11**(5): p. 1411.

15.Khan, X., et al., *Fijian Farmers' Attitude and Knowledge Towards Antimicrobial Use and Antimicrobial Resistance in Livestock Production Systems–A Qualitative Study.* Frontiers in Veterinary Science, 2022. **9**.

16.Zwetsloot, G.I.J.M., et al., *The case for research into the zero accident vision.* Safety science, 2013. **58**: p. 41-48.

17.Department of Agriculture Fisheries and Forestry, *Animal Welfare in Australia.* 2024, Australian Government.

18.European Commission, *EU animal welfare legislation.* 2024, European Commission.

19.Animal Legal Defense Fund, *Laws that Protect Animals.* 2024, Animal Legal Defense Fund.

20.Animal Welfare Standards, *The Australian Animal Welfare Standards and Guidelines for Cattle.* 2024, Animal Welfare Standards.

21.Government of Canada, *Provincial and territorial legislation concerning animal welfare.* 2024, Government of Canada.

22.Affairs., D.f.E.F.R., *Farm animals: looking after their welfare* 2023, Department for Environment, Food & Rural Affairs.

23. Animal Control New Zealand, *5 essential laws and regulations for the management and care of farm animals in New Zealand*. 2023, Animal Control New Zealand.

24. Clayton, L.A., *Overview of Brazil's Legal Structure for Animal Issues*. 2011, Michigan State University College of Law.

25. Cassuto, D.N. and C. Eckhardt, *Don't be cruel (anymore): A look at the animal cruelty regimes of the United States and Brazil with a call for a new animal welfare agency*. BC Envtl. Aff. L. Rev., 2016. **43**: p. 1.

26. World Animal Protection, *South Africa*. 2020, World Animal Protection.

27. Australian Pork, *Stockperson competency*. 2024, Australian Pork.

28. Laine, M. and E. Vinnari, *The Transformative Potential of Counter Accounts: A Case Study of Animal Rights Activism*. Accounting Auditing & Accountability Journal, 2017. **30**(7): p. 1481-1510.

29. Meat and Livestock Australia, *A national guide for smallholder livestock producers*. 2020.

30. Department of Primary Industries, *Welfare scoring nutritionally deprived beef cattle, dairy cattle and their crosses, sheep and horses*. 2019.

31. Henneke, D., et al., *Relationship between condition score, physical measurements and body fat percentage in mares*. Equine Vet J, 1983. **15**(4): p. 371-372.

32. Budsiness Queensland, *Assess the condition of pigs*. 2022, Queensland Government.

33. RSPCA, *What is animal sentience and why is it important?* 2024, RSPCA.

34. Department of Primary Industries and Regional Development's Agriculture and Food, *Preparing for animal welfare before emergencies*. 2020, Department of Primary Industries and Regional Development's Agriculture and Food.

35. Business Queensland, *Duty of care to livestock*. 2020, Queensland Governmnet.

36. Signature Staff, *The importance of inducting new employees into your workplace*. 2012, Signature Staff.

37. Meat & Livestock Australia Limited, *Diseases*. 2024, Meat & Livestock Australia Limited.

38. Andrews, T., *Live cattle assessment*. 2019, NSW Department of Primary Industries and Regional Development.

39. US Food & Deug Administration, *African Swine Fever*. 2024, US Food & Deug Administration.

40.Carlson, C.J., et al., *The Global Distribution of Bacillus Anthracis and Associated Anthrax Risk to Humans, Livestock and Wildlife.* Nature Microbiology, 2019. **4**(8): p. 1337-1343.

41.Rao, S., et al., *Risk Factors Associated With the Occurrence of Anthrax Outbreaks in Livestock in the Country of Georgia: A Case-Control Investigation 2013-2015.* Plos One, 2019. **14**(5): p. e0215228.

42.Hart, R., *Anthrax Outbreak In Wyoming Sparks Health Warning—Here's What To Know.* 2024, Forbes.

43.Reichel, M.P., S.R. Lanyon, and F.W. Hill, *Perspectives on Current Challenges and Opportunities for Bovine Viral Diarrhoea Virus Eradication in Australia and New Zealand.* Pathogens, 2018. **7**(1): p. 14.

44.Yue, X., et al., *The Effect of Bovine Viral Diarrhea Virus Introduction on Milk Production of Dutch Dairy Herds.* Journal of Dairy Science, 2021. **104**(2): p. 2074-2086.

45.Leemput, E.S.-V.D., et al., *Comparison of Milk Production of Dairy Cows Vaccinated With a Live Double Deleted BVDV Vaccine and Non-Vaccinated Dairy Cows Cohabitating in Commercial Herds Endemically Infected With BVD Virus.* Plos One, 2020. **15**(10): p. e0240113.

46.Baumbach, L.F., et al., *HoBi-like Pestivirus Is Highly Prevalent in Cattle Herds in the Amazon Region (Northern Brazil).* Viruses, 2023. **15**(2): p. 453.

47.Meat & Livestock Australia Limited, *1-Disease Prevention.* 2024, Meat & Livestock Australia Limited.

48.Future Beef, *Vaccinations for beef cattle.* 2011, Future Beef.

49.Meat & Livestock Australia Limited, *Chemical resistance and residues.* 2024, Meat & Livestock Australia Limited.

50.CDC, *Controlling Antimicrobial Resistance: Livestock and Poultry Producers.* 2024, CDC.

51.US Department of Agriculture, *Antimicrobial Resistance Overview (AMR).* 2024, US Department of Agriculture.

52.Livestock Solutions, *Drench Resistance.* 2024, Livestock Solutions,.

53.Meat & Livestock Australia Limited, *Parasites.* 2024, Meat & Livestock Australia Limited.

54.Moreira, G.M.S.G., et al., *Immunogenicity of a Trivalent Recombinant Vaccine Against Clostridium Perfringens Alpha, Beta, and Epsilon Toxins in Farm Ruminants.* Scientific Reports, 2016. **6**(1).

55.Jiang, Z., et al., *Induction of Potential Protective Immunity Against Enterotoxemia in Calves by Single or Multiple Recombinant <i>Clostridium Perfringens</I> Toxoids.* Microbiology and Immunology, 2014. **58**(11): p. 621-627.

56.Salvarani, F.M. and E.V. Vieira, *Clostridial Infections in Cattle: A Comprehensive Review With Emphasis on Current Data Gaps in Brazil.* 2024.

57.Galvão, C.C., et al., *Measurement Over 1 Year of Neutralizing Antibodies in Cattle Immunized With Trivalent Vaccines Recombinant Alpha, Beta and Epsilon of Clostridium Perfringens.* Toxins, 2021. **13**(9): p. 594.

58.Maréchal, C.L., et al., *A Case Report of a Botulism Outbreak in Beef Cattle Due to the Contamination of Wheat by a Roaming Cat Carcass: From the Suspicion to the Management of the Outbreak.* Animals, 2019. **9**(12): p. 1025.

59.Gil, L.A.F., et al., *Production and Evaluation of a Recombinant Chimeric Vaccine Against Clostridium Botulinum Neurotoxin Types C and D.* Plos One, 2013. **8**(7): p. e69692.

60.Mbhele, Z., et al., *Evaluation of Aluminium Hydroxide Nanoparticles as an Efficient Adjuvant to Potentiate the Immune Response Against Clostridium Botulinum Serotypes C and D Toxoid Vaccines.* Vaccines, 2023. **11**(9): p. 1473.

61.Anniballi, F., et al., *Management of Animal Botulism Outbreaks: From Clinical Suspicion to Practical Countermeasures to Prevent or Minimize Outbreaks.* Biosecurity and Bioterrorism Biodefense Strategy Practice and Science, 2013. **11**(S1): p. S191-S199.

62.Moreira, C., et al., *Protective Efficacy of Recombinant Bacterin Vaccine Against Botulism in Cattle.* Vaccine, 2020. **38**(11): p. 2519-2526.

63.State of Queensland (Queensland Health), *Animal contact guidelines – reducing the risk to human health, 2014.* 2014.

64.Department of Agriculture and Fisheries, *Livestock vaccination.* 2016, The State of Queensland

65.Blackmon, A., *How to Vaccinate Cattle: Methods, Techniques, and Tips.* 2024, Redd Summit Advisors.

66.Meat & Livestock Australia Limited, *Matching the grazing system with livestock requirements.* 2024, Meat & Livestock Australia Limited,.

67.Watts, R.D.a.P., *4. Water requirements.* 2024.

68.Agriculture Victoria, *Water supply in stock containment areas.* Department of Energy, Environment and Climate Action.

Index

C

D

E

F

G

H